AF564702

TEXT BOOK
OF
VECTOR DYNAMICS

DPH PHYSICS SERIES

TEXT BOOK OF VECTOR DYNAMICS

By

D.K. Jha

DISCOVERY PUBLISHING HOUSE
NEW DELHI-110002

First Published-2005

ISBN 81-8356-001-6

Published by

DISCOVERY PUBLISHING HOUSE
4831/24, Ansari Road, Prahlad Street,
Darya Ganj, New Delhi-110002 (India)
Phone: 23279245 • Fax: 91-11-23253475
E-mail:dphtemp@indiatimes.com

Printed at
Arora Offset Press
Laxmi Nagar, Delhi–92

Preface

There are number of books on Vector Dynamics in the market for the use of degree students in various universities in India. It is the experience of author that the average students need the treatment of theory in a way that should be easily comprehensive to him. Therefore an effort has been made in this book to put the matter in a very lucid and simple way to that even a beginner has no difficulty in grasping the subject. Each chapter for this book contains complete theory and a fairly large number of solved examples sufficient problems have also been selected from various university examinations paper. At the end of each chapter an exercise containing objective questions only has been given. The answer to almost all unsolved problems have been checked and every care has been taken to avoid printing and other mistakes. It is sincerely hoped that this book will satisfy the needs of the students and if it gives them even part of pleasure that the author had in its preparations he will consider his labour amply rewarded.

The author will feel amply rewarded if the book serve the purpose for which it is means suggested for the importance of this book are always welcome.

I am very thankful to Mr. Tilak Wasan (Managing Director), Discovery Publishing House, for their valuable effort to complete this book.

D.K. Jha

Contents

Vectors

SCALAR AND VECTOR QUANTITIES

Physical quantities may, in general, be divided into two main classes:

(i) Scalar quantities or scalars, and

(ii) Vector quantifies or vectors.

(1) *Scalar quantities or scalars* are those physical quantities which possess *only magnitude and no direction* in space as, for example, *mass, volume, temperature, time, speed of light* etc.

A scalar quantity can, therefore, be completely specified simply by its magnitude or numerical value, indicating how many times its unit is contained in it. Thus, for example, when we speak of a mass of 50 g, we mean that the unit of mass, the *gram* (g), is contained 50 times in it.

Obviously, having no direction, the magnitude or measure of a scalar quantity is quite independent of any coordinate system. All such quantities are, however, subject to the ordinary algebraical laws of addition and multiplication, *viz.*, the following:

(i) *Law of commutation*, according to which the result of addition or multiplication of a number of scalars is quite independent of the order in which they may be taken. For example,

$$A + B \equiv B + A \text{ and } A \times B \equiv B \times A.$$

(ii) *Law of association*, according to which the sum or the product of a finite number of scalars is quite independent of the manner in which they may be grouped or associated. Thus,

$$A + B + C \equiv (A + B) + C \equiv A + (B + C) \equiv (A + C) + B.$$

And, similarly, $A\times (B \times C) \equiv (A \times B) \times C \equiv (A \times C) \times B$,

clearly indicating that in the case of continuous sums and products the brackets are really superfluous.

(iii) *Law of distribution*, which states that in expressions involving both addition and multiplication the result in the same as the sum of the individual term-wise products. For example,

$$A \times (B + C) \equiv A \times B + A \times C$$

and $$(A + B) \times C \equiv A \times C + B \times C.$$

(2) *Vector quantities or vectors* are those physical quantities which *possess magnitude as well as, direction and are subject, to the parallelogram law of addition*. This latter condition is important because a quantity may have magnitude as well as direction and may still not be a vector quantity. For instance the finite rotation of a rigid body about a given axis has magnitude (*viz.*, the angle of rotation), also direction (*viz.*, the direction of the axis) but it is not a vector quantity. This is so for the simple reason that two finite rotations of the body do not add up in accordance with the vector law of addition, *i.e.*, the resultant rotation is not the vector sum of two finite rotations.

However, if the rotation be small or infinitesimal, it may be regarded as a vector quantity; for, then, the arc described by the body in a, small interval of time is more or less a straight line and is thus representable as a vector, and so also the transverse and angular velocities of the body. Thus, whereas small rotations, angular velocity and angular momentum are vector quantities, large or finite ones are not.

It follows, therefore, that to qualify as a vector, a physical quantity must not only possess magnitude and direction but must also satisfy the parallelogram law of vector addition.

Vectors associated with a linear or directional effect are called *polar vectors* or, usually, simply *as vectors*, and those associated with rotation about an axis are referred to as *axial vectors*. Thus, force, linear velocity and linear momentum are *polar vectors* and a couple, angular velocity and angular momentum are axial vectors.

VECTOR NOTATION

In ordinary writing, whether on a black board or a paper, a *vector quantity* A, say, is represented by putting a wavy line underneath the letter A or a small arrowhead above it (indicating its direction), as $\vec{A}$. In print, however, it is represented by the bold-faced type letter A.

The *magnitude* or *modulus* of a vector, being a scalar quantity, is denoted by a light-faced italic letter. Thus, the magnitude of a vector A is denoted by A. It may also be denoted by | A | or *mod* A.

UNIT AND ZERO VECTORS

A vector of unit magnitude is called a *unit vector* and the notation for it in the direction of A, is $\hat{A}$ read as 'A, hat or A, caret'. It is also often denoted by bold-faced lower case type letter a. Thus, we have

$$A = \hat{A}A \text{ or } aA,$$

A unit vector, as will be readily seen, merely indicates direction.

A *vector of zero magnitude is called* a zero or a null vector, *denoted by 0.*

All zero or null vectors are taken to be equal and their directions are quite arbitrary and, indeed, quite immaterial.

Vectors other than null vectors are referred to as *proper vectors.*

GRAPHICAL REPRESENTATION OF A VECTOR

Graphically a vector is represented by an arrow drawn to a chosen scale, parallel to the direction of the vector. The length and the direction of the arrow thus represent; the magnitude and the direction of the vector respectively. Thus, the in Fig. 1.1(a) represents a vector A, parallel to the x-y plane and making an angle θ with the axis of x or a line parallel to it. If the magnitude of vector A be 7 units *i.e.*, if A = 7, the arrow representing it is drawn 7 times (the length of the arrow representing unit vector A, as shown in Fig. 1.1(b).

The *negative of vector* A, is the vector –A, having the *same magnitude* as A bat being oppositely directed to it. It is therefore,

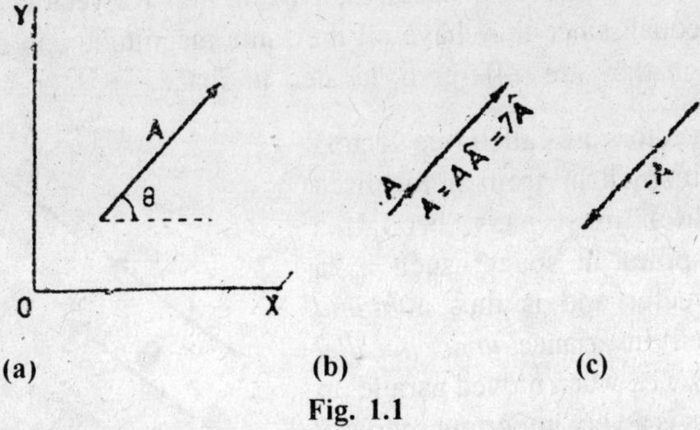

Fig. 1.1

represented by an arrow equal in length to that representing A but, antiparallel to it, as shown in Fig. 11(c).

MULTIPLICATION AND DIVISION OF VECTORS BY SCALARS

From the law of addition of vectors it follows that the sum A+ A + A + ... to *m terms* gives a vector of magnitude m times that of A in the same direction with it, *i.e.*, equal to mA. So that, the product of a vector A and a scalar m is a vector mA whose magnitude is m times the magnitude of A and which is similarly or oppositely directed to A according as the scalar m is positive or negative. Thus,

$$| m A | = mA.$$

Further, if m and n be two scalars and A and B, two vectors, we have

$$(m + n)A = mA + nA \text{ and } m(nA) = n(mA) = (mn) A,$$

showing clearly that *the multiplication of a vector by a scalar is distributive.*

The division of vector A by a non-zero scalar as is defined as the multiplication of vector A by 1/w. Thus, if $A = \hat{A}A$, we have $\hat{A} = A/A$.

EQUALITY OF VECTORS

Although, a vector may refer to a physical quantity defined at a particular point, it does not necessarily have any particular location. So that, we can compare two vectors even though they may measure physical quantities defined at quite different points of space and time. It follows, therefore, that *all vectors, with the same magnitude and direction, are equal despite their entirely different locations in space and remain so even if moved parallel to themselves*. This, in Fig. 1.2 vectors A, B and C are all equal, since-they have all the same magnitude and direction, even though they are differently located in space.

Such vectors are called free vectors to distinguish them from a localised vector which must pass through a specified point in space, such as a position vector and is thus a *bound vector*. Their 'invariance' *under parallel translation, i.e.*, when moved parallel to themselves, is a very important property of free vectors.

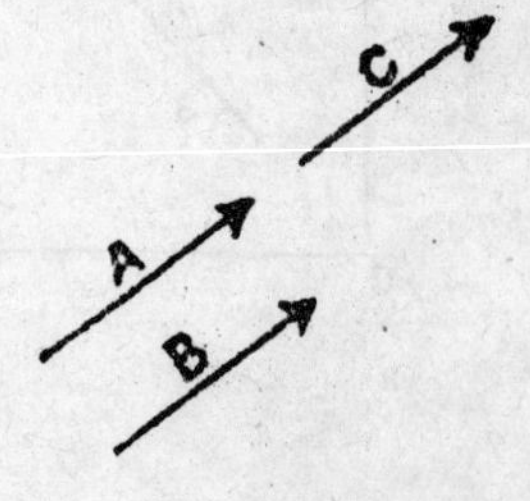

Fig. 1.2

COLLINEAR VECTORS

The term 'collinear' is used as being synonymous with parallel. So that, vectors having the same (as in the case of A and C) *or parallel lines of action* (as in the case of A and B) in Fig. 1.2, *are said to be collinear*. Further, *if the directions of the two parallel vectors be the same, they are referred to as like vectors.* An important property of such (*i.e., like*) vectors is the following:

If A *and* B *be two collinear or like vectors, there exists a scalar k such that* B = kA, *the absolute value of k being the ratio of the lengths of the two collinear or like vectors.*

ADDITION AND SUBTRACTION OF TWO VECTORS

Addition : (i) two vectors may be effected with the help of the *parallelogram law of vector addition*, according to which the sum or *the resultant* R of *two vectors* A *and* B *is the diagonal of the- parallelogram of which* A *and* B *are the adjacent sides, as shown in Fig.* 1.3(a), *i.e.,* R = A + B. As will be seen, A, B and R are all concurrent.

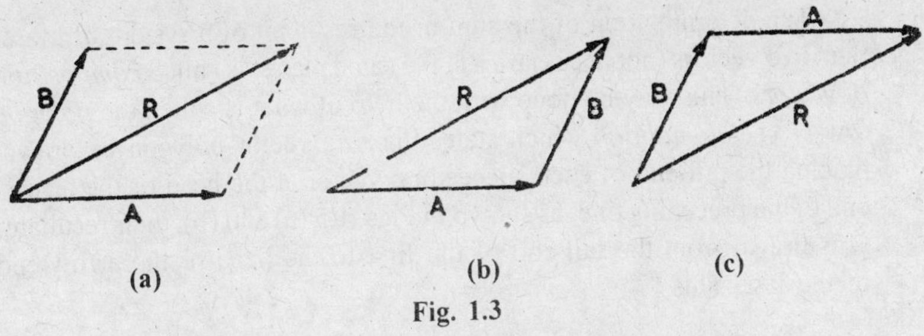

Fig. 1.3

(ii) *The triangle law* : The triangle law of vector addition follows from the parallelogram law and states that *if the tail-end of* one vector be placed at the head or the arrow-end of the other, their sum or resultant R is drawn from the tail-end, of the first to the head-end of the other. As is evident from Figs. 1.3 (b) and (c), the resultant R is the same irrespective of the order in which the vectors (A and B) are taken. So that, we have

$$R = A + B = B + A,$$

showing that vector addition too is commutative.

It may be noted that *vectors need not lie in the same plane for the laws of vector addition to be applicable to them.*

Subtraction : Clearly, if the sum of two vectors A and B be equal to a vector of zero magnitude, we have A + B = 0, showing that vectors A and B are equal in magnitude but opposite in direction. So that, we have B = –A, indicating that –A *is a vector of the same magnitude as vector A but pointing in a direction opposite to that of*

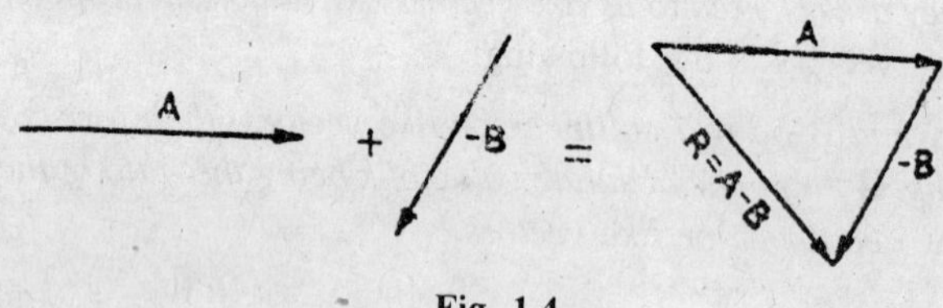

Fig. 1.4

A. And this at once gives us a method of subtracting a vector B from a vector A namely, the addition of the vector A to the vector –B. 50 that, A + (–B) = A – B, as shown in Fig 1.4.

ADDITION OF MORE THAN TWO VECTORS—(COMPOSITION OF VECTORS)

The determination of the sum or the resultant of a system of more-than two vectors (not necessarily in the same plane) is called *composition of vectors*. This may be done with the help of what is called the *polygon law of vector addition* which states that if a vector polygon be drawn, placing the tail-end of each succeeding vector at the head or the arrow-end of the preceding one, as shown in Figs. 1.5 (a) and (b), their resultant; R is drawn from the tail-end of the first to the head or the arrow-end of the last. This

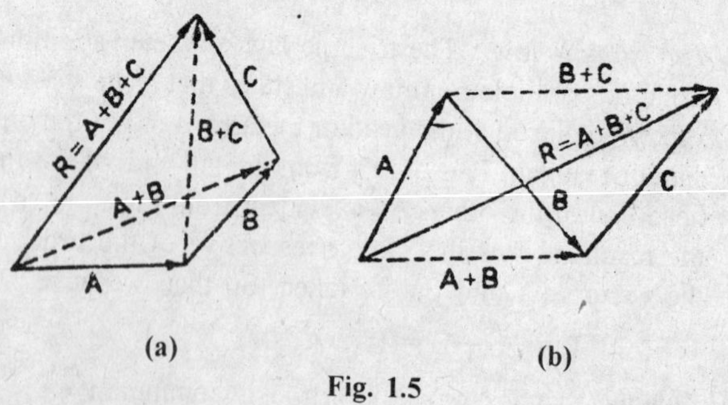

(a) (b)

Fig. 1.5

follows directly from the *Triangle law*. For, what we do, in effect, is to first obtain the resultant (A + B) of the first two vectors (A and B) by the *triangle law* and then the resultant R of (A + B) and C by the, same law. Or, we could fiat obtain the resultant (B + C) of B and C and then that of A and (B + C) which give R, the resultant of A, B and C. We thus have

$$R = A + B + C = (A + B) + C = A + (B + C),$$

showing that *vector addition too, like scalar addition, is associative* and may be extended to any number of rectors.

VECTORS IN A COORDINATE SYSTEM

Although, it is quite possible to deal with physical law in terms of vectors, it is found most helpful in actual practice to represent them in a coordinate system, and the one most widely favoured is the familiar *Cartesian coordinate system* which may be right-handed or left-handed. The former is shown in Fig. 1.6 (a) with its three mutually perpendicular axes OX, OY and OZ represented by the extended thumb, fore-finger and middle finger respectively of the right hand. Its mirror image [Fig 1.6 (b)] forms the left-handed system which is, however, not quite so convenient to me. We shall, therefore, stick to the right-handed system only, in which to an observer at O, right-handed rotations about the axes OX, OF and OZ appear to be from X to Y, Y to Z and Z to X respectively.

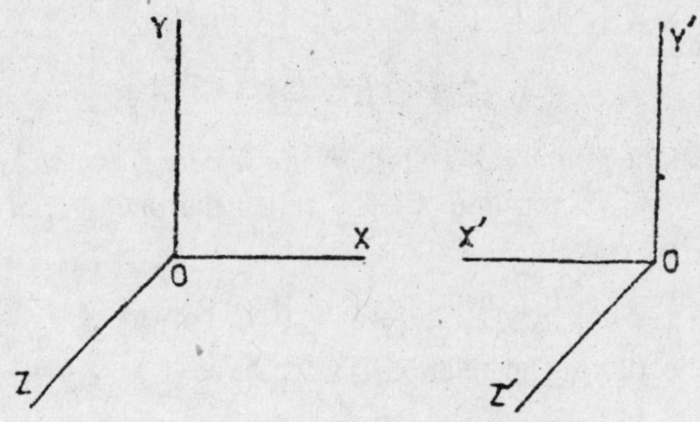

Fig. 1.6

RECTANGULAR COMPONENT OF VECTOR

Two or more vectors which, when compounded in accordance with the parallelogram law or the polygon law of vectors-give a vector A, are said to be the *components* of vector A. The most important components with which we are concerned are mutually perpendicular or rectangular ones along the three coordinate axes OX, OY and OZ respectively. Let us deduce their values.

With O as the origin, let vector OD be R. Then on OD as diagonal, let us construct a *rectangular parallelopiped* with its three edges along the three coordinate axes, as shown.

Let the vector intercepts of the three components of R (also spoken of as the *resolutes*, or the *resolved parts,* of R) along the three axes respectively be R_x, R_y and R_z. Then, we have

$$R = R_x + R_y + R_z \qquad \text{...(i)}$$

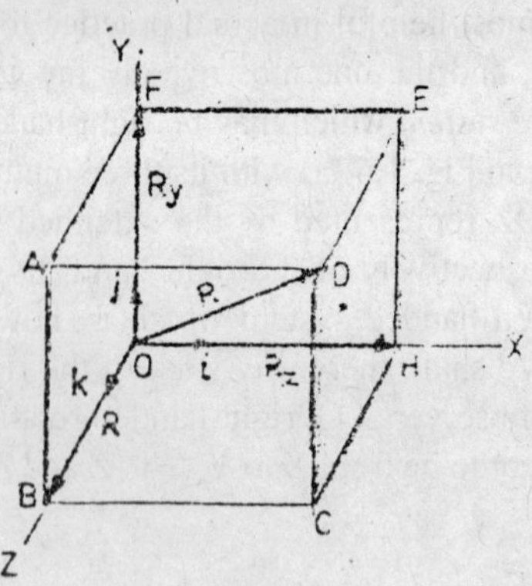

Fig. 1.7

Let i, j and k be *unit vectors along the three coordinate axes,* chosen as the *base vectors* or forming what is called the *orthogonal triad of vectors.* Then, we have

$$R_x = R_x i,\ R_y = R_y j \text{ and } R_z = R_z K.$$

So that, substituting in relation (i), we have

$$R = R_x i + R_y j + R_z k. \qquad \text{...(ii)}$$

Here, $R_x i$, $R_y j$ and $R_z k$ are the orthogonal projections of R on the directions of i, j and k respectively.

From the geometry of the figure, we have

$$OD^2 = OH^2 + OF^2 + OB^2$$

i.e. $$R^2 = R_x^2 + R_y^2 + R_z^2, \qquad \text{...(iii)}$$

where R is the *magnitude* (or *modulus*) of vector R.

Thus, *the square of the magnitude* (or *modulus*) *of a vector is equal to the sum of the squares of its rectangular components*

$$\therefore \qquad R = \sqrt{R_x^2 + R_y^2 + R_z^2} \qquad \text{...(iv)}$$

And, clearly, $R_x = R\ cos(R, x)$ $R_y = R\ cos(R, y)$ and $R_z = R \cos (R, z)$, where the three cosines, *viz., cos*(R, x) = R_x/R, *cos* (R, y)=R_y/R and cos(R, z) = R_z/R are referred to as the *direction cosines* of vector R, since they help determine its direction. They are quite often represented by the letters *l*, m and n respectively.

Dividing relation (iii) by R^2, we have

$$1 = R_x^2/R^2 + R_y^2/R^2 + R_x^2/R_y^2$$
$$= \cos^2 (R, x) + \cos^2 (R, y) + \cos^2 (R, z)$$
$$= l^2 + m^2 + n^2, \qquad \text{...(v)}$$

i.e., the sum of the squares of the three direction cosines of a vector is equal to unity.

Clearly, the unit vector along R is given by

$$R = \frac{R}{R} = \frac{R_x}{R} i + \frac{R_y}{R} j + \frac{R_z}{R} k,$$

where $R_x/R = \cos (R, x)$, $R_y/R = \cos (R, y)$ and $R_z/R = \cos (R, z)$ are the *direction cosines* of R. We therefore have

unit vector R = cos (R, x) i + cos (R, y)j + cos (R, z)k, ...(vi)

Further, if we have a number of vectors R_1, R_2, R_3 etc., their sum may be expressed in terms of their rectangular components in the form

$$\sum R = \left(\sum R_x\right)i + \left(\sum R_y\right)j + \left(\sum R_z\right)k.$$

So that, $\sum R_x$, is the resolute of $\sum R$ in the direction of i and $\sum R_y$ and $\sum R_z$, its resolutes in the directions of j and k respectively.

Since the directions of i, j and k may be chosen arbitrarily, we have the result that the *resolute of a sum of vectors in any direction is equal to the sum of the resolutes of the individual vectors in that direction*

POSITION VECTOR

The position of a point P from any assigned point, such as the origin O of the Cartesian coordinate system, for example, is uniquely specified by the vector $\overrightarrow{OP} = r$ (Fig. 1.8 .), called the *position vector of point* P *relative to* O.

The coordinates of point P being (x, y, z), we have

$$r = xi + yj + zk, \qquad ...(vii)$$

$$\text{where } r = \sqrt{x^2 + y^2 + z^2}.$$

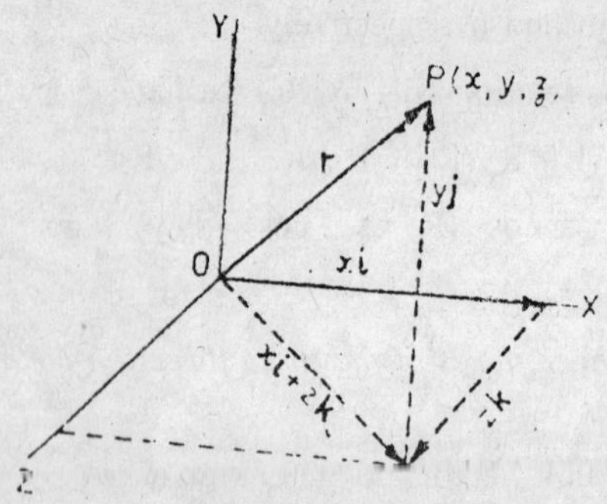

Fig. 1.8

The direction cosines of r are obviously x/r, y/r and z/r.

Such a point, with a position vector r, is often spoken of as the point r.

It follows from the above that if there be two points P_1 and P_2, with (x_1, y_1, z_1) and (x_2, y_2, z_2) (as their respective coordinates and $\overrightarrow{OP_1} = r_1$ and $\overrightarrow{OP_2} = r_2$ as their respective position vectors (Fig. 1.9), we have

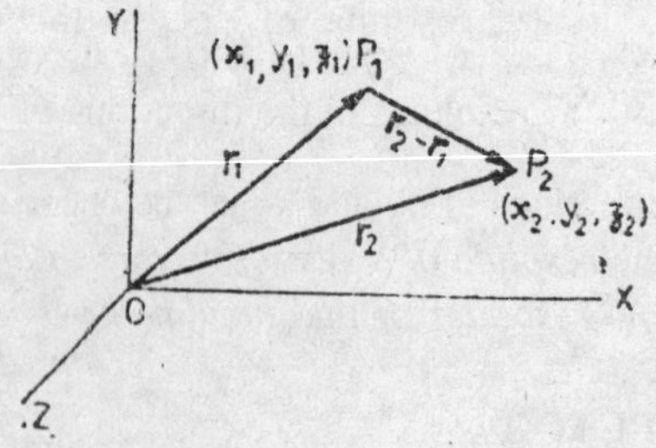

Fig. 1.9

$$r_1 = x_1 i + y_1 j + z_1 k$$

and $$r_2 = x_2 i + y_2 j + z_2 k$$

and therefore, $\overrightarrow{P_1P_2} = \overrightarrow{P_1O} + \overrightarrow{OP_2} = -r_1 + r_2 = r_2 - r_1$

$$= (x_2 - x_1)i + (y_2 - y_1)j + (z_2 - z_1)k.$$

Now, keeping the direction of A unchanged, *i.e.*, keeping it fixed in its position (with respect to any of the so called fixed objects) imagine the reference frame to be rigidly rotated about the origin, say, in the anticlockwise direction, with the coordinate axes now taking up the positions OX', OY' and OZ'. Then, if the new orthogonal triad of unit vectors be formed by i', j' and k' and the components of A along the new coordinate axes be A_x', A_y' and A_z', we have

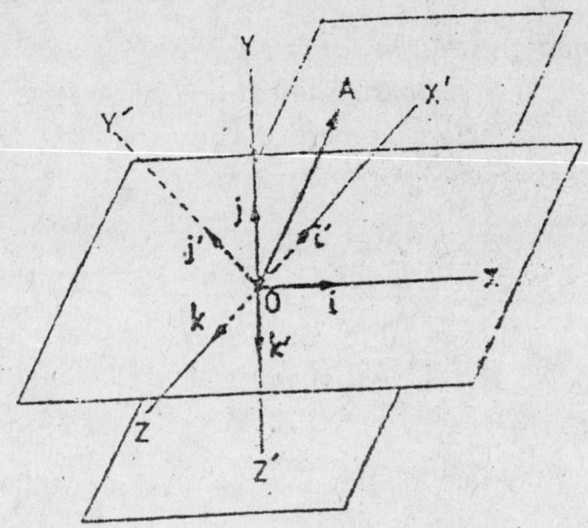

Fig. 1.10

$$A = A_x'i' + A_y'j' + A_y'k'. \qquad ...(ii)$$

Since the length of vector A must obviously be quite independent of the orientation of the coordinate system, or its frame of reference, we have

$$A^2 = Ax^2 + Ay^2 + Az^2 = Ax'^2 + Ay'^2 + Az'^2.$$

Or, $$A = \sqrt{A_x^2 + A_y^2 + A_z^2} = \sqrt{A_x'^2 + A_y'^2 + A_z'^2}$$

In other words rigid rotation of the Cartesian system (or the frame of reference) of a vector brings about no change in its form of magnitude.

PRODUCT OF TWO VECTORS

From the manner in which two vectors enter into combination in physics we come across two distinct kinds of products: (i) *a number or a scalar*, called the *scalar product,* and (ii) a vector, called the *vector product.* The process by which we obtain a product is called *multiplication* and each of the two vectors is called a *factor* of the product.

The scalar and the vector products are also called in *inner* and the *outer* products, after Grassmann.

SCALAR PRODUCT

The scalar product of two vectors A and B is denoted by A.B and is read as A dot B. It is also, therefore, known as the *dot Product* of the two vectors (also sometimes referred to as the *direct product*).

It is defined as the product of the magnitudes of the two vectors A *and* B *and the cosine of their included angle* θ (Fig. 1.11), *irrespective of the coordinate system used.* Thus,

A.B = AB cos θ or AB cos (A, B).

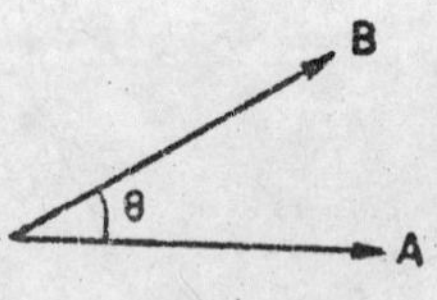

Fig. 1.11

Since cos (A, B) is equal to cos (B, A), *the scalars product is clearly commutative and* A.B = B.A

The order of the factors may thus be reversed without in any way affecting the value of the product.

Further, since B cos θ is the projection (or the resolute) of B in the direction of A and A cos θ, the projection (or the resolute) of A in the direction of B, as shown in Figs. 1.12 (a) and (b) respectively, we have

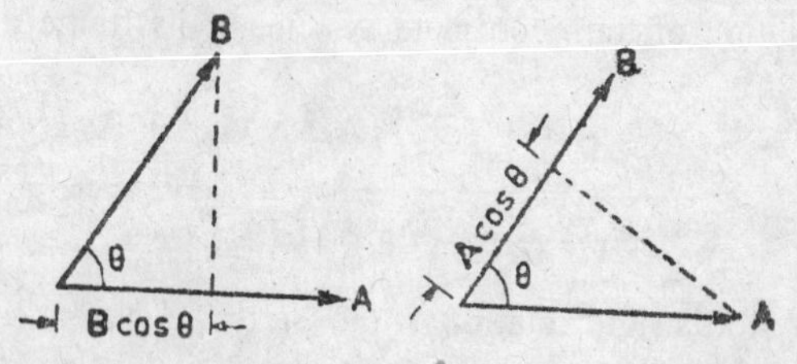

Fig. 1.12

A.B = A.B cos θ = B.A cos θ, *i.e., the scalar product of two vectors* (A *and* B) is the *product of the magnitude of either vector and the projection (or the resolute) of the other in its direction.*

It also follows from the above that the angle between the two vectors, *i.e.*, θ, is given by the relation cos θ = A.B/AB.

IMPORTANT POINTS ABOUT SCALAR PRODUCT

The following points in regard to scalar product of two vectors may usefully be remembered :

(i) *if the two vectors* (A *and* B) *be perpendicular to each other, i.e.,* θ = π/2 *and, therefore,* cos θ = cos (A, B) = 0, *i.e., the scalar product* A.B = AB cos θ = AB cos (A, B) = 0. *This is, therefore, the condition for two vectors to be perpendicular (or orthogonal) to each other.*

(ii) *If the two vectors have the same direction, clearly* θ = 0 and cos θ = cos (A, B) = 1. *So that, the scalar product* A.B = AB cos θ = AB cos (A, B) = AB cos 0° = AB, *i.e., the scalar product is equal to the product of the magnitudes of the two vectors.*

And, if A = B, *i.e., if the two vectors be equal*, we have

$$A.A = A.A = A^2. \text{ Or, } A^2 = A^2,$$

i.e., the square of a vector is equal to the square of its magnitude (or modulus).

(iii) *if the two vectors have opposite direction,* θ = π *and* ∴ cos θ = cos π = – 1. *So that,* A. B = – AB, *i.e., the scalar product is equal to the negative product of their magnitudes.*

(iv) *it follows from (ii) above that square of any unit vector, like* i, j or k *is unity. Thus,* $i^2 = j^2 = k^2$ = (1) (1) cos 0° = 1. or,

$$i.\,i = j.\,j = k.\,k = 1.$$

(v) *Since* i, j *and* k *are orthogonal (or perpendicular to each other), we have, from (i) above,*

$$i.\,j = j.\,k = k.\,i = (1)\,(1) \cos 90° = 0.$$

(vi) *If either vector* (A or B) *is multiplied by a scalar or a number m, the scalar product too is multiplied by that number. Thus,*

$$(mA).B = A.(mB) = m\,(A.B) = mAB \cos \theta = mAB \cos (A.\,B).$$

(vii) *If* A. B = 0, *it means that either* A or B = 0 or A *and* B *are perpendicular to each other.*

(viii) *The scalar product obeys the distributive law.* This may be seen from the following:

In Fig. 1.13, the scalar product of vector A and the resultant (B + C) of B and C is given by A. (B + C) = A. (*projection or resolute of* (B + C) *in the direction of* A), *i.e.*,

$$A.(B + C) = A \times ON = A(OM + MN)$$

$$= A.OM + A.MN. \quad ...(i)$$

Since OM is the projection (or resolute) of B in the direction of A and MN, the projection (or resolute) of C in tne direction of A, we have

$$A.OM = A.\ B$$

and $$A.MN = A.C.$$

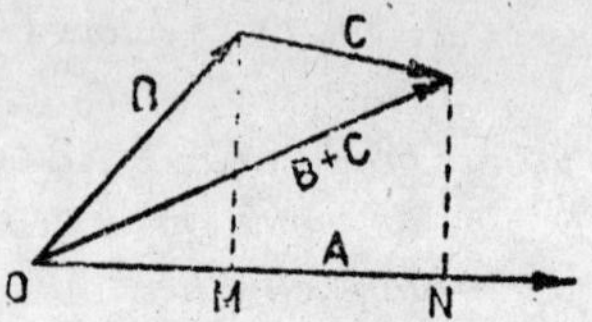

Fig. 1.13

Substituting these values in relation (i) above, we have

$$A.(B + C) = A.\ B + A.\ C,$$

Clearly showing that the distributive law holds good.

It thus follows that (A + B). (C + D) = A.(C + D) + B.(C + D)

$$= A.C + A.D = B.C + B.D$$

(ix) Repeated application of the result obtained in (viii) above shows that the *scalar product of two sums of vectors may be expanded as in ordinary algebra.* Thus.

$(A + B)^2 = A^2 + 2AB + B^2$ and $(A + B)(A - B) = A^2 - B^2$, *etc., etc.*

(x) *The distributive law also enables us to obtain an expression for the scalar product of two vectors in term of their rectangular components.* For, if $A = A_x i + A_y j + A_z k$ and $B = B_x i + B_y j + B_z k$, we have, remembering that i, j and k are mutually perpendicular.

$$A.\ B = A_x B_x + A_y B_y + A_z B_z,$$

i.e., the scalar product of two vectors is equal to the sum of the products of their corresponding rectangular components.

And, since A. B = AB cos θ, we have angle θ between the two vectors given by the relation cos

$$\theta = \frac{A_xB_x + A_yB_y + A_zB_z}{\sqrt{A_x^2 + A_y^2 + A_z^2}\,\sqrt{B_x^2 + B_y^2 + B_z^2}}$$

Further if A = B, we have

$$A.A = A^2 = A^2 = A_xA_x + A_yA_y + A_zA_z = A_x^2 + A_y^2 + A_z^2,$$

i.e., the square of a vector is equal to the sum of the squares of its rectangular components.

(xi) The magnitudes of the components of vector A along the three coordinate axes may be written as $A_x = A.i$, $A_y = A.j$, $A_z = A.k$, for, as we know,

$$A.i = (A_xi + A_yj + A_zk)i = A_xi.i + A_yi.j + A_zi.k.$$
$$= A_x + 0 + 0 = A_x,$$

Similarly $A_y = A.j$ and $A_z = A.k$.

And, if U*l* be the unit vector along a line L, the component of vector A along this line can be shown to be $A.U_l$.

(xii) *A scalar product, being a number, can occur as a numerical coefficient of a vector.* Thus, for example.

(i) (A. B) C is a vector of magnitude (A. B) C in the direction of C.

(ii) (A. B) (C. D) is a number obtained by the multiplication of two numbers (A.B and (C. D).

SOME ILLUSTRATIVE APPLICATIONS OF SCALAR PRODUCT

(i) *Deduction of the Cosine law:* Let two vectors A and B be represented by the two sides of a triangle, as shown in Fig 1.14. Then, the third side represents a vector C = – B + A = A – B.

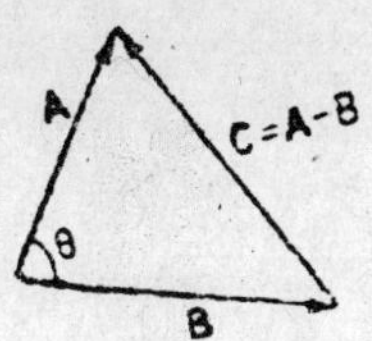

Fig. 1.14

Taking the scalar product of each side with itself, we have

$$C.C = (A - B).(A - B).$$

Or, $C^2 = A^2 + B^2 - 2AB$, because A.B = B.A.

Or, $C^2 = A^2 + B^2 - 2AB \cos (AB)$

$= A^2 + B^2 - 2AB \cos \theta.$

This is the well known trigonometric relation called the *cosine law*.

(ii) *Obtaining the equation of a plane:* Let ABCD be the plane (Fig. 1.15) whose equation is required to be obtained. Drop a normal N on to the plane from an origin O, lying quite outside the plane and let r be the position vector (with respect to O) of any point P in the plane.

Then, since the projection of r on normal N must be equal to the magnitude N or N, we have $N = r \cos \theta$ and $\therefore$ $N^2 = Nr \cos \theta = Nr$, which is the required equation of the plano.

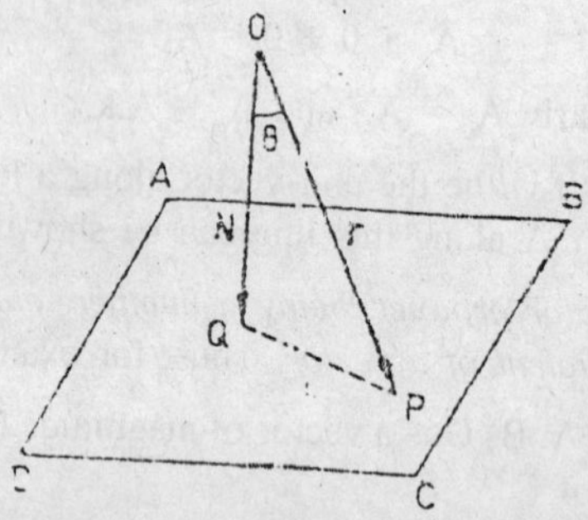

Fig. 1.15

Or, if we write N and r in terms of their rectangular components N_x, N_y, N_z and x, y, z respectively, along the three Cartesian coordinate axes, with i, j, k as the orthogonal triad of unit vectors along these, we have

$$N = N_x i + N_y j + N_z k \text{ and } r = xi + yj + zk.$$

So that, $N^2 = (N_x i + N_y j + N_z k)(xi + yj + zk), = xN_x + yN_y + zN_z.$

which is the usual form of the equation met with in Analytical Geometry.

(iii) *Deduction of Expressions for Work and Power*: As we know, work is said to be done by a force acting on a particle when the particle is displaced in a direction other than the one normal to the force. *It is a scalar quantity* proportional to the force and the resolved part of displacement in the direction of the force or to the displacement and resolved part of the force in the direction of displacement. If, therefore, vectors F and d represent force and displacement respectively, inclined to each other at an angle θ we have *work done*, W = (F cos θ) d = F. d, its value being zero only when F and d are at right angles to each other.

Further, if there be more than one force acting on the particle, say, F_1, F_2, F_3 etc., then, work done by them for a displacement d of the particle is respectively F_1 d, F_2. d. F_n. d. We, therefore, have *total work done*,

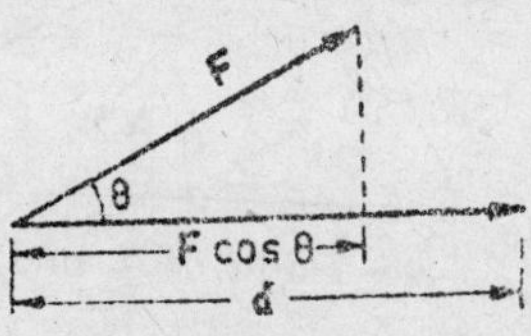

$$W = \sum_{1}^{n} F.d = d.\sum F = d.R,$$

Fig. 1.16

i.e., the effect off all the forces taken together is the same as that of one single force, R *equal to their resultant.*

Now, *power is the rate of doing work.* So that,

$$\text{power } P = \frac{dW}{dt} = \frac{d}{dt}(F.d) = F.\frac{d}{dt}(d).$$

Since $\frac{d}{dt}$ (d) is the velocity v of the particle, we have P = F.v.

VECTOR PRODUCT

The vector product (or the *outer product*) of two vectors A and B is denoted by A × B, read as 'A cross B'. It is, therefore, also called the *cross product* of the two vectors.

It may also be denoted simply as [A, B]. *It is defined as a vector* R *whose magnitude (or modulus) is equal to the product of the magnitudes of the two vectors* A *and* B *and the sine of their included angle* θ. This vector R is normal to the plane of A and B and points in the direction in which a right-handed screw would advance when rotated about an axis perpendicular to the plane of the two vectors in the direction from A to B through the smaller angle θ between them (Fig. 1.17).

Or, alternatively, we might state the rule thus:

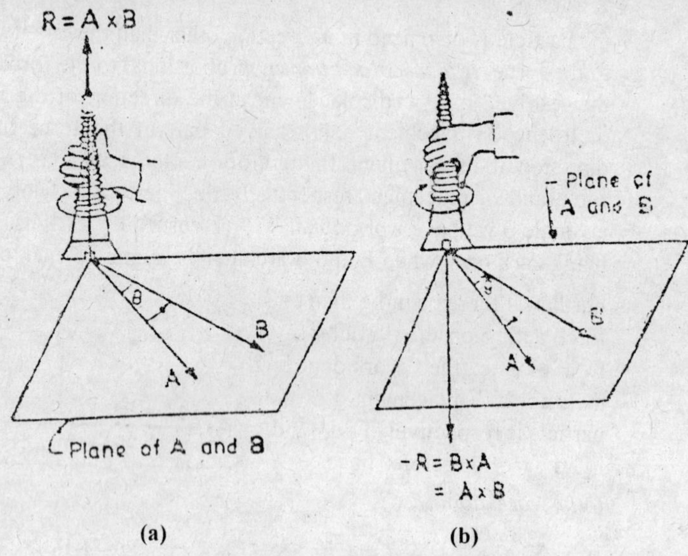

(a) (b)

Fig. 1.17

If the fingers of the right hand be curled in the direction in which vector A *must be turned through the smaller included angle* θ *to coincide with the direction of vectors* B, *the thumb points in the direction of* R, as shown in Fig. 1.17 (a).

Either of these rules is referred to as the *right handed screw rule.*

Thus, if $\hat{n}$ be the unit vector which gives the direction of vector R, we have

$$R = A \times B = AB \sin\theta\, \hat{n}, \qquad (0 \leq \theta \leq 180^\circ \text{ or } \pi,)$$

IMPORTANT POINTS ABOUT VECTOR PRODUCT

(i) Since $0 \geq \theta \geq \pi$, sin θ *and, therefore,* $|R|$ or R *cannot be negative.*

And, clearly,
$$\sin\theta = \frac{A \times B}{|A|\,|B|}$$

(ii) *A change in the order of factors* A *and* B *in the cross product obviously reverses its sign on account of the reversed sense of rotation of the screw,* [Fig. 1.20 (b)]. So that, the *cross product* $B \times A = -A \times B = -R = -AB \sin\theta\, \hat{n}$, indicating that *the vector product is not commutative.*

(iii) *If either vector* (A *or* B) *be multiplied by a number or a scalar m, their vector product too is multiplied by the same number. Thus*

$$(mA) \times B = A \times (mB) = (mAB) \sin \theta \ \hat{n}.$$

Again, $(mA) \times (nB) = (mA \ nB \ \sin \theta)\hat{n} = nA \times mB$

$= mnA \times B = A \times (mnB)$, where n is another scalar.

Thus, we see that the *vector product is associative.*

(iv) $\hat{A}$ and $\hat{B}$ be unit vectors, we have A = B = 1 and, therefore,

$$\hat{A} \times \hat{B} = AB \sin \theta \hat{n} = \sin \theta,$$

i.e., the magnitude of $\hat{A} \times \hat{B}$ *is the sine of the angle of inclination of the two.*

(v) *If the two vectors* (A *and* B) *be orthogonal or perpendicular to each other, we have* $\theta = 90°$ or $\pi/2$ *and therefore,* $\sin \theta = 1$. *So that.* $A \times B = AB\hat{n}$.

The vectors A, B *and* (A × B) *thus form a right-handed system of mutually perpendicular vectors.*

It follows at once from the above that in the *case of the orthogonal triad of unit vectors,* i, j and k (*each perpendicular to the other*), *we have*

$$\mathbf{i} \times \mathbf{j} = -\mathbf{j} \times \mathbf{i} = \mathbf{k};$$

$$\mathbf{j} \times \mathbf{k} = -\mathbf{k} \times \mathbf{j} = \mathbf{i}$$

and $$\mathbf{k} \times \mathbf{i} = -\mathbf{i} \times \mathbf{k} = \mathbf{j}.$$

(vi) *If the two vectors* (A *and* B) *be collinear or parallel,* $\theta = 0$ or π *and, therefore, sin* $\theta = 0$.

So that, $\mathbf{A} \times \mathbf{B} = \mathbf{AB} \ \mathbf{sin} \ \theta\hat{n} = \mathbf{0}$.

It follows at once, therefore, that A × A = 0, indicating that *the vector product of two parallel or equal vectors is zero* and , clearly therefore, *the vector product of a vector with its ownself is zero.*

Obviously, therefore, i*n the case of orthogonal triad of unit vectors, we have*

$$\mathbf{i} \times \mathbf{i} = \mathbf{j} \times \mathbf{j} = \mathbf{k} \times \mathbf{k} = \mathbf{0}$$

Conversely, *if the vector product* A × B = 0, we have AB sin $\theta\hat{n}$ = 0. So that, either A = 0 or B = 0 or sin θ and , therefore, θ = 0 or π, *i.e., either one of the vectors is a null (or zero) vector or the two vectors are parallel to each other.*

(vii) *The distributive law holds good for the cross product of vectors.* Thus,

$$A \times (B + C) = A \times B + A \times C.$$

This may be seen from the following: Putting U = A × (B +C) – A × B – A × C and taking the scalar product with an arbitrary vector V, we have

V. U = V.[A × (B + C) – A × B – A × C] = V. A × (B + C) – V. (A × B) – V. (A × C).

As we shall see under § 1.22, A. (B × C) = (A × B).C. Therefore, we have

V. U = (V × A). (B + C) – (V × A).B – (V × A).C.

Or, because scalar products are distributive, we have

V. U = (V × A).B + (V × A).C – (V × A).B – (V × A).C=0

Obviously, therefore, either V = 0 or perpendicular to U or U=0.

Since V is an arbitrary vector, it can be chosen to be *non-zero* and *non-perpendicular* to U. Hence, U = 0 and we, therefore, have

$$\mathbf{A \times (B + C) = A \times B + A \times C}$$

in accordance with the distributive law.

(viii) *A vector product can be expressed in terms of rectangular components of the two vectors and put in the determinant form,* as may be seen from the following:

Writing A and B in terms of their rectangular components, we have

$$A = A_x i + A_y j + A_z k \text{ and } B = B_x i + B_y j + B_z k$$

And ∴ $A \times B = (A_x i + A_y j + A_z k) \times (B_x i + B_y j\ B_z k)$

$$= (A_x B_x) i \times i + (A_x B_y) i \times j + (A_x B_z) i \times k$$

$$+ (A_y B_x)\, j \times i + (A_y B_y)\, j \times j + (A_y B_z) j \times k$$

$$+ (A_z B_x)\, k \times i + (A_z B_y)\, k \times j + (A_z B_z) k \times k.$$

Since i × i = j × j = k × k = 0 and i × j = k = – j × i etc.,[see (v) above], we have

$A \times B = (A_x B_y)k - (A_z B_z)j - (A_y B_x)k + (A_y B_z)i + (A_z B_x)j - (A_z B_y)$.

This may be grouped together as

$A \times B = (A_y B_z - A_z B_y)i + (A_z B_x - A_x B_z)j + (A_x B_y - A_y B_x)k$.

Or, putting it in the easily remembered *determinant form*, we have

$$A \times B = \begin{vmatrix} i & j & k \\ A_x & A_y & A_z \\ B_x & B_y & B_z \end{vmatrix}.$$

It may be noted that the *scalar components of the first vector A occupy the middle row of the determinant.*

(ix) if B = C + nA, where n is a number or a scalar, we have

$$A \times B = A \times (C + nA) = A \times C.$$

Conversely, however, if A × B = A × C, it *does not necessarily mean that* B = C. *All it implies is that* B *differs from* C *by some vector parallel to* A *which may or may not be a zero vector.*

SOME ILLUSTRATIVE APPLICATIONS OF VECTOR PRODUCT

(i) *Torque of moment of a force:* Let a force F be acting on a body free to rotate about O (Fig. 1.18) and let r be the *position vector* of any point P on the line of action of the force. Then, since torque = *force × perpendicular distance of its line of action from 0,* we have *torque (or moment) of force F about O i.e.,*

$$T = Fr \sin \theta.$$

Now, as we know, Fr sin θ is the magnitude of the cross or vector product r × F. So that, in vector notation, we have

$$\text{torque } \mathbf{T} = \mathbf{r} \times \mathbf{F}$$

Its direction being perpendicular to the plane containing F *and* r.

If we draw a set of three coordinate axes through O, as shown, we shall have r = xi + yj + zk,

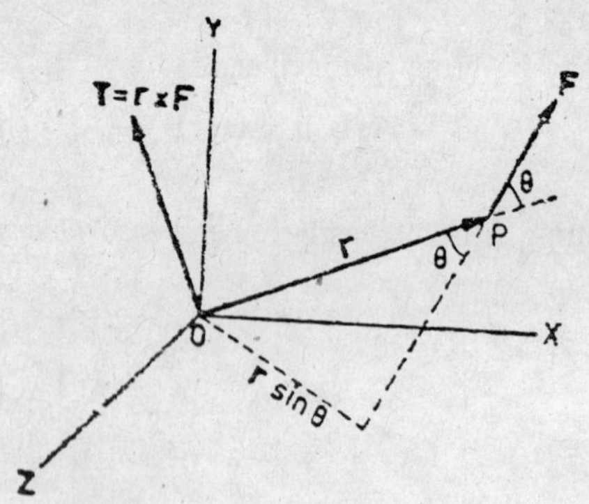

Fig. 1.18

Where xi, yj and zk are the rectangular components of r along the three axes respectively.

Similarly, $F = F_x i + F_y j + F_z k$ and $T = T_x i + T_y j + T_z k$,..(i)

the symbols having their usual meanings.

$$\text{Now,} \qquad T = r \times F = \begin{vmatrix} i & j & k \\ x & y & z \\ F_x & F_y & F_z \end{vmatrix}$$

which, on expansion, gives

$$T = i\,(F_z y - F_y z) + j\,(F_x z - F_z x) + k\,(F_y x - F_x y). \qquad \text{...(ii)}$$

Comparing expressions (i) and (ii), we have

$T_x = F_z y - F_y z$, $T_y = F_x z - F_z x$ and $T_z = F_y x - F_z y$,

where T_x, T_y and T_z are the respective scalar components of T along the three coordinate axes through O.

It will be easily seen that the scalar components T_x, T_y and T_z of torque Tare given by the dot or the scalar products of T and the respective unit vectors along the three coordinate axes. Thus,

$T_x = T.i$, $T_y = T.j$ and $T_z = T.k$. For, $T.i = (T_x i + T_y j + T_z k).i$

$$= T_x i.i + T_y j.i + T_z k.i$$

And since i. i = 1, j. i = 0 and k. i = 0 we have T_x = T.i

And, Similarly, T_y = T. j and T_z = T. k.

(ii) *Couple:* A couple, as we know, is a combination of two equal, opposite and parallel forces. Let F and –F be two such forces acting at points P and Q (Fig. 1.19) and let the position vectors of P and Q with respect to O be r_1 and r_2 respectively.

Then, since *moment of the couple* C with respect to O is equal to the sum of the moments, (with respect to O) of the two forces constituting the couple, we have

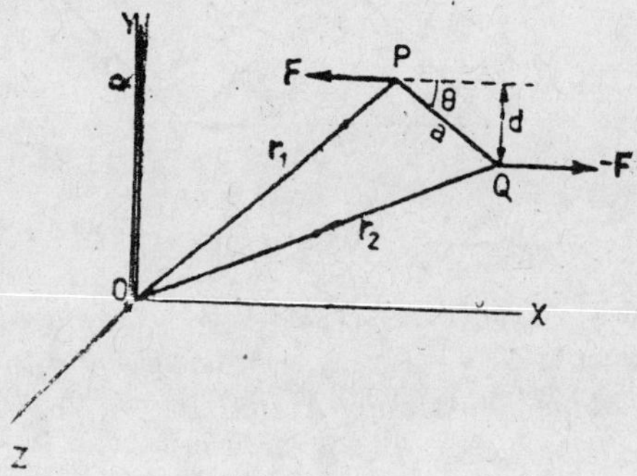

Fig. 1.19

$$C = r_1 \times F + r_2 \times (-F) = (r_1 - r_2) \times F.$$

Since $r_1 - r_2$ = a, where a lies in the same plane with F, we have

$$\mathbf{C = a \times F}$$

Thus, *the moment of the couple, C, is a vector lying in a plane perpendicular to that containing the two forces.*

The *magnitude of couple* C $= |a \times F|$ or a F sin θ.

And since a sin θ = d, *the perpendicular distance between the two forces, we have magnitude of the moment of the couple, i.e., C = Fd =* (*one of the forces*) (*perpendicular distance between the forces*).

(iii) *Area of a parallelogram*: Let vectors A and B form the adjacent sides of a parallelogram OPQR, inclined to each other at an angle θ. Then, if OD be the perpendicular dropped from O on to PQ, we have *area of the parallelogram*

$$= P\,Q\,(O\,D) = B\,(A \sin \theta).$$
$$= AB \sin \theta = AB \sin (A, B),$$

which is obviously twice the area of the triangle OPQ with the same adjacent sides A and B.

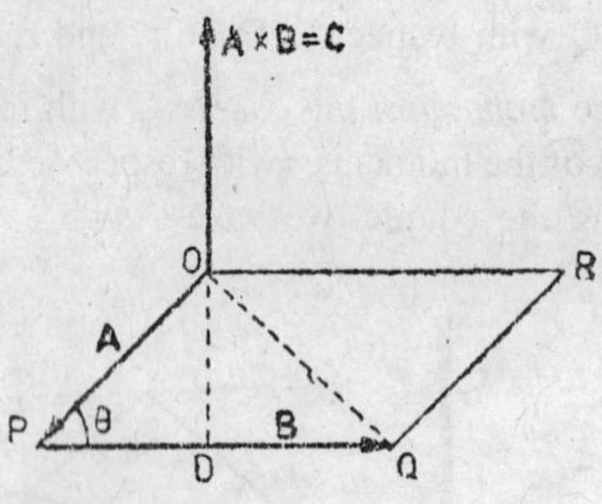

Fig. 1.20

Clearly, AB sin θ or AB sin (A, B) is *the magnitude* of the vector product A × B = C, say, *whose direction is perpendicular to the plane containing* A *and* B, *i.e., to the plane of the parallelogram.*

Now, an area, by itself, has no sign but may be regarded positive or negative in relation to the direction in which its boundary is described. A × B = C, *therefore, represents a vector area which gives both the magnitude and the orientation of the area of the parallelogram.*

The direction of vector C, drawn normal to the plane of the figure (*i.e.*, the parallelogram here) bears the same relation to the direction in which the boundary of the figure is described as the direction of advance of a right-handed screw does to its direction of rotation. Thus, with the parallelogram described as shown (A being taken first and B next), the normal vector C = A × B is drawn pointing upwards, the area of the parallelogram being regarded as positive in relation to this direction of C.

With this convention, therefore, any vector area (*i.e.*, the area of any plane figure) may be represented by a vector drawn normal to the plane of the figure in a direction relative to which the area of the figure is regarded as positive.

(iv) *The law of sines in a triangle*: Consider a triangle vector A, B and C such that A + B = C.

Taking vector product of both sides of the relation with A, we have

$$A \times A + A \times B = A \times C.$$

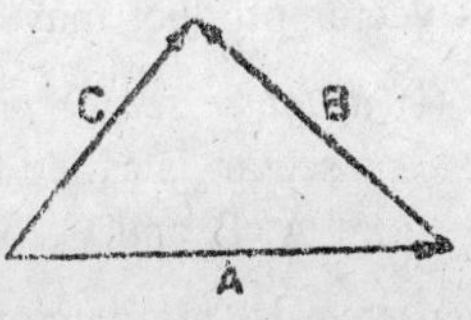

Fig. 1.21

Or, since $A \times A = 0$. We have $A \times B = A \times C$. So that, taking magnitudes of the two sides, we have

$$AB \sin (A, B) = AC \sin (A, C).$$

Or, $B \sin (A, B) = C \sin (A, C)$

whence, $$\frac{B}{\sin(A,C)} = \frac{C}{\sin(A,B)}$$

which is the familiar *law of sines in a triangle.*

(v) *Force on a moving charge in a magnetic field*: Imagine a charge q to be moving with velocity v at an angle θ with a magnetic field B at any given instant (Fig. 1.22). Then, the force acting on it in a direction perpendicular to B as well as v is $F = qv B \sin \theta$, where F, v and B are the *magnitudes of the force, velocity and the magnetic field* respectively. In vector form, therefore, we may put it as

$$F = qv \times B$$ *emu* or SI *units.*

This is called *Lorentz force law*, with the force itself referred to as the *Lorentz force.*

In most cases, q is taken in *esu* and B in *emu* (*i.e., gauss*). Since 1 *esu* charge = (1/c) *emn*, where c is the velocity of light in vacuo, we have q *esu* = q/c *emu* and, therefore,

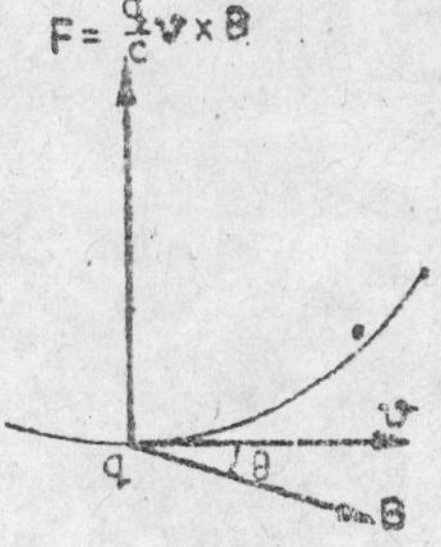

Fig. 1.22

$$F = \frac{q}{c} v \times B.$$

In case the charge also simultaneously passes through an electric field E, an additional force qE acts upon it and the Lorentz force law then takes the form $F = qE + \frac{q}{c} v \times B.$

TRIPLE PRODUCTS OF VECTORS

Let us consider three vector A, B and C. As we know, the cross or vector product of any two of them, say, B and C, *i.e.*, B × C is a vector. This vector product may have with the third vector A either

(i) *a dot or scalar product* A. (B × C) which, being a number or a scalar, is called the *scalar triple product* (or a *mixed product*) of A, B and C; or,

(ii) a cross or a *vector product* A × (B × C) which, being a vector, is referred to as the *vector triple product*.

Let us deal with each in a little more detail.

SCALAR TRIPLE PRODUCT

Let the three vectors A, B and C form a parallelopiped, as shown in Fig. 1.23. Then, if B and C be the *magnitudes* of vectors B and C, we have *area of the base of the parallelopiped* = BC sinϕ = B × C.

Thus, the *magnitude of the vector product* B × C *is the area of the base of the parallelopiped, and its direction perpendicular to this area,* as shown.

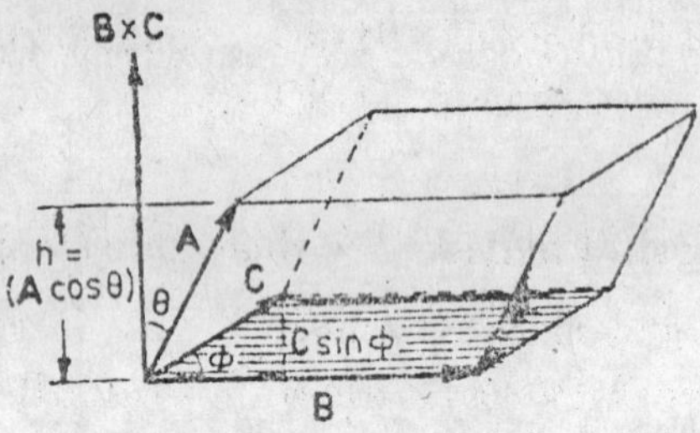

Fig. 1.23

If θ be the angle that the direction of the vector (B × C) makes with vector A (forming one of the edges of the parallelopiped), we have *altitude of the parallelopiped*, h = A cos θ, where A is the magnitude of vector A.

And, therefore, A. (B × C) = A cos θ (B × C) = *vertical height of parallelopiped* × *area of its base* = *volume of the parallelopiped,* V, the sign of V being positive if θ be an acute angle, *i.e.*, if A, B and C form a right-handed system of vectors.

Thus, we see that *the scalar triple product* A.(B × C) *represents the volume of a parallelopiped, with the three vectors forming its three edges.*

Obviously, since any one of the faces of the parallelopiped may be taken to be its base, its volume V is also given by B (C × A) and C. (A × B), *with the cyclic order of* A, B *and* C *maintained.*

In case this cyclic order be altered, the sign of the product is changed since, as we know, B × C = – C × B.

Further, since the order of the terms in a scalar product is quite immaterial, we have

$$V = A.(B \times C) = (B \times C).A = -A.(C \times B) = -(C \times B).A$$
$$= B.(C \times A) = (C \times A).B = -B.(A \times C) = -(A \times C).B$$
$$= C.(A \times B) = (A \times B).C = -C.(B \times A) = -(B \times A).C.$$

It will thus be seen that *the value of a scalar triple product is depends on the cyclic order of the vectors and is quite independent of the positions of the dots and the crosses, which may be interchanged as desired.* It is, therefore, usual to denote a scalar triple product of vectors A, B and C by [A, B, C], or [A, B, C] putting the three vectors in their cyclic order but without any dots or crosses.

Further Points to be Noted About a Scalar Triple Product

(i) *In case the three vectors be coplanar, their scalar triple product is zero,* because the vector (B × C) being then perpendicular to vector A, their scalar product is zero.

Thus, the *condition for coplanarity of three vectors* is that *their scalar triple product* [ABC] *should vanish. This condition is satisfied when two of the vectors are parallel.*

(ii) *If two of the vectors be equal, the scalar triple product is zero.* For, if the three vectors be A, A and B, we have [AAB] = (A × A)B. And, since A × A = 0 [the vector product of a vector with itself being zero. We have [AAB] = 0.

(iii) *If two of the vectors be parallel, the scalar triple product is zero.* For, of the three vectors A, B and C, if A and B be parallel,, B = kA, where k is some scalar, We therefore, have

$$[ABC] = (A \times B).C = (A \times kA).C$$
$$= k\,(A \times A).C = k\,[AAC] = 0.$$

(iv) In terms of the Cartesian components of the three vectors, we have

$B \times C = (B_y C_z - B_z C_y)i + (B_z C_x - B_x C_z)j + (B_x C_y - B_y C_x)k.$

And ∴ $A.(B \times C) = [ABC] = (B_y C_z - B_z C_y)i + (B_z C_x - B_x C_z)j + (B_x C_y - B_y C_x)k]. (A_x i + A_y j + A_z k)$

$= A_x (B_y C_z - B_z C_y) + A_y (B_z C_x - B_x C_z) + A_z(B_x C_y - B_y C_x).$

Or,
$$[ABC] = \begin{vmatrix} A_x & A_y & A_z \\ B_x & B_y & B_z \\ C_x & C_y & C_z \end{vmatrix}$$

which is the familiar expression for the volume of a parallelopiped with one of its corners at the origin (Fig. 1.23)

(v) *The scalar triple product of the orthogonal vector triad is unity.* For; [i j k] = (i × j).k = k.k = 1.

(vi) *Since the distributive law holds good for both scalar and vector products, it also holds good for scalar triple products.* Thus, for instance,

[A, B + D, C + E] = [ABC] + [ABE] +[ADC] + [ADE],

the cyclic order of the vectors being maintained in each term.

VECTOR TRIPLE PRODUCTS

The cross product of two vectors, one of which is itself the cross product of two vectors, is called a vector triple product (for the simple reason that it is a vector). Thus, if A, B and C be three vectors, we have A × (B × C), B × (C × A) and C × (A × B) as examples of *vector products*.

The vector triple product A × (B × C) is a vector lying in a plane perpendicular to that A and (B × C). But, as we know, vector (B × C) lies in the plane normal to that of B and C. It follows, therefore, that vector A × (B × C) lies in the plane of vectors B and C and is perpendicular to that of

Similarly, vector B × (A × C) lies in the plane of A and C and is perpendicular of B. And vector C × (A × B) lies in the plane of A and B and is perpendicular to C. So that, in general,

A × (B × C) = (A × B) × C.

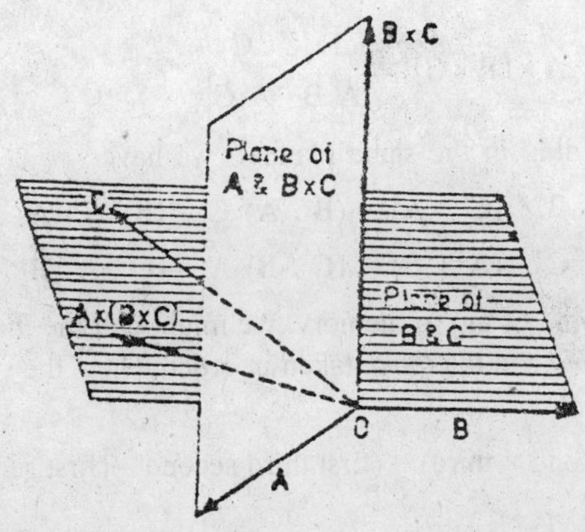

Fig. 1.24

EVALUATION OF THE VECTOR TRIPLE PRODUCT A × (B × C)

Let, i, j, k be the right handed *orthogonal tracts of unit vectors* with unit vectors i along B, j perpendicular to it in the plane of B and C, and k perpendicular to i and j. Then, we have $B_y = 0$, $B_z = 0$ and $C_z = 0$. So that.

$A = A_x\, i + A_y\, j + A_z\, k$, $B = B_x\, i + 0 + 0 = B_x\, i$ and $C = C_x\, i + C_y\, j + 0 = C_x\, i + C_y\, j$.

Hence $\quad B \times C = B_x\, i \times (C_x\, i + C_y\, j) = B_x\, C_y\, k$

$[\because\ i \times i = 0$ and $i \times j = k$.

And $\therefore \quad A \times (B \times C) = A_y\, B_x\, C_y\, i - A_x\, B_x\, C_y\, j.$

$[\because\ i \times k = -j,\ j \times k = i$ and $k \times k = 0$.

Adding and subtracting $A_x\, B_x\, C_x i$ to the right-hand side, we have

$$A \times (B \times C) = A_x\, B_x\, C_x i + A_y\, B_x\, C_y i - A_z\, B_x\, C_z i - A_x\, B_x\, C_y j$$

$$= (A_x\, C_x + A_y\, C_y)\, B_x i - A_x\, B_x\, (C_x i + C_y j)$$

Now, $\quad A_x\, C_x + A_y\, C_y = A.C$, $A_x\, B_x = A.B$,

$[\because\ i.i = j.j = 1$ and $i.j = j.k = 0$.

$B_x i = B$ and $C_x i + C_y j = C$.

we, therefore, have $\quad$ **A × (B × C) = (A . C) B – (A. B) C.**

Or, to put in the determinant form, we have

$$A \times (B \times C) = \begin{vmatrix} B & C \\ A.B & A.C \end{vmatrix}$$

And, proceeding in the same manner, we have

$$\mathbf{B \times (C \times A) = (B \,.\, A)\, C - (B \,.\, C)\, A}$$

and $$\mathbf{C \times (A \times B) = (C \,.\, B)\, A - (C \,.\, A)\, B}$$

As a mnemonic or aid to memory, we might call the three vectors, the *first*, the *second* and the *third*, taken in order. Then, the vector triple product

= first × (second × third) = (first.third second – (first.second) third.

VECTOR DERIVATIVES-VELOCITY—ACCELERATION

Let r be a *single-valued function* of a scalar variable t such that for every value of t there exists only one value of r. Then, as t varies continuously, r also does so.

We are particularly interested in the case in which t represents the time variable and r stands for the position vector of a moving particle with respect to a fixed origin O (Fig. 1.25). Then, as t varies continuously, the point moves along a continuous curve in space. So that, if r and r + δr be the position vectors of the point in positions P and P' relative to origin O for the values t and t + δt of the scalar variable, we have

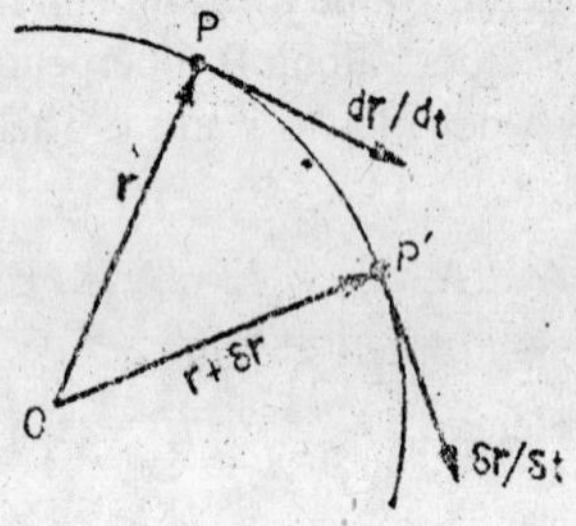

Fig. 1.25

change in the value of r = δr.

The quotient δr/δt (of vector δr by the scalar or a number δt) is also a vector. As δt → 0, point P' approaches P and the chord PP' tends to coincide with the tangent of the curve at P.

The limiting value of δr/δt as δt tends to zero is dr/dt and is a vector *whose direction is that of the tangent at* P *in the sense in which t increases*. It is called the *time derivative of* r or *the differential coefficient* of r with respect to t. We thus have

$$\frac{dr}{dt} = \underset{\delta t \to 0}{Lt} \frac{\delta r}{\delta t}.$$

When this limit exists, the function r is said to be differentiable.

The second and third derivatives of r are respectively d^2r/dt^2 and d^3r/dt^3.

Clearly, δr represents the displacement of the particle in time interval δt and, therefore, δr/δt gives its *average velocity* during interval δt. The limiting value of this average velocity, as δt → 0, is the *instantaneous velocity* v of the particle. Thus, we have

$$\mathbf{v = dr/dt}$$

along the tangent to the path of the particle.

Proceeding in the same manner if, δv be the increase in the velocity v of the particle during the time-interval δt, the *rate of change of velocity or the average acceleration during the interval* = δv/δt *and therefore, instantaneous acceleration* a of the particle is the limiting value dv/dt of δv/δt as δt → 0. Thus,

$$a = \frac{dv}{dt} = \frac{d^2r}{dt^2}$$

Now, since r = xi + yj + zk and since x, y and z are functions of time, we also have

$$v = \frac{dr}{dt} = \frac{dx}{dt}i + \frac{dy}{dt}j + \frac{dz}{dt}k$$

and $$a = \frac{d^2r}{dt^2} = \frac{d^2x}{dt^2}i + \frac{d^2y}{dt^2}j + \frac{d^2z}{dt^2}k.$$

RADIAL AND TRANSVERSE COMPONENTS OF VELOCITY

As we have just seen above, v = dr/dt,

where $r = r\,\hat{r}$. with r standing for the magnitude of r and r for the *unit vector along its direction.* We, therefore, have

$$v = \frac{dr}{dt} = \frac{dr}{dt}\hat{r} + r\frac{d\hat{r}}{dt}.$$

The vector $\frac{dr}{dt}\hat{r}$ is called the *radial component of velocity* v of the particle, its magnitude being obviously dr/dt.

And the vector $r\frac{d\hat{r}}{dt}$ is called the *transverse component of velocity* v *of the particle, because it is perpendicular to* r, as may be seen from the following:

Since $\hat{r}$ is a *unit vector*, its scalar product with itself is equal to 1, *i.e.*, $\hat{r}\,\hat{r} = 1$. So that, differentiating with respect to t, we have

$$\frac{d\hat{r}}{dt}r + \hat{r}\frac{d\hat{r}}{dt}\ 2\hat{r}\frac{d\hat{r}}{dt} = 0. \text{ Or, } \hat{r}\frac{d\hat{r}}{dt} = 0,$$

indicating that $\hat{r}$ and $d\hat{r}/dt$ are perpendicular to each other.

And, since $\hat{r}$ lies along r, it follows that $d\hat{r}/dt$ and hence $r\,d\hat{r}/dt$ is perpendicular to r, its magnitude being obviously $r\left|\frac{dr}{dt}\right|$.

RADIAL AND TRANSVERSE VELOCITY AND ACCELERATION OF A PARTICLE MOVING IN A PLANE—CASE OF MOVING AXES

This is really a case of the axes themselves moving in the plane of motion of the particle, so that we obtain components of velocity and acceleration *along these moving axes.*

Let OX and OY be the reference system of rectangular axes in the plane of motion of the particle and OX' and OY' another system of rectangular coordinate axes moving relative to the former, with its position at any instant given by that of the moving particle at the time.

Components of Velocity

Let the position vector $r = \overrightarrow{OP}$ of the particle along OX' make an angle θ with OX (Fig. 1.26). Then, if i and j be the unit vectors along the axes OX and OY and i' and j' along the axes OX' and OY' respectively, we have r = xi + yj. Or, r = r cos θi + r sin θ'j.

Since r = ri', where r is the *magnitude* or vector r, we have

$$ri' = r\cos\theta i + r\sin\theta j,$$

$$\text{whence, } i' = \cos\theta i + r\sin\theta j. \qquad \text{...(i)}$$

Similarly, since vector $n = \overrightarrow{OQ} = nn'$, where n is the magnitude of n, we have $j' = -\cos(\pi/2 - \theta)i + \sin(\pi/2 - \theta)j$

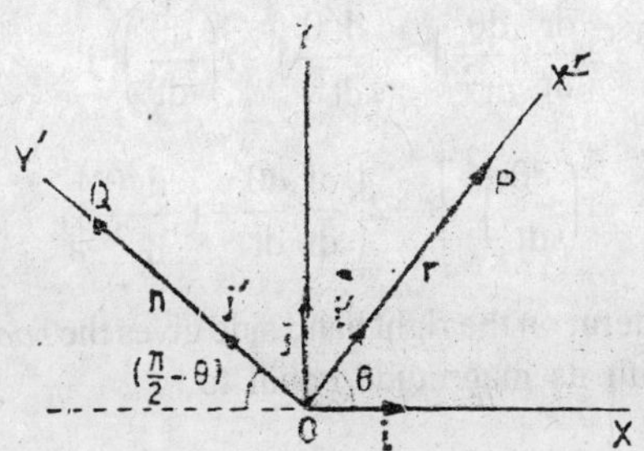

Fig. 1.26

$$= -\sin\theta\ i + \cos\theta\ j \qquad \text{...(ii)}$$

As we know, *velocity* $v = \frac{dr}{dt} = \frac{d}{dt}(r\cos\theta i + r\sin\theta j)$

$$= \left(\frac{dr}{dt}\cos\theta - r\sin\theta\frac{d\theta}{dt}\right)i + \left(\frac{dr}{dt}\sin\theta + r\cos\theta\frac{d\theta}{dt}\right)j$$

$$= \frac{dr}{dt}(\cos\theta i + \sin\theta j) + \frac{rd\theta}{dt}(-\sin\theta i + \cos\theta j).$$

Or, $$v = \frac{dr}{dt}i + r\frac{d\theta}{di}j. \qquad \text{...(iii)}$$

Thus, *radial component of* v is $\frac{dr}{dt}i'$, having a magnitude $\frac{dr}{dt}$ and transverse *component of* v is $r\frac{d\theta}{dt}j'$, having a magnitude $r\frac{d\theta}{dt}$.

Components of acceleration. As we know,

acceleration $$a = \frac{dv}{dt} = \frac{d}{dt}\left(\frac{dr}{dt}i' + r\frac{d\theta}{dt}j'\right)$$

$$= \frac{d^2r}{dt^2}i' + \frac{dr}{dt}\frac{di}{dt} + \frac{dr}{dt}\frac{d\theta}{dt}j' + r\frac{d^2\theta}{dt^2}j' + r\frac{d\theta}{dt}\frac{dj}{dt} \qquad \text{...(iv)}$$

Since from relations (i) and (ii) above, $\frac{di}{dt} = (\sin\theta i + \cos\theta j)\frac{d\theta}{dt}$ $= j\frac{d\theta}{dt}$ and $\frac{dj'}{dt} = (-\cos\theta i - \sin\theta j)\frac{d\theta}{dt} = -i'\frac{d\theta}{dt}$, we have from relation (iv),

$$a = \frac{d^2r}{dt^2}i' + 2\frac{dr}{dt}\frac{d\theta}{dt^2}j' + r\frac{d^2\theta}{dt^2}j' - r\left(\frac{d\theta}{dt}\right)^2 i'$$

Or $$a = \left[\frac{d^2r}{dt^2} - r\left(\frac{d\theta}{dt}\right)^2\right]i' + 2\left(\frac{dr}{dt}\frac{d\theta}{dt} + r\frac{d^2\theta}{dt^2}\right)j'. \qquad \text{...(v)}$$

Thus, the first term on the right hand side gives the *radial acceleration* of the particle, with its magnitude equal to

$$\frac{d^2r}{dt^2} - r\left(\frac{d\theta}{dt}\right)^2.$$

And, the second term gives the *transverse acceleration* of the particle, with its magnitude equal to

$$2\frac{dr}{dt}\frac{d\theta}{dt} + r\frac{d^2\theta}{dt^2} = \frac{1}{r}\frac{d}{dt}\left(r^2\frac{d\theta}{dt}\right).$$

DIFFERENTIAL COEFFICIENTS OF SUMS OF VECTORS

The differential coefficient of the sum A + B of two differentiable vectors, both of which are functions of t, is equal to the sum of their individual differential coefficients, as show below:

If, for an increase δt in the value of t, the corresponding increases in the values of A and B be δA and δB respectively, we have

$$\delta(A + B) = (A + \delta A + B + \delta B) - (A + B) = \delta A + \delta B$$

and hence the quotient $\dfrac{\delta(A+B)}{\delta t} = \dfrac{\delta A}{\delta t} + \dfrac{\delta B}{\delta t}.$

Or, taking the limiting values on either side as $\delta t \to 0$, we have

$$\frac{d}{dt}(A+B) = \frac{dA}{dt} + \frac{dB}{dt}.$$

What applies to the sum of two vectors applies equally well to the sum of any number of them.

DIFFERENTIAL COEFFICIENTS OF PRODUCTS OF VECTORS

As in the case of an algebraic product, so also here, the differential coefficient of a product of vectors in the sum of the quantities obtained by differentiating one single vector at a time, leaving the other unchanged.

(i) *Product of a scalar and a vector*: Let us consider the product ur of a vector r with a scalar u, both being differentiable functions of the variable t.

If, for the increase in the value of t from t to t + dt, the corresponding increases in the values of u and r be du and dr, we have

$\delta(ur) = (u + \delta u)(r + \delta r) - ur = \delta u r + u\delta r + \delta u \delta r.$

Dividing throughout by dt, we have

$$-\frac{\delta(ur)}{\delta t} = \frac{\delta u}{\delta t} r + u\frac{\delta r}{\delta t} + \frac{\delta u}{dt}\delta r.$$

Or, in the limit, $\frac{d}{dt}(ur) = \frac{du}{dt} r + u\frac{dr}{dt}.$ ...(i)

From relation (i) coupled with the results obtained earlier, it follows that *the components of the differential coefficient* (*or derivative*) *of a vector are the derivatives of its components for fixed directions.*

Thus, if vector r be expressed in terms of its rectangular component vectors, we have r = xi + yj + zk (where i, j and k are, as we know, constant vectors and x, y, z functions of time), and the differential coefficient or r is given by

$$v = \frac{dr}{dt} = \frac{dx}{dt} i + \frac{dy}{dt} j + \frac{dz}{dt} k,$$

with dx/dt, dy/dt and dz/dt respectively representing the magnitudes of the components of vector v, *here representing velocity.*

And, in the same manner, we have

acceleration $$a = \frac{dv}{dt} = \frac{d^2r}{dt^2} = \frac{d^2x}{dt^2} i + \frac{d^2y}{dt^2} j + \frac{d^2z}{dt^2} k,$$

where d^2x/dt^2, d^2y/dt^2 and d^2z/dt^2 represent respectively the magnitudes of the components of acceleration a.

(ii) *Scalar product of two vectors:* Let us now try to obtain the differential coefficient of a scalar product A.B. Proceeding as in case (i), if δA and δB be the increments in A and B respectively corresponding to an increase δt in t, we have increase in the scalar product given by

$\delta(A.B) = (A + \delta A).(B + \delta B) - A.B = \delta A.B + A.\delta B + \delta A.\delta B$
$= \delta A.B + \delta B.A,$

neglecting $\delta A.\delta B$ compared with other terms.

So that, dividing throughout by dt and proceeding to the limit $\delta t \to 0$, we have

$$\frac{d}{dt}(A.B) = \frac{dA}{dt}.B + A.\frac{dB}{dt},$$

where the order of the factors in any of the terms is quite immaterial.

(iii) Vector product of two vectors: Let us now consider the vector product $A \times B$. Proceeding exactly as above, we have

$\delta(A \times B) = (A + \delta A) \times (B + \delta B) - A \times B = \delta A \times B + A \times \delta B + \delta A \times \delta B.$

$= \delta A \times B + A \times \delta B,$

whence, as before, $\frac{d}{dt}(A \times B) = \frac{dA}{dt} \times B + A \times \frac{dB}{dt}$, *where the order of the factors in each term must* not *be changed unless accompanied also by change of sign.*

(iv) *Triple Products:* In the case of triple products too, we proceed as before to obtain the differential coefficients. Thus,

(a) in the case of a *scalar triple product* [ABC], we have

$$\frac{d}{dt}[ABC] = \left[\frac{dA}{dt}BC\right] + \left[A\frac{dB}{dt}C\right] + \left[AB\frac{dC}{dt}\right],$$

where the cyclic order in each term must be maintained.

And (b) in the case of a *vector triple product* $A \times (B \times C)$, we have

$$\frac{d}{dt}[A \times (B \times C)] = \frac{dA}{dt} \times (B \times C) + A\left(\frac{dB}{dt} \times C\right) + A \times \left(B \times \frac{dC}{dt}\right),$$

where, again, the order of the factors in each term must be maintained.

CIRCULAR MOTION

Let us try to obtain expressions for the velocity and acceleration of a particle P moving at a constant speed along a angular path in the x.y plane, say, of constant radius r (Fig. 1.27).

The position vector r, a function of the scalar variable t, moving with constant angular velocity (or angular frequency) of magnitude ω may, at any given instant t, be expressed in terms of its components along the axes of x and y. Thus, if i and j be the *unit vectors* along the two axes respectively, we have

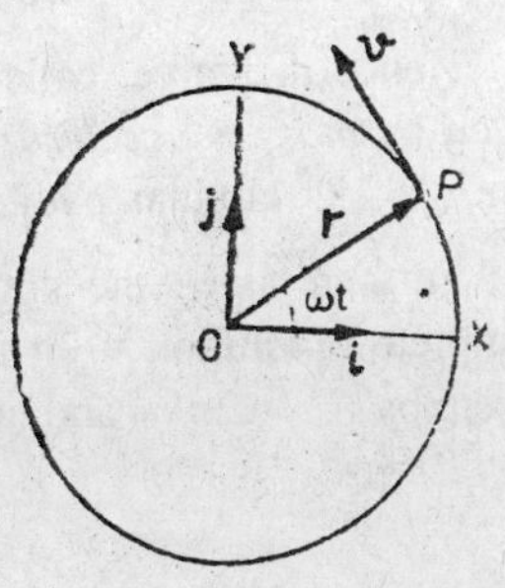

Fig. 1.27

$$r = xi + yj,$$

i.e., $r = r\cos\omega t i + r\sin\omega t j$

$$= r(\cos\omega t\ i + \sin\omega t\ j)$$

The velocity v of partiale P is thus given by

$$v = \frac{dr}{dt} = r\frac{dr}{dt} + \hat{r}\frac{dr}{dt}$$

where $\hat{r}$ is the unit vector along the direction or r.

Now, r being constant for the circular path, dr/dt = 0. So that, we have

$$v = r\frac{dr}{dt} = r\left(\frac{d}{dt}\cos\omega t\, i + \frac{d}{dt}\sin\omega t\, j\right)$$

Or, $v = r(-\omega\sin\omega t\ i + \omega\cos wt\ j)$

$$= \omega r(-\sin\omega t\ i + \cos\omega t\ j).$$

Therefore, with v as the magnitude of the velocity of P, we have

$$v^3 = v\,.v = \omega r(-\sin\omega t\ i + \cos\omega t\ j).\ \omega r(-\sin\omega t\ i + \cos\omega t\ j),$$

which, from our knowledge of scalar products and the fact that i.j = 0, gives

$$v^2 = \omega^2 r^3(\sin^2\omega t + \cos^2\omega t) = \omega^2 r^3, \text{ whence, } v = \omega r.$$

Now, since acceleration a is the rate of change of velocity, we have

$$a = dv/dt = \omega^2 r(-\cos\omega t\ i - \sin\omega t\ j) = -\omega^2 r(\cos\omega t\ i + \sin\omega t\ j)$$

And since $r(\cos\omega t\ i + \sin\omega t\ j) = r$, we have

$$a = -\omega^2 r.$$

Or, the *magnitude of the acceleration of particle* P, *i.e.*,, $a = \omega^2 r$, its direction being—r or *towards the centre of the, circle*—hence the name centripetal (*i.e.*, centre-seeking) acceleration given to it.

Or, since $\omega r = v$, we may also put $a = r\ (v/r)^2 = v^2/r$.

Hence the force, called *centripetal force*, acting on the particle is given by *mass* × *acceleration* = mv^2/r and is also directed towards the centre of its circular path.

Further, since the angle swept out by the radius vector in one complete revolution = 2π radians, we have *time period* of the particle given by $T = 2\pi/\omega$ and its frequency by $n = 1/T = \omega/2\pi$, whence, $\omega = 2\pi n$.

ANGULAR VELOCITY VECTOR

The angular velocity of a particle, having magnitude as well as direction, is obviously a *vector.* To obtain its value, we note that the linear velocity v of the particle is, at any instant, perpendicular to both the angular velocity ω and the radius vector r. So that, the vector equation corresponding to the relation $v = \omega r$ comes out to be $v = \omega \times r$, showing that *the linear velocity of the particle is the cross product of the angular velocity vector and the position vector with respect to a fixed point on the axis of rotation.*

The cross product of this with r gives

$$r \times v = r \times (\omega \times r) = \omega\ (r.r) - r\ (r.\omega) = r^2\omega,$$

(because $r.\omega = 0$, r and ω being perpendicular to each other),

whence, $\omega = r \times v/r^2$ and can be easily evaluated, its direction being parallel to the axis of rotation in the positive sense relative to the direction of rotation (*i.e.*, in accordance with the right-handed screw rule).

SCALAR AND VECTOR FIELDS

A physical quantity, expressed as a continuous function of the position of a point in a region of space is referred to as *point function* (or *function of position*). The region of space concerned is known as a *scalar field* or a *vector field* according as the physical quantity in question is a scalar or a vector one, expressed as a continuous single-valued function, ϕ (x, y, z) or F (x, y, z) respectively, of position in that region. In actual practice, *the two functions of positions themselves are spoken of as the scalar and vector fields respectively.*

As examples of scalar fields may be mentioned the distribution of temperature or magnetic and electrostatic potentials in a given region of space. The field here consists of a series of surfaces, (isothermal or equipotential, as the case may be), each having a fixed and definite value of temperature or magnetic or electrostatic potential, *i.e.*, of the function ϕ (x, y, z). Such surfaces are referred to as *level surfaces*, each level surface having its own constant value of ϕ (x, y, z) all over.

And, as examples of vector fields, we have distribution of magnetic and electric intensity or the distribution of velocity in a moving (continuous) fluid. At any given point, in this case, the single-valued vector function F (x, y, z) is specified by a vector having a definite magnitude and direction, both of which continuously change from point to point throughout the field.

Starting from any desired point in the field and proceeding through infinitesimal distances from point to point in the direction of the field, we obtain, in general, a curved line, called the *line of flow*, the *flux line* or the *vector line,* the tangent to which at any point gives the direction of the vector at that point. The measure or magnitude of the vector is given by the number of flux lines passing per unit area of a surface normal to their direction.

PARTIAL DERIVATIVES—GRADIENT

Partial derivatives. The differentiation of a function such as $\phi(x_1, x_2, x_3...)$ of two or more independent variables, x_1, x_2, x_3 etc., with respect to one of them, keeping the others constant, is called *partial differentiation.* The derivatives thus obtained are referred to as *partial derivatives* and are denoted by the symbol ∂ instead of d.

Thus, if we have a continuously differentiable scalar point function ϕ (x, y, z), *i.e.*, a function of the coordinates x, y, z its partial derivatives along the three coordinate axes are

$$\frac{\partial\phi}{\partial x}, \frac{\partial\phi}{\partial y} \text{ and } \frac{\partial\phi}{\partial z}$$

Gradient. The vector function $i\frac{\partial\phi}{\partial x} + j\frac{\partial\phi}{\partial y} + k\frac{\partial\phi}{\partial z}$ is called the *gradient* of the scalar point function ϕ and is denoted by grad ϕ.

Thus,

$$\text{grad}\,\phi = i\frac{\partial\phi}{\partial x} + j\frac{\partial\phi}{\partial y} + k\frac{\partial\phi}{\partial z}.$$

$$\therefore \quad \text{differential } d\phi = \frac{\partial\phi}{\partial x}dx + \frac{\partial\phi}{\partial y}dy + \frac{\partial\phi}{\partial z}dz$$

$$= (idx + jdy + kdz)\left(i\frac{\partial\phi}{\partial x} + j\frac{\partial\phi}{\partial y} + k\frac{\partial\phi}{\partial z}\right)$$

Or, $$d\phi = dr \text{ grad } \phi.$$

It can easily be show that grad ϕ at any point is quite independent of the choice of the coordinate axes. It follows, therefore, that the *gradient of a scalar point function is a vector point function.*

If ϕ be a constant, obviously the partial derivatives $\partial\phi/\partial x$, $\partial\phi/\partial y$ and $\partial\phi/\partial z$ will all be zero and hence grad $\phi = 0$. And, converse, if grad $\phi = 0$, the partial derivatives are all zero and hence function ϕ is a constant. Thus, *grad $\phi = 0$ only if ϕ be constant.*

THE OPERATOR ∇

We have in vector algebra a *differential operator,* denoted by ∇ (*i.e.*, inverted Δ). It is called 'del' or 'nabla'. *It operates distributively and is formally assumed to have the character of a vector.* And since it is assumed to have the character of a vector, *its product $\nabla\phi$ with a scalar ϕ is a vector.*

If we use Cartesian orthogonal coordinates, we may write

$$\nabla = i\frac{\partial}{\partial x} + j\frac{\partial}{\partial y} + k\frac{\partial}{\partial z}.$$

So that, $$\text{grad}\phi = i\frac{\partial\phi}{\partial x} + j\frac{\partial\phi}{\partial y} + k\frac{\partial\phi}{\partial z}$$

$$= \left(i\frac{\partial}{\partial x} + j\frac{\partial}{\partial y} + k\frac{\partial}{\partial z}\right)\phi = \nabla\phi, \text{a vector.}$$

MAGNITUDE AND DIRECTION OF $\nabla\phi$

Let us consider two level surface L_1 and L_2 through two close points A and C, distance dr apart (Fig. 1.28), with the values of the scalar function ϕ and $(\phi + d\phi)$ respectively. And let AB = dn be the normal to the surface L_1 at A.

Since L_2 is a level surface, the value of the scalar function is the same [*viz.*, $(\phi + d\phi)$] at B as at C. Obviously, therefore, the *rate of change of* ϕ along the normal AB is the highest (this being the shortest distance between the two surfaces) and is equal to $\partial\phi/\partial n$.

Now, we may put $dn = dr\cos\theta = \hat{n}\,dr$, where $\hat{n}$ is the unit vector normal to the surface L_1 at A. So that,

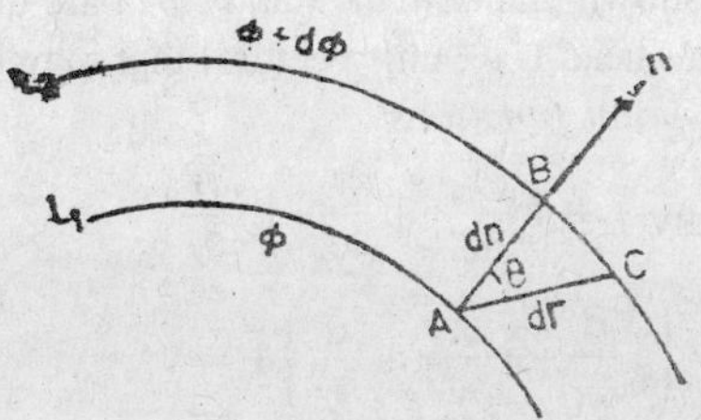

Fig. 1.28

$$d\phi = \frac{\partial \phi}{\partial n} dn = \frac{\partial \phi}{\partial n} \hat{n}.dr.$$

As we know, $\nabla\phi.dr = \left(i\frac{\partial \phi}{\partial x} + j\frac{\partial \phi}{\partial y} + k\frac{\partial \phi}{\partial z} \right) (idx + jdy + kdz)$

$$= \frac{\partial \phi}{\partial x} dx + \frac{\partial \phi}{\partial y} dy + \frac{\partial \phi}{\partial z} dz = d\phi.$$

Or, $$d\phi = \nabla\phi.dr = \frac{\partial \phi}{\partial n} \hat{n} dr,$$

whence, $$\nabla\phi = \left(\frac{\partial \phi}{\partial n} \right) \hat{n},$$

i.e., vector $\nabla\phi$ or grad ϕ has a magnitude equal to the maximum rate of change of ϕ and is directed along this maximum rate of change, *i.e.*, along the normal to the level surface L_1 (with ϕ constant).

DIVERGENCE AND CURL

A scalar point function ϕ we can obtain a vector point function grad ϕ. Similarly, from a vector point function f we can obtain two more point function: a scalar one, called *divergence* and a vector one, called *curl* (or *rotation*).

Thus, if f represents a vector field, *i.e.,* if f be a *continuously differentiable vector point function*, the function

$$i.\frac{\partial f}{\partial x} + j.\frac{\partial f}{\partial y} + k.\frac{\partial f}{\partial z},$$

which is a *scalar,* is called *divergence of* f and is written as div f. And the function

$$i\times\frac{\partial f}{\partial x} + j\times\frac{\partial f}{\partial y} + k\times\frac{\partial f}{\partial z},$$

which is a *vector*, is called *curl of* f and is written as *curl* f.

It can easily be shown that both *div* f and *curl* f are quite independent of the orthogonal unit triad (*i.e.*, unit vectors) that may be chosen. They are thus *essentially point functions.*

Now, clearly, div $f = i.\frac{\partial f}{\partial x} + j.\frac{\partial f}{\partial y} + k.\frac{\partial f}{\partial z}$,

$$= \left(i\frac{\partial}{\partial x} + j\frac{\partial}{\partial y} + k\frac{\partial}{\partial z}\right).f = \nabla f.$$

So that, *divergence of a vector point function* f *is the scalar product of the Del operator with* f. *And,*

$$\text{Curl } f = i\times\frac{\partial f}{\partial x} + j\times\frac{\partial f}{\partial y} + k\times\frac{\partial f}{\partial z} = \left(i\frac{\partial}{\partial x} + j\frac{\partial}{\partial y} + k\frac{\partial}{\partial z}\right)\times f = \nabla\times f,$$

i.e., curl of a vector point function f *is the vector product of the Del operator with* f. The term $\nabla\times$, [read as *Del cross* (*or Nabla cross*)], is therefore also used in place of the term '*curl*'.

Both divergence and curl of f may be expressed in terms of the components of vector f, as shown below:

if f = i fx + j fy + k fz, we have

$$\text{div } f = \sum i.\frac{\partial f}{\partial x} = \sum i\left[i\frac{\partial fx}{\partial x} + j\frac{\partial fy}{\partial x} + k\frac{\partial fz}{\partial x}\right]$$

$$= \frac{\partial fx}{\partial x} + \frac{\partial fy}{\partial y} + \frac{\partial fz}{\partial z}.$$

$$\text{And curl } f = \sum i\times\frac{\partial f}{\partial x} \sum i\times\left[i\frac{\partial fx}{\partial x} + j\frac{\partial fy}{\partial y} + k\frac{\partial fz}{\partial z}\right]$$

$$= i\left(\frac{\partial fz}{\partial y} - \frac{\partial fy}{\partial z}\right) + j\left(\frac{\partial fx}{\partial z} - \frac{\partial fz}{\partial x}\right) + k\left(\frac{\partial fy}{\partial x} - \frac{\partial fx}{\partial y}\right)$$

$$= \begin{vmatrix} i & j & k \\ \frac{\partial}{\partial x} & \frac{\partial}{\partial y} & \frac{\partial}{\partial z} \\ f_x & f_y & f_z \end{vmatrix}$$

APPLICATIONS OF DIVERGENCE AND CURL

(i) *Rate of flow of flux* (*or a fluid*). Let v be a vector point function (or the value of the vector field) representing the *velocity* of a fluid at any given instant t at a point P inside a small parallelopiped, *with edges* δx, δy, and δz *parallel to the three coordinate axes*, as shown in Fig. 1.29.

Let $\quad v = v_x i + v_y j + v_z k.$

Then, *velocity component along the x-axis at any point of the face* ABCD *normal to this axis*

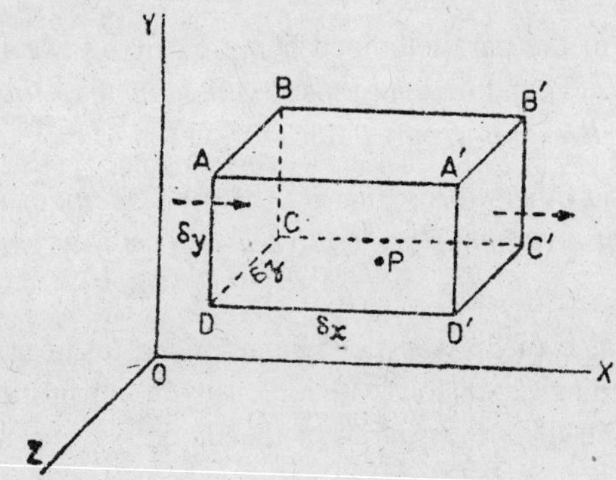

Fig. 1.29

$$= v_x - \frac{1}{2}\frac{\partial vx}{\partial x}\delta x,$$

neglecting the second and the higher powers of δx.

∴ *flux* (*or fluid*) *entering per second through face* ABCD *of the parallelopiped*

= (*component of velocity normal to the face*) (*area of the fact*)

$$= \left(Vx - \frac{1}{2}\frac{\partial V_x}{\partial x}\delta x\right)\delta y\delta z.$$

Similarly, *flux* (*or fluid*) *leaving per second through the opposite face* A' B' C' D' *of the parallelopiped*

$$= \left(xx + \frac{1}{2}\frac{\partial v_x}{\partial x}\delta x\right)\delta y\delta z.$$

Thus, *volume of the flux* (*or fluid*) *passing per second through faces* ABCD *and* A' B' C' D' *of the parallelopiped, i.e., along the direction of x*

$$= \left(vx + \frac{1}{2}\frac{\partial v_x}{\partial x}\delta x\right)\delta y\,\delta z - \left(v_z - \frac{1}{2}\frac{\partial v_z}{\partial x}\delta x\right)\delta y\,\delta z = \frac{\partial v_x}{\partial x}\partial x\,\partial y\,\partial z$$

Considering the other faces of the parallelopiped also, we have

total volume of flux (or fluid) moving out of the parallelopiped per second

$$= \left(\frac{\partial v_x}{\partial x} + \frac{\partial v_y}{\partial y} + \frac{\partial v_z}{\partial z} \right) \delta x \; \delta y \; \delta z$$

$$= (\nabla \; . \; v) \; \delta x \; \delta y \; \delta z \text{ or } (\text{div } v) \; \delta x \; \delta y \; \delta z.$$

The volume of the parallelopiped being $\delta x\ \delta y\ \delta z$, we find that in the limit when δx, δy, and δz tend to zero, *the volume of flux (or fluid) flowing past (or diverging from) point* (x, y, z) is ∇v or div v.

This means, in other words, that *if* v *be the velocity of the fluid at a point, its rate of divergence per unit volume from that point is given by div* v or ∇v.

A positive value of div v many be interpreted to mean either that the fluid is undergoing expansion (with its density falling proportionately) or that the point itself is the *source* of the fluid.

Similarly, a negative value of div v may be interpreted to mean the reverse, *viz.*, that either the fluid is undergoing contraction (with its density rising proportionately) or that the point serves as a sink (or a *negative source)* of the fluid.

If, however, div v = 0, the fluid entering and leaving the element (or the parallelopiped) is the same. *i.e.*, there is no change in the density of the fluid indicating that it is *incompressible.* In this case, therefore, if ρ be the density of the fluid, we have *mass of the fluid per unit volume that flows through a point* (x, y, z) = ρ div v.

(ii) *Solenoidal vector point function, A vector point function* F *is said to be solenoidal is a region if its flux across any closed surface in that region be zero,* which will happen only when div F is equal to zero.

This means that *for a vector point function to be solenoidal, either the lines of flow of its flux should form closed curves (like, for example, the lines of force in the magnetic field of an electric current) or extend to infinity*

It can be shown that every solenoidal vector function is the curl of some function. Further, as shall see in the next article, the divergence of every curl is zero. It follows, therefore, that *the curl of every function is necessarily solenoidal.*

SOME USEFUL RESULTS

The gradient of a scalar point function ϕ is a vector point function and so is the curl of a vector point function f. We may also, therefore, obtain their divergence and curl, as given below:

(i) div gradϕ = $\nabla.\nabla\phi = \frac{\partial^2\phi}{\partial x^2} + \frac{\partial^2\phi}{\partial y^2} + \frac{\partial^2\phi}{\partial z^2} = \nabla^2$.

where ∇^2 is the *Laplacian operator* (and a *scalar*).

(ii) curl grad $\phi = \nabla \times \nabla\ \phi = 0$, because the cross product of two equal vectors is zero, and here both the vectors or ∇'s are equal since they operate on the same function ϕ.

It follows from this that if *we have a vector field* f *such that its curl is zero, i.e.,* $\nabla \times \phi = 0$, f *is certainly the gradient of some scalar field* ϕ, *i.e.*, if $\nabla \times$ f = 0, we have $\phi = \nabla\phi$.

(iii) div *curl* f = $\nabla.\nabla \times \phi = 0$, because, as we know, it is a combination similar to A.A × B, where A × B is perpendicular to A and has no component in the direction of A.

It follows, therefore that *if the divergence of a vector field* D *is zero, then,* D *is clearly the curl of some vector field f, i.e., if* ∇ D = 0, we have D = $\nabla \times \phi$.

(iv) curl f = $\nabla \times$ (D × f) = $\nabla\nabla$.f – ∇^2f = grad div f – ∇^2f.

We may have two more combinations, *viz.*,

(v) $\nabla(\nabla$.f) and (vi) (∇'∇) f = ∇^2f. There is, however, nothing special about them. They represent just vector fields which we may possibly come across sometimes.

THE LAPLACIAN OPERATOR

If ∇ be the Del or Nabla operator, we have

$$\nabla^2 = \nabla\nabla = \left(\sum i\frac{\partial}{\partial x}\right).\left(\sum i\frac{\partial}{\partial x}\right) = \frac{\partial^2}{\partial x^2} + \frac{\partial^2}{\partial y^2} + \frac{\partial^2}{\partial z^2},$$

where ∇^2 is referred to as the *Laplacian operator*. Being the dot or scalar product of two vectors (∇), it is a *scalar operator* and is also called *del squared or Nabla squared* for the obvious reason that ∇ is a *del operator*. Above could also be spoken of as *del squared* (or *Nabla squared*) ϕ.

INTEGRATION OF VECTORS

(i) *Line integral*: Let dl be an element of length at a point on a smooth curve AB drawn in a vector field and F, a continuous vector point function, or vector, inclined at an angle θ to dl, as shown in Fig. 1.30. Such that it continuously varies in magnitude as well as direction as we proceed along the curve. Then, the integral

$$\int_A^B F.dl = \int_A^B F\cos\theta dl$$

is referred to as the line integral of vector F along the curve AB.

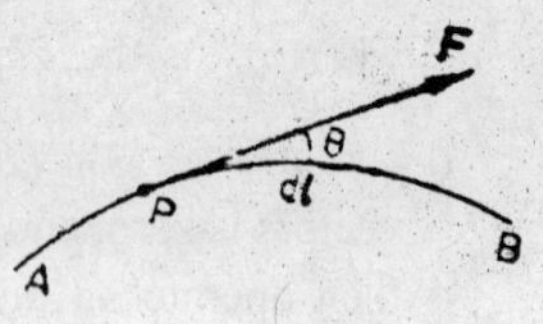

Fig. 1.30

In terms of the components of F along the three Cartesian coordinates, we have

$$\int_A^B F\,dl = \int_A^B \left(F_x\,dx + F_y dy + F_z dz\right)$$

Thus, if F represents the force acting on a particle moving along the curve from A to B, the line integral $\int_A^B F.dl$ represents the total work done by the force during the motion of the particle over its entire path from A to B.

If the value of the line integral depends only upon the location of the two points in the vector field and not upon the actual path taken between them, the vector field is referred to as a conservative field. Familiar examples of such fields are the electrostatic, magnetic and gravitational fields.

If, therefore, F represents the value of the electric (or magnetic) field intensity at the point P, the line integral $\int_A^B F.'dl$ represents the work done on unit charge (or unit pole) during its motion from A to B (*i.e.*, the potential difference between A and B) *irrespective of the path taken.*

In Hydrodynamics, the line integral of a continuous vector point function or vector (F) along a closed curve is called the circulation of F along the curve. And, if the circulation of a vector point function along every closed curve in a region be zero, it is said to be irrotational in that region.

Relationship between line integral and curl. There is a definite relationship between the curl of a vector field F at a point and its line integral along the boundary C of an infinitesimal plane area around and including that point. To clearly bring this out, let us, for convenience, calculate the line integral of F around an infinitesimal plane rectangular area ABCD, of sides AB = δx and BC = δy in the x = y plane (Fig. 1.31) surrounding and including a point P.

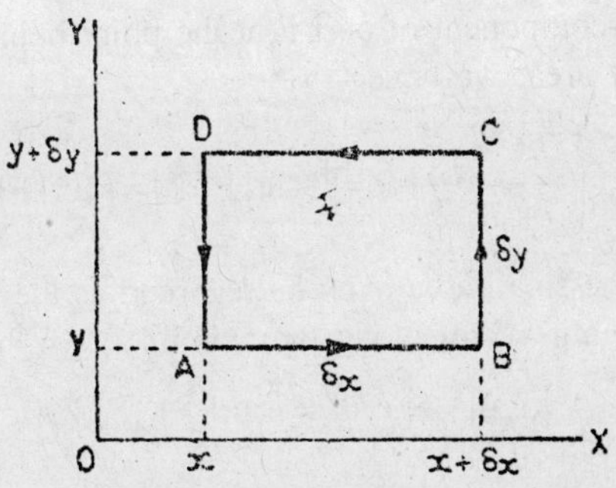

Fig. 1.31

Let the components of F along the axes of x and y be F_x and F_y respectively. Then, since the sides of the rectangle are small, the average values of these components along them may respectively be taken to be the same as those at their mid-points. So that, we have

average value of x-component along the path AB $= F_x + \frac{\partial F_x}{\partial x}.\frac{\delta x}{2}$

and average value of x-component along the path CD

$$= F_x + \frac{\partial F_x}{\partial x}.\frac{\delta x}{2} + \frac{\partial F_s}{\partial y}\delta y.$$

Remembering that the path CD is oppositely directed to the x-axis, we have *line integral of* F *along* AB + *line integral of* F *along* BC given by

$$\int_A^B F\,dl + \int_C^D F.dl = \left(F_x + \frac{\partial F}{\partial x}\partial x\right)\delta x - \left(F_x + \frac{\partial F_x}{\partial x}\frac{\delta x}{2} + \frac{\partial F_x}{\partial y}\partial y\right)\delta x$$

$$= \frac{F_x}{\partial y}\partial x\delta y \quad \text{...(i)}$$

Proceeding exactly in the same manner and remembering that DA is oppositely directed to the y-axis, we have

line integral of F *along* BC + *line integral of* F *along* DA *given by*

$$\int_B^C F.dl + \int_D^A F.dl = \frac{\partial F_y}{\partial y}\partial x\,\delta x\,\partial y \quad \text{...(ii)}$$ Thus,

adding relation (i) and (ii), we have

line integral over the entire boundary C the rectangle ABCD given by

$$\oint_{ABCD} F.dl = \oint_{C} F.dl = \left(\frac{\partial F_x}{\partial x} - \frac{\partial F_x}{\partial y}\right)\delta x \delta y \quad ...(iii)$$

Now, the z-component of curl F at the point being the above line integral per unit area, we have

$$\left(\frac{\partial F_y}{\partial x} - \frac{\partial F_x}{\partial y}\right) = \left[(\text{curl } F).k\right] = (\text{curl } F)_z$$

So that, if the small area $\delta x \delta y$ be regarded as the vector area $\delta_{zs} k$ (where k is the unit vector along the axis of z), we have

$$\oint_{ABCD} F.dl = \oint_{C} F.dl = (\text{curl } F)_z . \delta_{sz} k.$$

It follows, therefore, that if we have an infinitesimal plane area at the point in question given by $\partial s = \delta_{sx} i + \delta_{sz} j + \delta_{sz} k$, with a boundary C and ted in any direction, we have

$$\oint_{C} F.dl = \text{curl } F.\delta s,$$

the maximum of $\oint F.dl$ being $|\text{curl } F|\ \delta s$.

We thus see that *the magnitude of curl* F *at a point is the maximum value of the line integral of* F *per unit area along the boundary* C *of an infinitesimal plane area* (δs) *at that point.* And the direction of the curl is perpendicular to the plane of the infinitesimal area (when the value of the line integral is the line integral is the maximum) in accordance with the right-handed screw rule.

It has been mentioned earlier that the curl of a conservative vector field is zero at all points in space. This may easily be seen from the following:

The line integral between any two points P and Q in the case of a conservative vector field is, as we know, a constant property of those points, irrespective of the path taken between them. It follows, therefore that if we choose two different paths I and II between the two points, as indicated in Fig. 1.32, we shall have

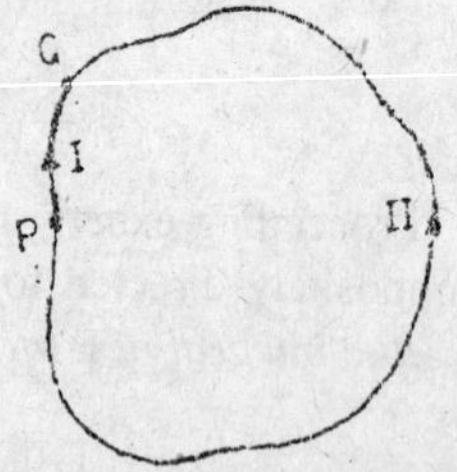

Fig. 1.32

$$\int_P^Q \mathbf{F}.d\mathbf{l} = \int_P^Q \mathbf{F}.d\mathbf{l}$$

Path I Path II

If Q be brought nearer and nearer to P, path I becomes shorter and in the limit when Q is made to coincide with P, path I is reduced to zero and hence the integral along this, or any other path too, is reduced to zero. Thus the line integral of F along any closed path (also called *circulation of field* F *around the closed path*), beginning and ending at the same point, is zero in the case of a conservative field, *i.e.*,

$$\oint_{\text{Closed path}} \mathbf{F}.d\mathbf{l} = 0,$$

Since this relation holds good for all conservative fields irrespective of the closed path chosen these field must have zero curl at all points in space. That is why they are referred to as curl-free or non-curl fields.

(ii) *Surface integral*: Imagine a smooth surface S (Fig. 1.33) drawn in a vector field and a continuously varying vector point function or vector F at a point P in a small element dS of the surface, at an angle θ with the normal to the surface, at the point (drawn outwards if the surface be closed and always towards the same side otherwise). Then, the integral $\iint_S \mathbf{F}.\mathbf{S}dS = \iint_S F \cos\theta \, dS$ over the entire surface is called the *normal surface integral,* or generally, simply the *surface integral* of vector F over the surface.

In terms of the Cartesian components of F, we have

$$\iint_S \mathbf{F}.d\mathbf{S} = \iint \left(F_x dS_x + F_y dS_y + F_z dS_z\right)$$

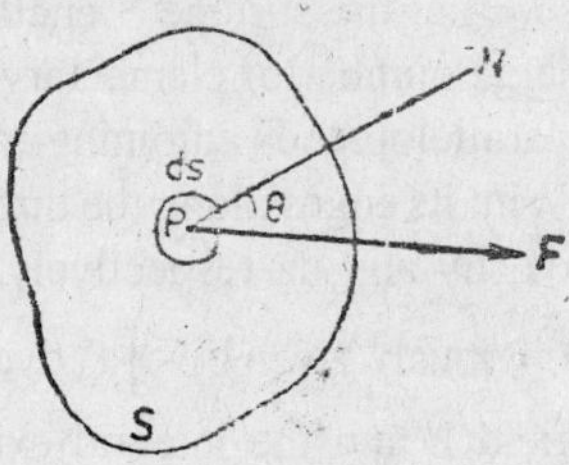

Fig. 1.33

The surface integral of the normal component of a continuous vector point function F over a closed surface S is called the *flux of* F across the surface. As we have seen before (ii)], if the flux of a vector point function across every closed surface in a region be zero, it is said to be *solenoidal* in that region.

Example:

(i) if F represents the electric or magnetic induction at the point P the surface integral $\iint_S \mathbf{F}.d\mathbf{S}$ *represents the total normal induction over the surface.*

(ii) If the surface S be drawn over a region of a moving or a flowing fluid such that its velocity v varies from point the surface integral of v. viz., $\iint_S \mathbf{v}.d\mathbf{S}$ *gives the volumes rate of the fluid cross the surface.*

(iii) Volume integral. Suppose we have a surface enclosing a region of volume V and that F is a vector point function at a point in a small element dV of the region. Then, the integral $\iiint_V \mathbf{F}\, dV$, *covering the entire region, is called the volume integral of vector F over the surface.*

In terms, of the Cartesian components, we have

volume integral $\iiint_V \mathbf{F}\, dV = \mathbf{i}\iiint \hat{V}\, F_x dxdydz + \mathbf{j}\iiint \hat{V}\, F_y dxdydz$

$$+ \mathbf{k}\iiint \hat{V}\, F_z dxdydz.$$

GAUSS'S THEOREM OF DIVERGENCE

The theorem states that the *normal surface integral of a function* F *over the boundary of a closed surface* S (*i.e., the flux across* S) *is equal to the volume integral of the divergence of the function over the volume* V *enclosed by the surface, i.e..,*

$$\text{flux} = \iint_S \mathbf{F}.d\mathbf{S} = \iiint \hat{V}\ \text{div}\ \mathbf{F}dV = \iiint \hat{V}\ (\nabla.\mathbf{F})dV.$$

This may be easily shown as follows:

Let the surface S enclosing a volume V be divided up into a very large number of elementary volumes in the form of cubes or rectangular parallelopipeds adjoining each other. Imagine one such cube (Fig. 1.34). With its edges along the three Cartesian coordinate axes and their lengths dx, dy and dz respectively. The flux outwards through the left face is obviously given by $-\int F_x dydz$, where F_x is the x-component of the vector field F and the integral extends over the axes of the face.

Since the cube considered is an infinitesimal one this integral-may be taken to be very nearly equal to the product of the x-component of F at the centre P_1 of the face and the area dydz of the face. So that, denoting the x-component of F at the centre of the face by $F_x(P_1)$, we have

flux through left face of the cube $= -F_x(P_1)\, dydz$

Similarly, *flux through right face of the cube* $= F_x(P_2)\, dydz$, the component $F_x(P_2)$ now being positive.

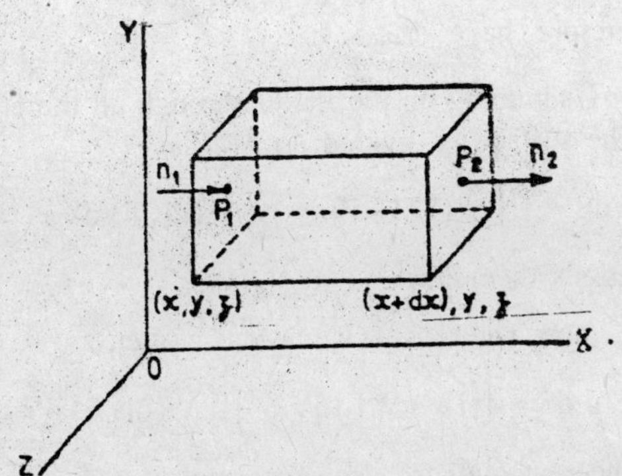

Fig. 1.34

Taking $F_x(P_2) = F_x(P_1) + \frac{\partial F_x}{\partial x}dx$, we have

flux through right face of the cube $= \left[F_x(P_1) + \frac{\partial F_x}{\partial x}dx\right]dydz.$

$\therefore$ *flux through the left and right faces*

$$= \left[F_x(P_1) + \frac{\partial F_x}{\partial x}dx\right]dydz - F_x(P_1)dydz = \frac{\partial F_x}{\partial x}dxdydz.$$

Similarly, *flux through top and bottom faces of the cube*

$$= \frac{\partial F_y}{\partial y}dxdydz$$

and *flux through the remaining faces of the cube*

$$= \frac{\partial F_z}{\partial z}dxdydz.$$

Therefore, flux through all the faces of the cube is given by the sum of all these. So that,

$$\iint\limits_{\substack{\text{Surface}\\\text{of cube}}} F\,dS = \left(\frac{\partial F_x}{\partial x} + \frac{\partial F_y}{\partial y} + \frac{\partial F_z}{\partial z}\right)dxdydz.$$

where dS is the area of the surface of the cube.

Now $\frac{\partial F_x}{\partial x} + \frac{\partial F_y}{\partial y} + \frac{\partial F_z}{\partial z} = \nabla.F$ or div F

and $dxdydz = dV$, volume of the cube.

We, therefore, have $\iint_{\substack{\text{Surface}\\ \text{of cube}}} F\,dS = (\nabla.F)\,dV.$

Therefore, summing up the fluxes through all the elemental cubes into which the surface is divided up, we have

$$\iint_S F dS = \iiint \hat{V}\,(\nabla.F) dV = \iiint \hat{V}\,(\text{div}\,F)\,dV,$$

which is Gauss's theorem.

Perhaps, more rigorously, it may be written as

$$\iint_S F.n\ dS = \iiint \hat{V}\,(\nabla.F) dV = \iiint \hat{V}\,(\text{div}\,F)\,dV,$$

where n is the *unit outward drawn normal vector.*

It may be noted that $\iint_S F\,dS$ gives the surface integral of F over the area of the bounding surface S enclosing volume V, because surfaces of the elemental cubes other than those forming part of surface S itself are common to two adjacent cubes and their surface integrals thus all cancel out in view of their oppositely directed normals.

Obviously, since $F.dS = F_x dS_x + F_y dS_y + F_z dS_z = F_x dydz + F_y dzdx + F_z dxdy$, we can express *Gauss's* theorem in terms of the Cartesian coordinates as

$$\iint_S \left(F_x dS_x + F_y dS_y + F_z dS_z\right) = \iiint \hat{V} \left(\frac{\partial F_x}{\partial x} + \frac{\partial F_y}{\partial y} + \frac{\partial F_z}{\partial z}\right) dxdydz.$$

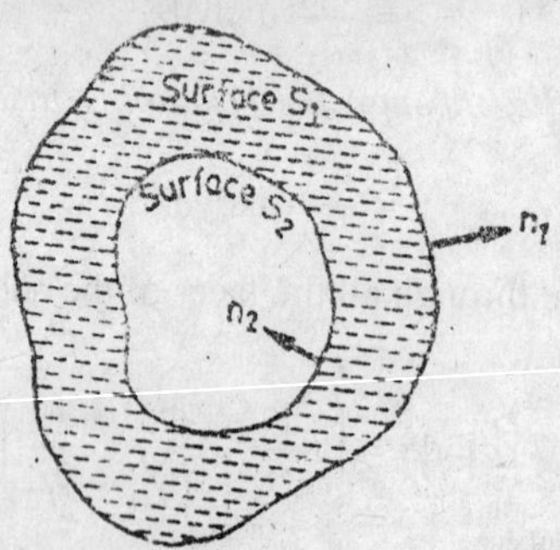

Fig. 1.35

It may as well be mentioned here that Gauss's theorem applies also to a volume bounded by two closed surface (and, indeed by any number of closed surfaces). Only, it must be remembered that the outward drawn normals (n_1 and n_2) at any points on the surfaces concerned must point away from the enclosed volume, as shown.

STOKES' THEOREM

This theorem states that *the line integral (or the circulation) of a vector field* F *around any closed curve* C *is equal to the normal surface integral of the curl of* F *taken over any surface* S *of which curve* C *forms the contour or the boundary, i.e.,*

$$\oint_C F dl = \iint_S \text{curl } F.dS = \iint_S (\nabla \times F).dS.$$

This may be shown as follows:

Let a closed curve forming the boundary of a surface S of whatever shape (Fig. 1.36) be divided up into a very large number of small loops, infinitesimal in size and squarish in shape, all lying on the surface. These loops enclose small areas dS_1, dS_2 etc. inside them which are all flat, in view of the small size of the loops. Considering one such small squarish loop enclosing area dS_1, we have

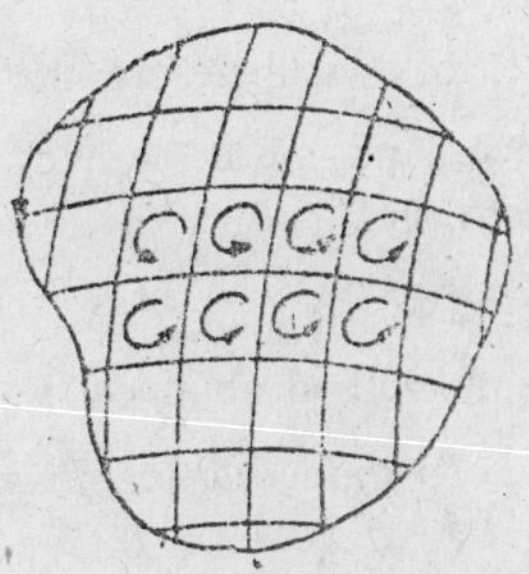

Fig. 1.36

line integral (or circulation) of vector F *around it, i.e.,*

$\oint F dl = \text{curl } F dS_1$, the direction of the curl being perpendicular to the plane of area dS in accordance with the right-handed screw rule.

Now, if we sum up the left hand side, *i.e.*, the circulation around all these small loops into which the curve has been divided, we find that the contributions due to all the bounding lines of the loops, other than those that form part of the curve itself, cancel out since these contributions are all in the form of equal and oppositely directed pairs. So that, the summation simply gives the circulation or the line integral of the vector around the original closed curve alone.

And, summation on the right hand side of the expression obviously gives the surface integral of *curl* F over the entire surface S closed by the curve. So that, we have

$\oint F dl = \iint_S \text{curl } F.dS$, which is the required theorem.

APPLICATIONS OF GAUSS'S AND STOCKES' THEOREMS

1. *To show that the curl of the gradient of a scalar function ϕ is equal to zero, i.e.,* $\nabla \times \Delta \phi = 0$. We know that the line integral of the gradient $\nabla \phi$ of the function along a closed loop is zero, *i.e.*,

$$\oint_{loop} \nabla\phi . \, dl = 0.$$

Now in accordance with Stokes' theorem, this line integral of the vector $\nabla \phi$ along a closed loop is equal to the surface integral of the curl of the vector over the surface B bounded by the loop, *i.e.*,

$$\oint \nabla\phi \, dl = \iint \hat{S} \, \nabla \times (\nabla\phi) \, dS.$$

It follows, therefore, that the integral $\iint \hat{S} \, \nabla \times (\nabla\phi) dS = 0$ over any surface. And, since the integral is zero, the integrand must-necessarily be zero, *i.e.*,

$$\nabla \times \nabla\phi = 0,$$

a result we had obtained by vector algebra earlier.

2. *To show that the divergence of the curl of a vector function f is zero. i.e.,* $\nabla . \nabla \times f = 0$. We know that according to Stocke' theorem, the normal surface integral of the curl of a vector f over a surface is equal to the line integral of the vector over the closed loop that forms the boundary of the surface. It follows, therefore that as the boundary, or the loop, decreases, *i.e.*, as the surface becomes more and more closed, (Fig. 1.37), the line integral of the vector around the loop, and hence the normal surface integral of the curl of the vector over the surface. Must progressively decrease until, when the loop decreases to a mere point, *i.e.*, the surface becomes closed, the line and the surface integrals become zero. *Thus, for a closed surface.*

$$\iint_S (\nabla \times f) \, dS = 0.$$

But, in accordance with Gauss's theorem, this normal surface integral of a function over the boundary of a closed surface S is equal to the volume integral of the divergence of the function over the volume V enclosed by the surface. We therefore, have

$$\iint_S (\nabla \times f) dS \times \iiint v \, \Delta.(\nabla \times f) dV.$$

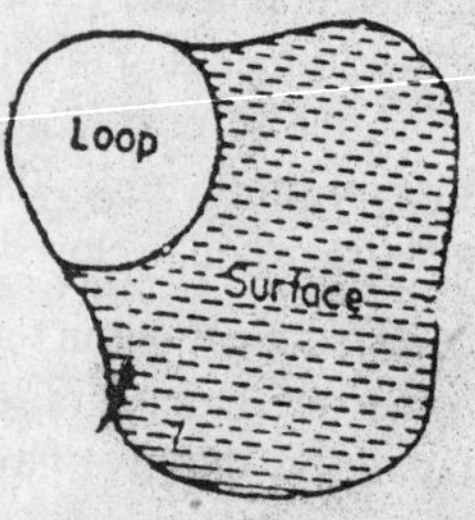

Fig. 1.37

Obviously, therefore, $\iiint \hat{V}\nabla.(\nabla\times f)dV = 0$ for any vector field f. And, since this is so for any volume, it follows that at every point in space the

$$\text{integrand } \nabla.(\nabla\times f) = 0,$$

which, again, is a result we had arrived at earlier (iii) by vector algebra.

RECIPROCAL SYSTEM OF VECTORS

If vectors a', b' and c' be defined by the equations

$$a' = \frac{b\times c}{[abc]},\ b' = \frac{c\times a}{[abc]} \text{ and } c' = \frac{a\times b}{[abc]},$$

they are said to be *reciprocal vectors* to a, b and c which are assumed to be *non-coplanar* so that the product [abc] = 0. The planes of a', b' and c' are respectively perpendicular to those of b, c; c, a and a, b.

Since, obviously, the dot products a.a' = b.b' = c.c' = 1, the name *reciprocal vectors* for a', b' and c' is only natural.

The dot product of any other pair; one from each system, is zero. Thus, a.b' = 0 because a.b' = aca/abc and the numerator, being a triple product containing two equal or identical vectors (a and a), is equal to zero. Similarly, c b' = b.a' = b.c' = c.a' = a.c' = 0.

It can be shown that

$$a = \frac{b'\times c'}{[a'b'c']},\ b = \frac{c'\times a'}{[a'b'c']} \text{ and } c = \frac{a'\times b'}{[a'b'c']},$$

So that, a, b, c, and a', b', c' are really reciprocal system of vectors to each other. They, therefore, possess the same sign. *i.e.*, they are either both right-handed or both left-handed. The unit vectors i, j and k along the three coordinate axes are, however, their own reciprocals.

SOLVED EXAMPLES

Example 1:

A motor boat, with its engine on running river and blown over by a horizontal wind is observed to travel at 20 Km/hour in a direction 53° east of north. The velocity of the boat with its engine on in still water and blown over by the horizontal wind is 4 Km/hour eastward and the velocity of the boat with its engine on over the running river, in the absence of wind is 8 Km/hour due south. Calculate

(a) *the velocity of the boat in magnitude and direction, over still water in the absence of wind,*

(b) *the velocity of the wind in magnitude and direction.*

Solution:

Suppose $\vec{v}_B$ = velocity of boat over still water in the absence of wind.

$\vec{v}_W$ = velocity of wind with respect to ground

$\vec{v}_R$ = velocity of river with respect to ground

Now, $\vec{v}_B + \vec{v}_W + \vec{v}_R$ = 20 Km/hour, East of north = $\hat{i}(20\sin 53^\circ) + \hat{j}(20\cos 53^\circ)$ ...(1)

$\vec{v}_B + \vec{v}_W$ = 4 Km/hour, due east

$= 4\hat{i} + 0\hat{j}$ ...(2)

and $\vec{v}_B + \vec{v}_R$ = 8 Km/hour, due south

$= \hat{i}\times 0 + 8(-\hat{j})$ ...(3)

(i) $(\vec{v}_B + \vec{v}_W) + (\vec{v}_B + \vec{v}_R) - (\vec{v}_B + \vec{v}_W + \vec{v}_R) = \vec{v}_B$

$= -12\,\hat{i} - 20\,\hat{j}$

i.e., $|\vec{v}_B| = \sqrt{[(12)^2 + (20)^2]}$ = **23.32 Km/hour**

Its direction is given by

$$\tan\theta = \frac{-20}{-12} = \frac{5}{3}$$

or $\theta = \tan^{-1}\left(\frac{5}{3}\right)$ = **59° south of west**

(ii) $\vec{v}_W = \vec{v}_B + \vec{v}_W - \vec{v}_B = \hat{i}\times 16 + \hat{j}\times 20$

$\therefore |\vec{v}_W| = \sqrt{[(16)^2 + (20)^2]}$ = **25.61 Km/hour**

Its direction is given by

$$\tan\beta = \tan^{-1}\left(\frac{20}{16}\right)$$

$= \tan^{-1}\left(\frac{5}{4}\right)$ = **51.34° North of East**

Example 2:

If $\vec{A} \times \vec{B} = 0$ and $\vec{A} . \vec{B} = 0$ does it imply that one of the vectors $\vec{A}$ or $\vec{B}$ must be a null vector?

Solution:

$$\vec{A} \times \vec{B} = (AB \sin \theta)\, \hat{n}$$

where θ is the angle between vectors $\vec{A}$ and $\vec{B}$ and $\hat{n}$ is a unit vector perpendicular to both $\vec{A}$ and $\vec{B}$

Hence $\vec{A} \times \vec{B} = 0$ means that

$(AB \sin \theta)\ \hat{n} = 0$

i.e. either $A = 0$ or $B = 0$ or $\sin \theta = 0$ ...(1)

Now, $\vec{A} . \vec{B} = AB \cos \theta$

Hence, $\vec{A} . \vec{B} = 0$ implies that

$$AB \cos \theta = 0 \qquad ...(2)$$

If $\vec{A}$ and $\vec{B}$ are not null vectors then from the conditions (1) and (2), it follows that sin θ and cos θ both should be zero simultaneously. But it can not be possible. Therefore, **it is essential that either $\vec{A} = 0$ or $\vec{B} = 0$ *i.e.* one of the vectors $\vec{A}$ or $\vec{B}$ must be a null vector.**

Example 3:

A particle moves in the x – y plane under the action of a force $\vec{F}$ such that the value of its linear momentum $\vec{p}$ at any time t is P_x = 2 cos t and P_y = 2 sin t. What is the angle θ between $\vec{F}$ and $\vec{p}$ at a given time t ?

Solution:

As $\vec{p} = \hat{i}\, p_x + \hat{j}\, p_y = \hat{i}\,(2 \cos t) + \hat{j}\,(2 \sin t)$

$\therefore\ |\vec{p}| = \sqrt{[2 \cos t)^2 + (2 \sin t)^2]} = 2\sqrt{2}.$

Now, according to Newton's second law of motion

$\vec{F} = d\vec{p}/dt$

Hence $\vec{F} = d/dt\, [\,\hat{i}\,(2 \cos t) + \hat{j}\,(2 \sin t) = \hat{i}\,(-2\sin t) + \hat{j}\,(2\cos t)$

with $|\vec{F}| = \sqrt{[(-2 \sin t)^2 + (2 \cos t)^2]} = 2\sqrt{2}$

Now as $\vec{F}.\vec{p} = F_x p_x + F_y p_y$

$= (-2 \sin t)(2 \cos t) + (2 \cos t)(2 \sin t) = 0$

$$\therefore \quad \theta = \cos^{-1}\left[\frac{\vec{F}.\vec{p}}{Fp}\right]$$

$$= \cos^{-1}\left[\frac{0}{2\sqrt{2}\times 2\sqrt{2}}\right] = \cos^{-1}(0) \quad i.e.\ \theta = \mathbf{90^o}$$

Example 4:

The x and y components of vector $\vec{A}$ are 4 and 6 m respectively. The x and y components of vector $\vec{A}+\vec{B}$ are 10 and 9 m respectively. Calculate for the vector $\vec{B}$ the following :

(a) its x and y components

(b) its length, and

(c) the angle its makes with x-axis.

Solution:

In terms of components

$\vec{A}+\vec{B} = (\hat{i}A_x + \hat{j}A_y) + (\hat{i}B_x + \hat{j}B_y)$

i.e. $\vec{A}+\vec{B} = \hat{i}(A_x + B_x) + \hat{j}(A_y + B_y)$

Given that : $A_x + B_x = 10$ and

$A_x + B_x = 9$ and

$A_x = 4$ and $A_y = 6$

Hence,

(a) $B_x = 10 - 4 = 6$ m and $B_y = 9 - 6 = 3$ m

(b) $B = \sqrt{(B_x^2 + B_y^2)} = \sqrt{(6^2 + 3^2)} = \sqrt{(45)}$ m $= 3\sqrt{(5)}$m

(c) $\theta = \tan^{-1}(B_y/B_x) = \tan^{-1}(3/6) = \tan^{-1}(1/2) = 26.6^o$.

Example 5:

If $|\vec{A}+\vec{B}| = |\vec{A}-\vec{B}|$, then find the angle between $\vec{A}$ and $\vec{B}$.

Solution :

Here, $|\vec{A}+\vec{B}| = |\vec{A}-\vec{B}|$

Suppose θ be the angle between $\vec{A}$ and $\vec{B}$. Dot product of a vector with itself is equal to square of its magnitude.

$|A + B|^2 = (\vec{A}+\vec{B}).(\vec{A}+\vec{B})$

$= \vec{A}.\vec{A}+\vec{A}.\vec{B}+\vec{B}.\vec{A}+\vec{B}.\vec{B}$

$= A^2 + 2AB \cos\theta + B^2$

Again, $|\vec{A}-\vec{B}|^2 = (\vec{A}-\vec{B}).(\vec{A}-\vec{B})$

$\vec{A}.\vec{A} - \vec{A}.\vec{B} - \vec{B}.\vec{A} + \vec{B}.\vec{B}$

$= A^2 - 2AB \cos\theta + B^2$

As $|\vec{A}+\vec{B}| = |\vec{A}-\vec{B}|$

hence, $|\vec{A}+\vec{B}|^2 = |\vec{A}-\vec{B}|^2$

or $A^2 + B^2 + 2AB \cos q = A^2 + B^2 - 2AB \cos\theta$

or $4 AB \cos\theta = 0$

If we assume that $\vec{A}$ and $\vec{B}$ are nonzero vectors, then 4 AB is nonzero and hence

$\cos\theta = 0$ *i.e.* $\theta = \mathbf{90^o}$

Example 6:

Find the components of vector $\vec{A} = 2\hat{i} + 3\hat{j}$ *along the directions of* $\hat{i} + \hat{j}$ *and* $\hat{i} - \hat{j}$

Solution:

Here $\vec{A} = 2\hat{i} + 3\hat{i}$

In order to find the component of $\vec{A}$ along the directions of $\hat{i} + \hat{j}$, let us find out the unit vector along $\hat{i} + \hat{j}$. If $\hat{a}$ is the unit vector along $\hat{i} + \hat{j}$, then

$$\hat{a} = \frac{\hat{i}+\hat{j}}{|\hat{i}+\hat{j}|} = \frac{\hat{i}+\hat{j}}{\sqrt{2}}$$

Hence, the magnitude of the component vector of $\vec{A}$ along $\hat{i}+\hat{j}$

$$= \vec{A}\cdot\hat{a} = (2\,\hat{i}+3\,\hat{j}).\frac{\hat{i}+\hat{j}}{\sqrt{2}} = \frac{1}{\sqrt{2}}(2+3) = \frac{5}{\sqrt{2}}$$

Therefore, component vector of A along $\hat{i}+\hat{j}$

$$= (\vec{A}.\hat{a})\,\hat{a} = \frac{5}{\sqrt{2}}\left(\frac{\hat{i}+\hat{j}}{\sqrt{2}}\right) = \frac{5}{2}(\hat{i}+\hat{j})$$

Similarly, if $\hat{b}$ is the unit vector along the direction of $\hat{i}-\hat{j}$, then magnitude of the component vector of $\vec{A}$ along $\hat{i}-\hat{j}$

$$= \vec{A}.\hat{b} = (2\,\hat{i}+3\,\hat{j}).\frac{(\hat{i}-\hat{j})}{|\hat{i}-\hat{j}|}$$

$$= (2\,\hat{i}+3\,\hat{j}).\frac{(\hat{i}-\hat{j})}{\sqrt{2}} = \frac{(2-3)}{\sqrt{2}} = -\frac{1}{\sqrt{2}}$$

$\therefore$ Component vector of $\vec{A}$ along $\hat{i}-\hat{j}$

$$= (\vec{A}.\hat{b})\,\hat{b} = -\frac{1}{\sqrt{2}}\left(\frac{\hat{i}-\hat{j}}{\sqrt{2}}\right) = -\frac{1}{2}(\hat{i}-\hat{j})$$

Example 46 :

If $|\vec{A}+\vec{B}| = |\vec{A}-\vec{B}|$, then find the angle between $\vec{A}$ and $\vec{B}$.

Solution:

Here, $|\vec{A}+\vec{B}| = |\vec{A}-\vec{B}|$

Suppose θ be the angle between $\vec{A}$ and $\vec{B}$. Dot product of a vector with itself is equal to square of its magnitude.

$$|A+B|^2 = (\vec{A}+\vec{B}).(\vec{A}+\vec{B})$$

$$= \vec{A}.\vec{A}+\vec{A}.\vec{B}+\vec{B}.\vec{A}+\vec{B}.\vec{B}$$

$$= A^2 + 2AB\cos\theta + B^2$$

Again, $|\vec{A}-\vec{B}|^2 = (\vec{A}-\vec{B}).(\vec{A}-\vec{B})$

$$\vec{A}.\vec{A} - \vec{A}.\vec{B} - \vec{B}.\vec{A} + \vec{B}.\vec{B}$$

$$= A^2 - 2AB\cos\theta + B^2$$

As $|\vec{A}+\vec{B}| = |\vec{A}-\vec{B}|$

hence, $|\vec{A}+\vec{B}|^2 = |\vec{A}-\vec{B}|^2$

or $A^2 + B^2 + 2AB \cos q = A^2 + B^2 - 2AB \cos\theta$

or $4\ AB \cos \theta = 0$

If we assume that $\vec{A}$ and $\vec{B}$ are nonzero vectors, then 4 AB is nonzero and hence

$\cos \theta = 0$ *i.e.* $\theta =$ **90°**

Example 7:

Find the components of vector $\vec{A} = 2\hat{i} + 3\hat{j}$ *along the directions of* $\hat{i} + \hat{j}$ *and* $\hat{i} - \hat{j}$

Solution:

Here $\vec{A} = 2\hat{i} + 3\hat{i}$

In order to find the component of $\vec{A}$ along the directions of $\hat{i} + \hat{j}$, let us find out the unit vector along $\hat{i} + \hat{j}$. If $\hat{a}$ is the unit vector along $\hat{i} + \hat{j}$, then

$$\hat{a} = \frac{\hat{i}+\hat{j}}{|\hat{i}+\hat{j}|} = \frac{\hat{i}+\hat{j}}{\sqrt{2}}$$

Hence, the magnitude of the component vector of $\vec{A}$ along $\hat{i} + \hat{j}$

$$= \vec{A} \cdot \hat{a} = (2\hat{i}+3\hat{j}).\frac{\hat{i}+\hat{j}}{\sqrt{2}} = \frac{1}{\sqrt{2}}(2+3) = \frac{5}{\sqrt{2}}$$

Therefore, component vector of A along $\hat{i}+\hat{j}$

$$= (\vec{A}.\hat{a})\ \hat{a} = \frac{5}{\sqrt{2}}\left(\frac{\hat{i}+\hat{j}}{\sqrt{2}}\right) = \mathbf{\frac{5}{2}(\hat{i}+\hat{j})}$$

Similarly, if $\hat{b}$ is the unit vector along the direction of $\hat{i}-\hat{j}$, then magnitude of the component vector of $\vec{A}$ along $\hat{i}-\hat{j}$

$$= \vec{A}.\hat{b} = (2\hat{i}+3\hat{j}).\frac{(\hat{i}-\hat{j})}{|\hat{i}-\hat{j}|}$$

$$= (2\hat{i}+3\hat{j}).\frac{(\hat{i}-\hat{j})}{\sqrt{2}} = \frac{(2-3)}{\sqrt{2}} = -\frac{1}{\sqrt{2}}$$

$\therefore$ Component vector of $\vec{A}$ along $\hat{i}-\hat{j}$

$$= (\vec{A}.\hat{b})\,\hat{b} = -\frac{1}{\sqrt{2}}\left(\frac{\hat{i}-\hat{j}}{\sqrt{2}}\right) = -\frac{1}{2}(\hat{i}-\hat{j}).$$

Example 8:

(a) Obtain the magnitudes and direction cosines of vectors (A+B) and (A–B), if A = 3i + 2j + k and B = i – 2j +3k.

(b) A force F = 2.5i + 4.5j – 5k newton acts through the origin. What is the magnitude of this force and what angles does it make with the three coordinate axes?

Solution:

(a) Obviously (A + B) = 4i + 4k and (A – B) = 2i + 4j – 2k. So that, their *magnitudes* are

$$|A+B| = \sqrt{4^2+4^2} = 4\sqrt{2} \text{ and } |A-B| = \sqrt{2^2+4^2+2^2} = \sqrt{24}.$$

Hence the *direction cosines* of (A + B) = are $\frac{4}{4\sqrt{2}} = \frac{1}{\sqrt{2}}, 0$ and $\frac{4}{4\sqrt{2}} = \frac{1}{\sqrt{2}}$ and the *direction cosines* of (A – B) are

$$\frac{2}{\sqrt{24}}, \frac{4}{\sqrt{24}} \text{ and } \frac{2}{\sqrt{24}}.$$

(b) Clearly, the *magnitude of the force is* given by

$|F| = \sqrt{(2.5)^2 + (4.5)^2 + (-5^2)} = 7.176$ *newton and its direction cosines* are

(i) $\cos\theta_x = 2.5/7.176 = 0.3433$ and $\therefore \theta_x = 69.6^\circ$,

(ii) $\cos\theta_j = 4.5/7.176 = 0.6270$ and $\therefore \theta_y = 51.2^\circ$ and

(iii) $\cos\theta_z = 5/7.176 = -0.6068$ and $\therefore \theta_z = 124.2^\circ$.

Example 9:

Deduce the necessary condition for two vectors $A = a_1 i + b_1 j + c_1 k$ and $B = a_2 i + b_2 j + c_2 k$ to be collinear.

Solution:

Obviously, to be collinear (or parallel), the direction cosines of vector A must be equal to the respective direction cosines of vector B, *i.e.*, we should have $a_1/A = a_2/B$, $b_1/A = b_2/B$ and $c_1/A = c_2/B$, which gives $a_1/a_2 = b_1/b_2 = c_1/c_2 = A/B$.

Thus, the condition for collinearity of the two vectors works out to be that *the ratio between the scalars* a_1 and a_2, b_1 and b_2 and c_1 and c_2 *must be the same, i.e.*, $a_1/a_2 = b_1/b_2 = c_1/c_2$.

Alternatively, we could obtain the same condition by taking the cross product of the two vectors and equating it to zero. Thus

$$A \times B = (b_1c_2 - c_1b_2)i + (c_1a_2 - a_1c_2)j + (a_1b_2 - b_1a_2)k = 0,$$

which gives $b_1c_2 - c_1b_2 = 0$, $c_1a_2 - a_1c_2 = 0$ and $a_1b_2 - b_1a_2 = 0$, whence, $a_1/a_2 = b_1/b_2 = c_1/c_2$, as before.

Example 10:

(a) *The position vectors of four points A, B, C and D are* $a = 2i + 3j + 4k$, $b = 3i + 5j + 7k$, $c = i + 2j + 3k$ *and* $d = 3i + 6j + 9k$ *respectively. Examine whether vectors* $\overrightarrow{AB}$ *and* $\overrightarrow{CD}$ *are collinear*

(b) *Define the condition that vectors* $5i + 7j - 3k$ *and* $2i - bj + ck$ *may be parallel.*

Solution:

(a) Clearly, vector $\overrightarrow{AB} = (b - a) = i + 2j + 3k$ and vector $\overrightarrow{CD} = (d - c) = 2i + 4j + 6k$. For collinearity of vectors, as we have seen in example 2 above, the condition is that $a_1/a_2 = b_1/b_2 = c_1/c_2$. Here, $a_1/a_2 = 1/2$, $b_1/b_2 = 2/4 = 1/2$ and $c_1/c_2 = 3/6 = 1/2$, *i.e.*, the condition for collinearity is satisfied. The two vectors are thus collinear.

(b) For the two given vectors to be parallel (or collinear), we must have $5/2 = -7/b = -3/c$, whence, $b = -14/5$ and $c = -6/5$, which is thus the required condition.

Example 11:

If a position vector $r = 2i + 2j + k$, *obtain (i) its direction cosines and (ii) its components in the xz and yz planes.*

Solution:

Clearly, *magnitude of* r is given by $r = \sqrt{2^2 + 2^2 + 1^2} = \sqrt{9} = 3$ *units*, and

(i) it *direction cosines* are cos (r, x) = 2/3, cos (r, y) = 2/3 and cos (r, z) = 1/3 and

(ii) its components in the xz and yz planes are $r_{xz} = 2i + k$ and $r_{yz} = 2j + k$, with their magnitudes equal to

$$\sqrt{2^2 + 1^2} = \sqrt{5} \text{ and } \sqrt{2^2 + 1^2} = \sqrt{5} \text{ respectively.}$$

Example 12:

Show that the vectors $a - 2b + 3c$, $-2a + 3b - 4c$ and $-b + 2c$ are coplanar, where a,b and c are unit (or any) vectors.

Solution:

We have the condition for coplanarity of three vectors, namely, that the

$$\text{determinant} \quad \begin{vmatrix} A_X & B_X & C_X \\ A_Y & B_Y & C_Y \\ A_Z & B_Z & C_Z \end{vmatrix} = 0.$$

$$\text{Here, the determinant is} \quad \begin{vmatrix} 1 & -2 & 3 \\ -2 & 3 & -4 \\ 0 & -1 & 2 \end{vmatrix}.$$

So that, its value is $1(6 - 4) + (-2)(+4) + 3(2) = 2 - 8 + 6 = 0$, indicating that the three vectors are coplanar.

Alternatively, we could express one of the vectors in terms of the other two, equate the coefficients of a, b and c, solve any two of the three equations thus obtained and see if the same solution satisfies the third equation too. If it does, the three vectors are coplanar.

Example 13:

The projection velocity of a rocket is expressed as $V = 5\hat{V}_x + 7\hat{V}_y + 9\hat{V}_z$, where $\hat{V}_x$ $\hat{V}_y$ and $\hat{V}_z$ are unit velocity vectors along east, north and vertical directions respectively. Calculate the magnitude of the horizontal and vertical components of the velocity. Also deduce the change in the angle of projection if the vertical component be doubled.

Solution:

Obviously, the magnitude of vector V, *i.e.*,

$V = \sqrt{5^2 + 7^2 + 9^2} = \sqrt{155}$, and *horizontal component of the vector* is $5\hat{V}_x + 7\hat{V}_y$, with *its magnitude* $\sqrt{5^2 + 7^2} = \sqrt{74}$ and *vertical component of the vector* is $9\hat{V}_x$, with *its magnitude* $= 9$.

If this latter (*i.e.*, the vertical) component be *doubled*, the new vector will be, say, V' = $5\hat{V}_x + 7\hat{V}_y + 18\hat{V}_x$, with magnitude

$V = \sqrt{5^2 + 7^2 + 18^2} = \sqrt{398}$. Then, clearly, *angle between vectors* V and V' *will directly give the required change in the direction of velocity*.

Now, V.V' = VV' cos (V, V'). whence, cos (V, V') = V.V'/VV'. Since V.V' = 5(5) + 7(7) + 9(18) = 236 and $VV' = \left(\sqrt{155}\right)\left(\sqrt{398}\right)$, we have

$$\cos\left(V, V'\right) = \frac{236}{\left(\sqrt{155}\right)\left(\sqrt{398}\right)} = 0.95.$$

Hence, angle (V, V') or *the change in the direction of velocity* = $\cos^{-1}(0.95) = 18.2°$.

Example 14:

A particle is displaced from the point whose position vector is 5i – 5j – 7k to the point 6i + 2j – 2k under the action of forces 10i – j + 11k, 4i + 5j + 6k, – 2i + j – 9k. Find the total work done.

Solution:

Clearly, *total work done* = F.d,

where F is the *resultant force on the particle* and d, its *displacement*

Here. F = (10i – j + 11k) + (4i + 5j + 6k) + (– 2i + j – 9k) = 12i + 5j + 8k, and *displacement* d = *position vector of second point—position vector of first point.*

= (6i + 2j – 2k) – (5i – 5j – 7k) = i + 7j + 5k.

∴ *total work done* = F.d = (12i + 5j + 8k) . (i + 7j + 5k)

= 12(1) + 5(7) + 8(5) = 12 + 35 + 40 = 87 *units.*

Example 15:

The methane molecule CH_4 may be fitted in a cube such that the C atom lies at the body centre and four H atoms at non adjacent corners of the cube as shown in Fig. 1.38. Prove that the angle between any two C—H bonds is $\cos^{-1}$ (1/3).

Solution:

Let Fig. 1.38. represent a cube of each edge 2a with O, as its body centre where the C atom of the methane molecule lies and let the H atoms be at the four *non-adjacent* corners P, Q, R, and S, as shown.

Consider the two C—H bonds OP and OQ, with an angle θ between them. We are required to show that $\theta = \cos^{-1}(-1/3)$.

Clearly, *vector* OP = ai – aj + ak and *vector* OQ = ai + aj – ak. So that, OP. OQ = (OP) (OQ) cos θ = (ai – aj + ak) (ai + aj – ak) = $a^2 - a^2 - a^2 = -a^2$.

$$\text{And } (OP)(OQ) = \left(a\sqrt{3}\right)\left(a\sqrt{3}\right) = 3a^2$$

We, therefore, have $\cos\theta = \dfrac{O\,Q}{(I),(OQ)} = \dfrac{-a^2}{3a^2} = -\dfrac{1}{3}$.

And, therefore, $\theta = \cos^{-1}(-1/3)$.

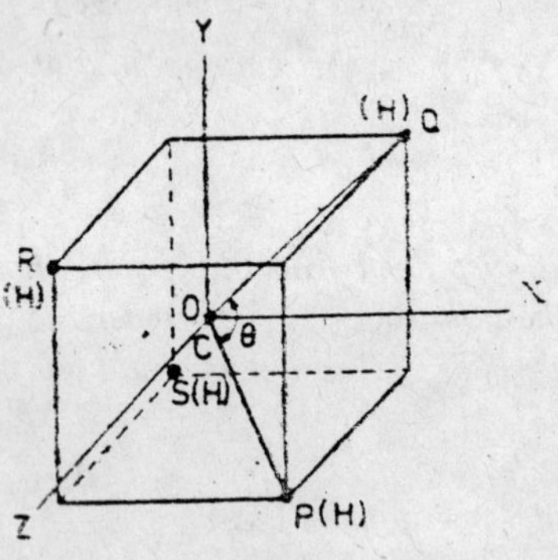

Fig. 1.38

The same can be shown to be true for any other pair of C—H bonds.

Example 16:

In a tetrahedron, if two pairs of opposite edges are perpendicular, show that the third pairs are also perpendicular to each other, and the sum of the squares on two opposite edges is the same for each pair.

Solution:

Let ABCD ba a tetrahedron (Fig 1.39) such that taking D as the origin of reference, the position vectors of points A, B and C are a,b and c respectively.

Then clearly, $\overrightarrow{AB} = (b - a)$. $\overrightarrow{AC} = (c - a)$ and $\overrightarrow{CB} = (b - c)$.

If the edge BD is perpendicular to CA, we have b. (c – a) = 0. *i.e.*,

$$b.c - b.a = 0, \text{ r, } b.c = a.b \qquad ...(i)$$

Similarly, if edge DA is perpendicular to BC, we have

$a.b\ (b - c) = 0$, *i.e.*,

$a.b - a.c = 0$.

or, $a.b = c.a$ ...(ii)

From relation (i) and (ii) therefore.

$a.\ b = b.\ c = c.\ a$ So that.

$b.\ c = c.\ a$

Or, $b.\ c - c.\ a = 0$.

Or, $c(b - a) = 0$,

showing that DC is perpendicular to BA.

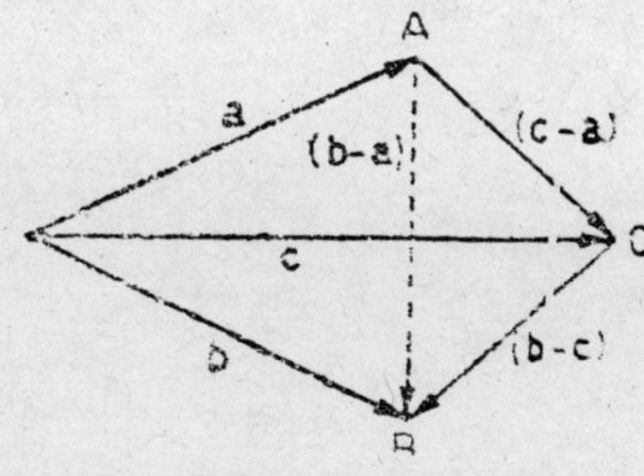

Fig. 1.39

Now, the sum of the squares on the opposite pair of edges BD and CA

$$= b^2 + (c - a)^2 = b^2 + c^2 + a^2 - 2ac$$

and since a. b = b. c = c. a, the sum of the square on the other two pairs of opposite edges (viz; DA, BC and DC. AB) will also be the same.

Example 17:

The position vectors of two particles ejected simultaneously from the same source are $r_1 = 3i + 4j + 5k$ *and* $r_2 = 2i + 6j + 8k$. *Obtain (i) the displacement r of particle two with respect to particle one, (ii) the magnitudes or* r_1. r_2 *and r, (iii) angles between* r_1 *and* r_2 *and* r_1 *and r and* r_2 *and r and (iv) projection of r on* r_1.

Solution:

(i) Here, obviously, *displacement of particle 2 with respect to particle* 1,

$$i.e., r = r_2 - r_1 = -i + 2j + 3k.$$

(ii) Magnitude of $r_1 = |r_1| = \sqrt{9+16+25} = 5\sqrt{2} = 7.07$,

Magnitude of $r_2 = |r_2| = \sqrt{4+36+64} = \sqrt{104} = 10.20$

and Magnitude of $r = |r| = \sqrt{1+4+9} = \sqrt{14} = 3.74$.

(iii) We have $r_1.r_2 = |r_1|\,|r_2| \cos\theta_{r_1r_2}$, whence, $\cos\theta_{r_1r_2} = \dfrac{r_1r_2}{|r_1||r_2|}$

$$= \frac{6+24+40}{(7.07)(10.20)} = 0.9706 \text{ and } \therefore \theta_{r_1 r_2} = 14.0^\circ.$$

Similarly,

$$\cos\theta_{r_1 r} = \frac{r_1 . r}{|r_1||r|} = \frac{-3+8+15}{(7.07)(3.74)} = 0.7583 \text{ and } \therefore \theta_{r_1 r} = 49.9^\circ.$$

and

$$\cos\theta^o_{r_2 r} = \frac{r_2 . r}{|r_2||r|} = \frac{-2+12+24}{(10.20)(3.74)} = 0.8913 \text{ and } \therefore \theta_{r_2 r} = 27.8^\circ.$$

(iv) And, projection of r on r_1 = r.r_1 (where r_1 is the unit vector in the direction of r_1) $\frac{r.r_1}{|r_1|} = \frac{20}{7.07} = 2.83.$

Example 18:

(a) Obtain the cross product of P = 3i – 2j + 8k and Q = – i – 2j – 3k.

(b) show that if a, b and c be three non-zero vectors, such that

$$a \times b = b \times c = 0, \text{ then, } a \times c = 0.$$

Solution:

(a) We have *cross product* $P \times Q = \begin{vmatrix} i & j & k \\ 3 & -2 & 8 \\ -1 & -2 & -3 \end{vmatrix}$

= i[(–2) (–3) – (–2) 8] + j [(–1)8 – (–3)3] + k [3(–2) – (–2) (–1)]

= 22i + j – 8k.

(b) Since a × b = 0, it is clear that a is parallel to b and, similarly, since b × c = 0, b is parallel to c. Obviously, therefore, a is parallel to c and hence

$$a \times c = 0$$

Example 19:

(a) Express the magnitude of a × b in terms of scalar products.

(b) If two vectors are 6i + 0 3j – 5k, i – 4.2j + 2.5k, find their vector product and by calculation prove that the new vector is perpendicular to the two.

Solution:

(a) Clearly, $(a \times b)^2 = (a \times b).(a \times b) = |a \times b|^2$

$$= |a|^2 |b|^2 \sin^2\theta = |a|^2 |b|^2 (1-\cos^2\theta)$$

$$= a^2b^2 - |a|^2 |b|^2 \cos^2\theta = a^2b^2 - (a.b)^2.$$

(b) Let the two given vectors be called a and b respectively. Them,

their *vector product c, say,* $= \begin{vmatrix} i & j & k \\ 6 & 0.3 & -5 \\ 1 & -4.2 & 2.5 \end{vmatrix}$

$= i\,[(0.3)(2.5) - (-5)(-4.2)] + [j(-5)\,1 - 6\,(2.5)] + k\,[6(-4.2) - (0.3)\,1]$

$= -20.25i - 20j - 25.5k.$

Now, taking the scalar product of the new vector c with vector a, we have

$$c.a = (-20.25i - 20j - 25.5k).(6i - 0.3j + 5k)$$

$$= -121.5 - 6 + 127.5 = 0$$

showing that c is perpendicular to a

similarly, $c.b = (-20.25i - 20j - 25.5k).(i - 4.2j + 2.5k)$

$$= -20.25 + 84 - 63.75 = 0,$$

showing that c is also perpendicular to b.

Example 20:

Find the unit vector perpendicular to each of the vector a = 3i + j + 2k and b = 2i – 2j + 2k. Also determine the sine of the angle between a and b. What will be the vector perpendicular to a and b and having a magnitude $4\sqrt{3}$?

Solution:

By its very definition, the vector product a × b is perpendicular to the plane of a and b. So that, we have

unit vector perpendicular to a and b, say, $\hat{n} = \dfrac{a \times b}{|a \times b|}.$

Now, $a \times b = (3i + j + 2k) \times (2i - 2j + 4K) = 8i - 8j - 8k$

and ∴ $|a \times b| = \sqrt{64+64+64} = 8\sqrt{3}.$

Hence unit vector $\hat{n}$, perpendicular to a and b

$$= \frac{8i-8j-8k}{8\sqrt{3}} = \frac{1}{\sqrt{3}}(i-j-k)$$

Finally, since the unit vector perpendicular to a and b is $\hat{n}$ equal to $\frac{1}{\sqrt{3}}$ (i – j – k), we have *vector of magnitude* $4\sqrt{3}$ and perpendicular to a and b = $4\sqrt{3}\,\hat{n}$

$$= 4\sqrt{3}\left[\frac{1}{\sqrt{3}}(i-j-k)\right] = 4i-4j-4k = 4(i-j-k).$$

Example 21:

(a) Find the area of the parallelogram determined by the vectors a = 3i + 2j and b = 2j – 4k.

(b) Find the area of the triangle whose vertices are (1, – 1, – 3), (4, – 3, 1) and (3, – 1, 2).

Solution:

(a) Here, *vector area of the parallelogram*

$$= a \times b = (3i + 2j) \times (2j - 4k)$$
$$= 8i - 12j + 6k.$$

∴ *magnitude of the area of the parallelogram*

$$= |8i+12j+6k|$$
$$= \sqrt{64+144+36} = 2\sqrt{61}.$$

(b) Clearly, the position vectors of the vertices A, B and C of the triangle ABC are a = (i – j – 3k) b = (4i – 3j + k) and c = (3i – j – 2k).

∴ $\overrightarrow{BA} = a - b = -3i + 2j - 4k$ and $\overrightarrow{BC} = c - b = -i + 2j + k.$

Hence *vector area of the triangle* $= \frac{1}{2}\overrightarrow{BA} \times \overrightarrow{BC}$

$$= 1/2\ (-3i + 2j - 4k) \times (-i + 2j + k)$$
$$= 1/2\ (10i + 7j - 4k)$$

and *magnitude of the area of the triangle* $= \frac{1}{2}\left|10i + 7j - 4k\right|$

$$= \frac{1}{2}\sqrt{100+49+16} = \frac{1}{2}\sqrt{165}.$$

Example 22:

Prove that the vector area of a triangle whose vertices are a, b, c is 1/2 (b × c + c × a + a × b).

What is the condition for the collinearity of the vertices ?

Solution:

As in example 18 above, let the triangle be ABC. Then, we have

$$\overrightarrow{BC} = (c - b) \text{ and } \overrightarrow{BA} = (a - b)$$

∴ *vector area of the triangle* $= 1/2\,\overrightarrow{BC} \times \overrightarrow{BA} = 1/2\,[(c - b) \times (a-b)]$

$$= 1/2\,(c \times a - c \times b - b \times a + b \times b]$$

Now, $b \times b = 0$, $- c \times b = b \times c$ and $- b \times a = a \times b$

So that, the *vector area of the triangle*

$$= 1/2\,(b \times c + c \times a + a \times b).$$

If the vertices of the triangle be collinear, we shall have area of the triangle equal to zero and, therefore vector area of the triangle too equal to zero. Thus, the *condition for collinearity of the vertices* is that 1/2 (b × c + c × a + a × b) = 0.

Example 23:

Show that if the vector area of each face of a tetrahedron has the direction of the outward normal, the sum of their vector areas is zero.

Solution:

Let DABC be the tetrahedron, such that taking D as the origin, the position vectors of A, B and C are a, b and c respectively. Then, we clearly have *vector area of face* DBC = 1/2 b × c, *of face* DAB = 1/2 a × b, *of face* DCA = 1/2 c × a and of face BAC = 1/2 (b – a) (b – c) = 1/2 (b × b – b × c – a × b + a × c) = – 1/2 (b × c + a × b + c × a), all directed along their respective outward normals.

∴ *sum of the vector areas of all the faces of the tetrahedron*

= 1/2 (b × c + a × b + c × a) – 1/2 (b × c + a × b + c × a) = 0.

Example 24:

A rigid body is rotating with angular velocity 3 radians/second about an axis passing through the point 2i – j – k and parallel to i – 2j + 2k. Find the magnitude of the velocity of the point (of the rigid body) whose position vector is 2i + 3j – 4k.

Solution:

Obviously, the unit vector, $\hat{n}$,say, in the direction of

$$i - 2j + 2k = \frac{i-2j+2k}{|i-2j+2k|}$$

$$= \frac{i-2j+2k}{\sqrt{1+4+4}} = \frac{1}{3}(i-2j+2k).$$

angular velocity of the rigid body, $\omega = \omega\hat{n} = 3\ [1/2\ (j - 2j + 2k)]$

$$= i - 2j + 2k.$$

Let the point whose velocity is desired to be determined be P. Then, its positive with respect to the point 2i – j – k on the axis given by

$$r = (2i + 3j - 4k) - (2i - j - k) = 4j - 3k.$$

Hence, if v be the velocity of point P, we have $v = \omega \times r$

$$= (i-2j+2k)\times(4j-3k) = \begin{vmatrix} i & j & k \\ 1 & -2 & 2 \\ 0 & 4 & -3 \end{vmatrix} = -2i+3j+4k.$$

So that, *magnitude of the velocity* $= \sqrt{(-2)^2 + 3^2 + 4^2} = \sqrt{29}$ units.

Example 25:

A force F = – 2i + 3j +4k is acting at a point 5i + 4j + 3k. Obtain the moment of the force about the origin.

Solution:

Here, r = 5i + 4j + 3k. Therefore, *moment of the force about the origin* $r \times F = (5i + 4j + 3k) \times (-2i + 3j + 4k)$

$$= i\ (16 - 9) + j\ (-6 - 20) + k\ (15 + 8) = 7i - 26j + 23k.$$

Example 26:

A = 4i – 5j + 3k, B = 2i – 10j – 7k and C = 5i + 7j – 4k calculate the following: (i) A × B . C, (ii) A × (B × C).

Solution:

We have

$$\text{(i) } A \times B.C \text{ or } [ABC] = \begin{vmatrix} 4 & -5 & 3 \\ 2 & -10 & -7 \\ 5 & 7 & -4 \end{vmatrix}$$

$$= 4(40+49)-5(-35+8)+3(14+50) = 356+135+192 = 683.$$

(ii) A × (B × C) = B (A.C) – C (A.B)

= (2i – 10j – 7k) 20 – 35 – 12) – (5i + 7j – 4k) (8 + 50 – 21)

= – 54i + 270j + 189k – 185i – 259j + 148k = – 239i + 11j + 337k.

Example 27:

(a) Prove that a × (b × c) + b × (c × a) + c × (a × b) = 0.

(i) Prove that the four points (4i + 5j + k), – (j + k), (3i + 9j + 4k) and (4 – i – j – k) are coplanar.

Solution:

(a) As we know a × (b × c) = (a.c)b – (a.b)c ...(i)

– b × (c × a) = (a.b)c– (b.c)a ...(ii)

and c × (a × b) = (b.c)a – (c.a) b. ...(iii)

So that, adding up the three, we have

a × (b × c) + b × (c × a) + c × (a × b) = 0.

(b) If 4i + 5j + k, – (j + k), 3i + 9j + 4k and 4 (– i + j + k) be the position vectors of four points A, B, C and D with reference to an origin O, clearly the points will be coplanar if the vectors $\overrightarrow{BA}$, $\overrightarrow{BC}$ and $\overrightarrow{CD}$ be coplanar.

Now, $\overrightarrow{BA} = \overrightarrow{OA} - \overrightarrow{OB}$ = (4i + 5j + k) – [– (j + k)]=4i + 6j + 2k,

$\overrightarrow{BC} = \overrightarrow{OC} - \overrightarrow{OB}$ = (3i + 9j + 4k) – [– (j + k)] = 3i + 10j + 5k

and $\overrightarrow{CD} = \overrightarrow{OD} - \overrightarrow{OC}$ = (– 4i + 4j + 4k) – (3i + 9j + 4k) – 7i – 5j.

And, these vectors ($\overrightarrow{BA}$, $\overrightarrow{BC}$ and $\overrightarrow{CD}$) will be coplanar if their scalar product is zero, Let us see if its is so.

The triple scalar product of the vectors is clearly

$$[\overrightarrow{BA}\ \overrightarrow{BC}\ \overrightarrow{CD}\ ] = \begin{vmatrix} 4 & 6 & 2 \\ 3 & 10 & 5 \\ -7 & -5 & 0 \end{vmatrix} = 4\,(25) + 6\,(-35) + 2\,(55) = 0$$

The three vectors $\overrightarrow{BA}$ $\overrightarrow{BC}$ and $\overrightarrow{CD}$, and hence the four points A, B, C and D, are thus coplanar.

Example 28:

The edge of a parallelopiped are given by the vectors i + 2j + 3k, 5j and 4j + mk. What should be the value of m in order that the volume of the parallelopiped be 20 units?

Solution:

The volume of a parallelopiped, as we know is given by the scalar triple product of the vectors representing its three edges. So that, we have:

$$\text{volume of the parallelopiped} = \begin{vmatrix} 1 & 2 & 3 \\ 0 & 5 & 0 \\ 0 & 4 & m \end{vmatrix} = 1(5m) + 2(0) + 3(0) = 5m.$$

Since the volume is given to be 20 units, we have 5m = 20, whence, m = 4.

Example 29:

Show tha. (i) (a × b) × (c × d) = [abd]c – [abc]d.

(ii) (a × b) × (c × d) = [acd]b – [bcd]a.

Solution:

(i) Let a × b = m. Then, (a × b) × (c × d) = m × (c × d)

$$= (m\ .\ d)\ c - (c\ .\ m)d$$

$$= [(a \times b).d]\ c - [c.\ (a \times b)]d.$$

Since in the triple products (a × b).d and c.(a × b), the positions of the dots and crosses are quite immaterial and may be interchanged or omitted, as desired, we have (a × b) × (c × d) = [abd]c – [abc]d.

(ii) Again, putting (c × d) = n, we have

$$(a \times b) \times (c \times d) = (a \times b) \times n = (a.n)b - (b.n)a$$

$$= [a.(c \times d)]\ b - [b.(c \times d)]a$$

$$= [acd]b - [bcd]a.$$

Example 30:

Obtain the value of the product $(A \times B).(C \times D)$ and show that $A \times [B \times (C \times D)] = (A \times C)(B.D) - (A \times D)(B.C)$

Solution:

Since in the case of scalar products, the positions of dots and crosses are freely interchangeable, we have

$$(A \times B).(C \times D) = A.[B \times (C \times D)] = A.[(B.D)C - (B.C)D]$$
$$= (A.C)(B.D) - (A.D)(B.C)$$

Again, $A \times [B \times (C \times D)] = A \times [(B.D)C - (B.C)D]$

$$= (A \times C)(B.D) - (A \times D)(B.C).$$

Example 31:

Establish the kinematic relation $v = u + at$, $S = ut + 1/2at^2$ and $v^2 - u^2 = 2aS$.

Solution:

We know that *acceleration* a = dv/dt. So that, dv = a.dt, which, on integration gives $v = at + C_1$ where C_1 is a constant of integration.

Since at t = 0, $v = C_1$, clearly C_1 stands for the initial velocity u.

We, therefore, have $v = at + u$

or $v = u + at.$

Now, $v = dS/dt = u + at.$

Or $dS = (u + at)dt.$

Integrating, we have $S = ut + 1/2at^2 + C_2$, where C_2 is another constant of integration.

Since at t = 0, S = 0 and $\therefore$ $C_2 = 0$, we have $S = ut + 1/2\, at^2$

Finally, taking the scalar product of a with v, we have

$$v.a = v.\frac{dv}{dt} \quad \text{Or,.} \quad \frac{dS}{dt}.a = v.\frac{dv}{dt}.\left[\therefore a = \frac{dv}{dt} \text{ and } v = \frac{dS}{dt}.\right.$$

Integration gives $S.a = 1/2\, v.v + C_2$ (a being a constant).

where C_3 is yet another constant of integration.

Since $v.v = v^2 = v^2$. we have $S.a = 1/2v^2 + C_3$.

Again, when S = 0, v = u and we have $0 = 1/2v^2 + C_3$ whence, $C_3 = -u^2/2$.

So that, $S.a = 1/2v^2 - 1/2u^2 = 1/2v^2 - 1/2u^2$, whence, $v^2 - u^2 = 2aS$

$[\because u^2 = v^2$ and $u^2 = u^2$.

Example 32:

If the acceleration of moving particle at any instant be given by $\mu r + v/r^3$ show that (i) it has a constant speed and that (ii) its angular momentum consists of two components, one constant in magnitude and direction and the other constant in magnitude in the direction of, r.

Solution:

(i) Clearly, the equation of motion of the particle is

$$\frac{d^2r}{dt^2} = \frac{\mu r \times v}{r^3} = \frac{\mu r}{r^3} \times \frac{dr}{dt}$$

Taking the scalar product of relation (i) with dr/dt, we have

$$\frac{dr}{dt}.\frac{d^2r}{dt^2} = \frac{\mu r}{r^3} \times \frac{dr}{dt}.\frac{dr}{dt} = 0.$$

This, on integration, given $(dr/dt)^3$ = constant,

indicating that *the speed* dr/dt *of the particle is constant.*

(ii) Obtaining the vector product of (i) with r, we have

$$r \times \frac{d^2r}{dt^2} = \frac{\mu}{r^3} r \times \left(r \times \frac{dr}{dt}\right) = \frac{\mu}{r^3}\left[\left(r.\frac{dr}{dt}\right)r - (r.r)\frac{dr}{dt}\right]$$

$$\frac{\mu}{r^3}\left[r\frac{dr}{dt}r - r^2\frac{dr}{dt}\right] = \mu\frac{d}{dt}\left(\frac{r}{r}\right),$$

which, on integration, gives $r \times \frac{dr}{dt} = -\mu\frac{r}{r} + a$, where a is a *constant vector.*

Clearly, $r \times \frac{dr}{dt}$ is the angular velocity of the particle and hence its

ular momentum is given by $m\left(r \times \frac{dr}{dt}\right) = -\frac{m\mu r}{r} + ma.$

clearly consists of two components, vis (i) a component—$m\mu r/$

e magnitude is the constant$|-m\mu|$ and whose direction is that of

i) the component ma, which is constant both in magnitude and

Example 33:

A rigid body is spinning with an angular velocity of 4 radians/about an axis parallel to 3j – k passing through the point i + 3j – k. Find the velocity of the particle at the point 4i – 2j +k.

Solution:

Let $\hat{n}$ be the unit vector in the direction of 3j – k. Then,

$$\hat{n} = \frac{3j-k}{\sqrt{0+9+1}} = \frac{1}{\sqrt{10}}(3j-k),$$

∴ angular velocity of the particle P, say, $\omega = \omega\hat{n} = \frac{4}{\sqrt{10}}(3j-k).$

The position vector of P with reference to point i + 3j – k, *i.e.*,

$$r = (4i - j + k) - (i + 3j - k) = 3i - 5j + 2k.$$

Hence *velocity (linear) of particle* P = r × ω = (3i – 5j + 2k) × $\frac{4}{\sqrt{10}}(3i-k)$

This may be seen from the following: Consider, a particle P in a rigid body, revolving around an axis with an angular velocity ω. Let r be the position vector of the particle with respect to any point O on the axis. Then, since the particle describes a circle of radius r sinθ, lying in the plane perpendicular to the axis or rotation, we have

magnitude of its linear velocity (or of the velocity vector) v = ω r sin θ.

But by the definition of cross product, ω × r has also the same magnitude ωr sin θ and is perpendicular to the plane containing r and ω.

We, therefore have v = r × ω.

$\frac{4}{\sqrt{10}}(i+3j+9k)$ and its magnitude $= \frac{4}{\sqrt{10}}\sqrt{1+9+81} = 4\sqrt{\frac{91}{10}}$

= 12 *units* very nearly.

Example 34:

A particle describes a circle in the i, j plane with a uniform speed in 12 seconds in the direction i to j. If the initial position of the radius vector relative to the centre of the circle be i, determine its position at the end of the 3rd second from the start. What will be its velocity vector then?

Solution:

Let OX and OY be the rectangular coordinates in the plane of the circle through its centre O and let i and j be the unit vectors along the two axes respectively. Then, taking the circle to be of unit radius, we have OA = OB =1, where A and B are the points in which the circle meets the two axes respectively.

Let A be the initial position of the particle and P, its position after t *second* from the start. So that, angular velocity of the particle being $2\pi/12$ or $\pi/6$ radian/s, angle XOP = $(\pi/6)t$.

If PM be the perpendicular dropped from P on to OA, we have

$$\overrightarrow{OP} = \overrightarrow{OM} + \overrightarrow{MP}.\ \text{Or,} \overrightarrow{OP} = \left(OP\cos\frac{\pi}{6}t\right)i + \left(OP\sin\frac{\pi}{6}t\right)j.$$

Or, since OP = 1 and t = 3, we have $\overrightarrow{OP} = \left(\cos\frac{3\pi}{6}\right)i + \left(\cos\frac{3\pi}{6}\right)j$

$$= (\cos\pi/2)\,i + (\sin\pi/2)j = j$$

Thus, the position of the radius vector will, after 3 seconds, be along OB. And, the velocity vector of the particle will, therefore, be $-\pi i/6$, the velocity v being tangential to the circle at B, as shown.

Example 35:

Find grad r^m where r is the distance of any point from the origin, Also show that differential $d\phi = dr$ grad ϕ.

Solution:

we have $\phi(x, y, z) = r^m = (x^2 + y^2 + z^2)^{m/2}$

$$\left[\begin{array}{l} \because r^2 = x^2+y^2+z^2 \\ \text{and} \therefore r = \left(x^2+y^2+z^2\right)^{1/2} \end{array}\right]$$

$$\therefore \quad \frac{\partial\phi}{\partial x} = mx\left(x^2+y^2+z^2\right)^{\frac{m}{2}-1} = mxr^{m-2}$$

Similarly, $\dfrac{\partial\phi}{\partial y} = myr^{m-2}$ and $\dfrac{\partial\phi}{\partial z} = mzr^{m-2}$

Hence grad

$$r^m\left(\text{or}\ \nabla r^m\right) = \sum i\frac{\partial\phi}{\partial x} = mr^{m-2}\sum ix = mr^{m-2}r, \qquad \text{...(i)}$$

where r is the *position vector* of the point

Or, since $\mathbf{r} = r\hat{\mathbf{r}}$, where r is the magnitude of r and r, the unit vector along r, we have $\hat{\mathbf{r}} = \mathbf{r}/r$.

So that, grad r^m also $= mr^{m-1}\,(\mathbf{r}/r) = mr^{m-1}\hat{\mathbf{r}}$ ...(ii)

Now, differential $d\phi = \frac{\partial \phi}{\partial x}dx + \frac{\partial \phi}{\partial y}dy + \frac{\partial \phi}{\partial z}dz$

$$= (\mathbf{i}dx + \mathbf{j}dy + \mathbf{k}dz)\left(\mathbf{i}\frac{\partial \phi}{\partial x} + \mathbf{j}\frac{\partial \phi}{\partial y} + \mathbf{k}\frac{\partial \phi}{\partial z}\right) = d\mathbf{r}\text{ grad }\phi.$$

Example 36:

Show that the curl of the velocity of any particle of a rigid body is equal to twice the angular velocity of the body.

Solution:

We have the relation $\mathbf{v} = \boldsymbol{\omega} \times \mathbf{r}$, where v is the velocity of any particle of the body, with r, the *position vector* relative to a fixed point and ω, the *angular velocity* of the body.

Obviously, v is a *vector point function* and vector ω is the same for all the particles.

Now,

$$\text{curl }\mathbf{v} = \text{curl}(\boldsymbol{\omega} \times \mathbf{r}) = \sum \mathbf{i}\,\frac{\partial}{\partial x}(\boldsymbol{\omega} \times \mathbf{r}) = \sum \mathbf{i}\left(\boldsymbol{\omega} \times \frac{\partial \mathbf{r}}{\partial x}\right) = \sum \mathbf{i} \times (\boldsymbol{\omega} \times \mathbf{i})$$

$$= \sum (\mathbf{i}.\mathbf{i})\,\boldsymbol{\omega} - \sum (\mathbf{i}.\boldsymbol{\omega})\mathbf{i} = 3\boldsymbol{\omega} - \boldsymbol{\omega} = 2\boldsymbol{\omega}.$$

Example 37:

Establish the results : (i) div (ϕ A) = ϕ div A + A. grad ϕ and (ii) div (A × B) = B. curl A – A. curl B.

Solution:

(i) We have div $(\phi \mathbf{A}) = \sum \mathbf{i}.\,\frac{\partial(\phi \mathbf{A})}{\partial x} = \sum \mathbf{i}\left(\phi\frac{\partial \mathbf{A}}{\partial x} + \frac{\partial \phi}{\partial x}\mathbf{A}\right)$

$$= \sum \mathbf{i}.\phi\frac{\partial \mathbf{A}}{\partial x} + \sum \mathbf{i}.\frac{\partial \phi}{\partial x}\mathbf{A} = \phi\sum \mathbf{i}.\,\frac{\partial \mathbf{A}}{\partial x} + \left(\sum \mathbf{i}\frac{\partial \phi}{\partial x}\right)\mathbf{A}.$$

$= \phi$ div A + A. grad. ϕ.

(ii) div $(\mathbf{A} \times \mathbf{B}) = \sum \mathbf{i}.\,\frac{\partial(\mathbf{A} \times \mathbf{B})}{\partial x} = \sum \mathbf{i}\left(\frac{\partial \mathbf{A}}{\partial x} \times \mathbf{B} + \mathbf{A} \times \frac{\partial \mathbf{B}}{\partial x}\right)$

$$= \sum \mathbf{i}.\,\frac{\partial \mathbf{A}}{\partial x} \times \mathbf{B} + \sum \mathbf{i}.\mathbf{A} \times \frac{\partial \mathbf{B}}{\partial x} = \sum \mathbf{i} \times \frac{\partial \mathbf{A}}{\partial x}\mathbf{B} - \sum \mathbf{i} \times \frac{\partial \mathbf{B}}{\partial x}\mathbf{A}$$

$$= \left(\sum i \times \frac{\partial A}{\partial x}\right).B - \left(\sum i \times \frac{\partial B}{\partial x}\right).A \text{ curl } A.B - \text{curl } B.A$$

= B. curl A – A. curl B.

Example 38:

Obtain div grad r^m. Also show that curl grad r^m = 0.

Solution:

We have

grad $r^m = \nabla r^m = mr^{m-2}\, r = mr^{m-2}\,(ix + jy + kz)$

$$\therefore \text{div grad } r^m \text{ or div} \Delta r^m = \frac{\partial}{\partial x}\left(mr^{m-2}x\right) + \frac{\partial}{\partial y}\left(mr^{m-2}y\right) + \frac{\partial}{\partial z}\left(mr^{m-2}z\right).$$

$$\text{Now, } \frac{\partial}{\partial x}\left(mr^{m-2}x\right) = m(m-2)r^{m-3}\frac{\partial r}{\partial x} + mr^{m-2}$$

$$= m(m-2)r^{m-1}x^2 + mr^{m-2},$$

because $\partial r/\partial x = x/r$.

Similarly, $\frac{\partial}{\partial y}(mr^{m-2}y) = m(m-2)\, r^{m-1}\, y^2 + mr.^{m-2}$

and $\frac{\partial}{\partial y}(mr^{m-2}z) = m(m-2)\, r^{m-1}\, z^2 + mr.^{m-2}$

$\therefore$ div grad r^m = div∇ $r^m = m(m-2)\, r^{m-1}\, r^2 + 3\, mr^{m-2}$

$= m\,(m+1)r^{m-2}$

$$\text{And, curl grad } r^m = \begin{vmatrix} i & j & k \\ \frac{\partial}{\partial x} & \frac{\partial}{\partial y} & \frac{\partial}{\partial z} \\ mr^{m-2}x & mr^{m-2}y & mr^{m-2}z \end{vmatrix}$$

$$\text{So that coefficient of i in curl grad } r^m = \frac{\partial}{\partial y}mr^{m-2}z - \frac{\partial}{\partial z}mr^{m-2}y$$

$$= m(m-2)r^{m-3}\frac{\partial r}{\partial y}z - m(m-2)r^{m-3}\frac{\partial r}{\partial z}y$$

$$= m(m-2)r^{m-3}\frac{yz}{r} - m(m-2)r^{m-3}\frac{zy}{r} = 0.$$

Hence *curl grad* $r^m = \nabla \times \nabla r^m = 0$.

Example 39:

Show that $\int_S F \times n dS = -\int_V \text{curl } F dV$.

Solution:

Let us put f = a × F, where f is a vector function and a, an *arbitrary constant vector*. Then, applying Gauss's theorem to f, we have

$$\int_S a \times F.ndS = \int_V div(a \times F)dV. \quad ...(i)$$

Now, $$a \times F.\ n = a.\ F \times n \quad ...(ii)$$

and div (a × F) = ∇. (a × F) = – a.∇ × F. ...(iii)

So that, using relations (ii) and (iii), relation (i) may be put as

$$\int_S a.F \times ndS = -\int_V a.\nabla + FdV.\ \text{Or,}\ a\int_S F \times ndS = -\ a\int_V \nabla \times dFV.$$

Or, $$a\left[\int_S F \times ndS + \int_V \nabla \times FdV\right] = 0.$$

Since a is just an arbitrary vector, we have

$$\int_S F \times ndS + \int_V \nabla \times FdV = 0.$$

Or, $$\int_S F \times ndS = -\int_V \nabla \times FdV = -\int_V \text{curl } FdV.$$

Example 40:

Verify Stokes' theorem for the function F = x (ix + jy), integrated round the square, in the plane z = 0, whose sides are along the lines x = 0, y = 0, x = a, y = a,

Solution:

Referring to the square shown in Fig . 1.42, we clearly have

$$\oint_C F.dl + \int_{OA} F.dl + \int_{AB} F.dl + \int_{BC} F.dl + \int_{CO} F.dl,$$

where, $$\int_{OA} F.dl = \int_0^a x(ix + jy).idx = \int_0^a x^2 dx = \frac{1}{3}a^3,$$

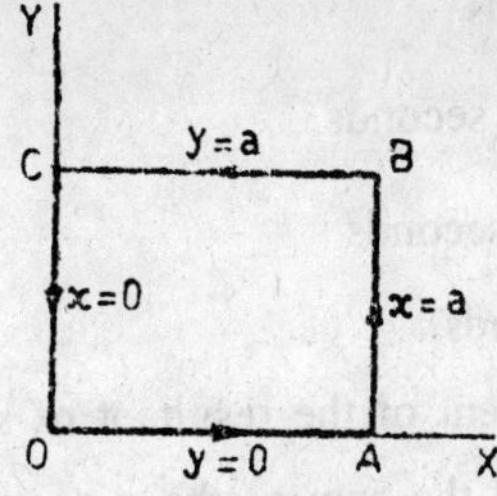

Fig. 1.40

$$\int_{AB} F.dl = \int_o^a x(ix + jy).jdy = \int_o^a aydy = \frac{1}{2}a^3,$$

$$\int_{BC} F.dl = \int_a^o x(ix + jy).idx = -\int_o^a x^2dx = -\frac{1}{3}a^3 \text{ and}$$

$$\int_{CO} F.dl = \int_a^o x(ix + jy).jdx = 0$$

So that, $\oint_c F.dl = \frac{1}{3}a^3 + \frac{1}{2}a^3 - \frac{1}{3}a^3 + 0 = \frac{1}{2}a^3.$

Now, in accordance with Stokes' theorem, $\oint c\ F.d\ l = \iint S$ curl FdS.

Since curl x (ix + jy) = ky, we have

$$\iint_S \text{curl FdS} = \iint_S \text{curl}\, x(ix + jy)dS = \int_o^a\int_o^a ky.kdxdy$$

$$= \int_o^a\int_o^a ydxdy = \frac{1}{2}a^3.$$

Stokes' theorem thus stands verified.

EXERCISES

1. What is (i) *a polar vector* (ii) *an axial vector* (iii) *unit vector* (iv) *a zero vector* ? What do you understand by the term *orthogonal unit vector triad* ?
2. What is meant by the *position vector* of a point ? Express the position vector of a point in terms of the orthogonal unit vector triad. What is the modulus of the vector r ?
3. Four persons K, L, M and N are initially at the corners of a square of side of length d. If every person starts moving, such that K is always headed towards L, L is always towards M, M is headed directly towards N and N towards K. Then the four persons will meet after
   (a) d/v seconds
   (b) $\sqrt{(2)}$ d/v seconds
   (c) d/ $\sqrt{(2)}$ v seconds
   (d) d/2v seconds
4. The x component of the resultant of several vectors
   (a) is equal to the sum of the x-components of the vectors
   (b) may be smaller than the sum of the magnitudes of the vectors

(c) may be greater than the sum of the magnitudes of the vectors

(d) may be equal to the sum of the magnitudes of vectors

5. An iron sphere and a wooden sphere of same mass are projected upwards in air of density r with a velocity v m/sec. Neglecting the effect of viscosity

(a) the maximum height attained by both of them is same

(b) iron sphere will rise higher

(c) both will reach the same maximum height but they will take different times

6. A motor car is going due north at a speed of 50 km/h. It makes a 90° left turn without changing the speed. The change in the velocity of the car is about

(a) 50 km/h towards west

(b) 70 km/h towards south-west

(c) 70 km/h towards north-west

(d) zero

7. The displacement of a particle is represented by

$$x = a_0 + \frac{a_1}{2} t + \frac{a_2}{3} t^2$$

then the acceleration of the particle will be

(a) a_0 (b) $a_1/2$

(c) $2a_2/3$ (d) none of the above

8. Wind is blowing west to east along two parallel tracks. Two trains moving with same speed in opposite directions have the steam track of one double than other. The speed of each train is

(a) equal to that of wind

(b) double that of wind

(c) three times that of wind

(d) half that of wind

9. A stone is released from an elevator going up with an acceleration a. The acceleration of the stone after the release is

(a) a upward (b) (g – a) upward

(c) (g – a) downward (d) g downward

10. A particle starting from rest travels a distance x in first 2 seconds and a distance y in next 2 seconds then

(a) y = x (b) y = 2x

(c) y = 3x (d) y = 4x

11. A particle has a velocity u towards east at t = 0. Its acceleration is towards west and is constant. Let x_A and x_B be the magnitude of displacements in the first 10 seconds and the next 10 seconds

(a) $x_A < x_B$

(b) $x_A = x_B$

(c) $x_A > x_B$

(d) the information is insufficient to decide the relation of x_A with x_B

12. The engine of a train passes an electric pole with a velocity u and the last compartment of the train crosses the same pole with a velocity x. Then the velocity with which the midpoint of the train passes the pole is

(a) u (b) v

(c) (u + v)/2 (d) $\sqrt{[(u^2 + v^2)/2]}$

13. Two projectiles A and B are projected with angle of projection 15° for the projectile. A and 45° for the projectile B. If R_A and R_B be the horizontal range for the two, then

(a) $R_A < R_B$

(b) $R_A = R_B$

(c) $R_A > R_B$

(d) the information is insufficient to decide the relation of R_A with R_B

14. A body is thrown vertically upwards in air when air resistance is taken into consideration, let the time of ascent be t_1 and the time of descent be t_2. Then

(a) $t_1 = t_2$ (b) $t_1 < t_2$

(c) $t_1 > t_2$ (d) $t_1 \leq t_2$

15. A person travelling on a straight line moves with a uniform velocity V_1 for some time and with uniform velocity V_2 for the next equal time. The average velocity V is given by

(a) $v = \frac{V_1 + V_2}{2}$ (b) $v = \sqrt{V_1 V_2}$

(c) $\frac{2}{v} = \frac{1}{V_1} + \frac{1}{V_2}$ (d) $\frac{1}{v} = \frac{1}{v_1} + \frac{1}{v_2}$

16. The speed of the car is reduced to one-third of its original value in travelling a distance S. Then the car is brought to rest in a minimum distance

(a) 9 s (b) (8/9) s

(c) (9/8) s (d) 3 s

17. In a projectile motion the velocity

(a) is always perpendicular to the acceleration

(b) is never perpendicular to the acceleration

(c) is perpendicular to the acceleration for one instant only

(d) is perpendicular to the acceleration for two instants

18. The range of a projectile fired at an angle of 15° is 50m. If it is fired with the same speed at an angle of 45°, its range will be

(a) 25 m (b) 37 m

(c) 50 m (d) 100 m

19. A stone is dropped into a well in which the level of water is h below the top of the well. If v is velocity of sound, the time T after which the splash is heard is given by

(a) T = 2h/v (b) $T = \sqrt{\left(\frac{2h}{g}\right)} + \frac{h}{v}$

(c) $T = \sqrt{\left(\frac{2h}{v}\right)} + \frac{h}{g}$ (d) $T = \sqrt{\left(\frac{h}{2g}\right)} + \frac{2h}{v}$

20. Two bullets are fixed simultaneously, horizontally and with different speeds from the same plane. Which bullet will hit the ground first ?

    (a) the faster one

    (b) the slower one

    (c) both will reach simultaneously

    (d) depends on the masses

21. The displacement is given by $x = 2t^2 + t + 5$ the acceleration at $t = 2s$ is

    (a) 4 m/s^2 (b) 8 m/s^2

    (c) 10 m/s^2 (d) 15 m/s^2

22. A person standing near the edge of the top of a building throws two balls A and B. The ball A is thrown vertically upward and B is thrown vertically downward with the same speed. The ball A hits the ground with a speed v_A and the ball B hits the ground with a speed v_B. We have

    (a) $v_A > v_B$

    (b) $v_A < v_B$

    (c) $v_A = v_B$

    (d) the relation between v_A and v_B depends on height of the building above the ground

23. A particle moves a long a straight line such that its displacement at any time t is given by

    $S = t^3 - 6t^2 + 3t + 4$ metres

    The velocity when the acceleration is zero is

    (a) 3 ms^{-1} (b) – 12 ms^{-1}

    (c) 42 ms^{-1} (d) – 9 ms^{-1}

24. A body sliding on a smooth inclined plane requires 4 seconds to reach the bottom starting from rest at the top. How much time does it take to cover one fourth the distance starting from the rest at the top?

    (a) 1 seconds (b) 2 seconds

    (c) 4 seconds (d) 16 seconds

25. A river is flowing from west to east at a speed of 5m/min. A man on the south bank of the river, capable of swimming at 10 metre/min, in still water, wants to swim across the river in the shortest time. He should swim in a direction

    (a) due north (b) 30° east of north

    (c) 30° north of west (d) 60° east of north

26. A particle moves along x-axis in such a way that its coordination (x) varies with time (t) according to the expression

    $x = (2 - 5t + 6t^2)$ metre

    The initial velocity of the particle is

    (a) – 5 m/sec (b) – 3 m/sec

    (c) 6 m/sec (d) 3 m/sec

27. A person travelling on a straight line moves with a uniform velocity v_1 for a distance x and with a uniform velocity v_2 for the next equal distance. The average velocity v is given by

    (a) $v = \frac{v_1 + v_2}{2}$ (b) $v = \sqrt{v_1 v_2}$

    (c) $\frac{2}{v} = \frac{1}{v} + \frac{1}{v_2}$ (d) $\frac{1}{v} = \frac{1}{v_1} + \frac{1}{v_2}$

28. A particle starts from rest and travels a distance s with uniform acceleration, then it travels a distance 2s with uniform speed, finally it travels a distance 3s with uniform retardation and comes to rest. If the complete motion of the particle is a straight line then the ratio of its average velocity to maximum velocity is

    (a) 6/7 (b) 4/5

    (c) 3/5 (d) 2/5

29. A particle at rest starts moving in a horizontal straight line with a uniform acceleration. The ratio of the distance covered during the fourth and the third second is

    (a) 4/3 (b) 26/9

    (c) 7/5 (d) 2

30. A particle moves in x – y plane according to equations. $x = 4t^2 + 5t + 16$, $y = 5t$. The acceleration of the particle must be

(a) 8 m/sec^2 (b) 13 m/sec^2

(c) 14 m/sec^2 (d) none of the above

31. The acceleration 'a' in m/sec^2, of a particle is given by $a = 3t^2 + 2t + 2$ where 't' is the time. If the particle starts out with a velocity V = 2m/sec at t = 0, then the velocity at the end of 2 seconds is

(a) 12 m/sec (b) 14 m/sec

(c) 16 m/sec (d) 18 m/sec

32. The motion of a body is given by the equation dv(t)/dt = 6.0 – 3v (t) where v (t) is the speed in m/s and t in sec. If the body was at t = 0

(a) the terminal speed is 2.0 m/s

(b) the speed varies with time at $v(t) = 2(1 - e^{-t})$ m/s

(c) the speed is 1.0 m/s when the acceleration is half the initial value

(d) the magnitude of the initial acceleration is 6.0 m/s^2

33. The acceleration of a particle is increasing linearly with time t as bt. The particle starts from the origin with an initial velocity V_0. The distance travelled by the particle in time t will be

(a) $V_0 t + 1/3\ bt^2$ (b) $V_0 t + 1/3\ bt^3$

(c) $V_0 t + 1/6\ bt^3$ (d) $V_0 t + 1/2\ bt^3$

34. The deceleration experienced by a moving motor boat after its engine is cut off, is given by $dv/dt = -kv^3$ where k is constant. If V_0 is the magnitude of the velocity at cut-off, the magnitude of the velocity at a time t after the cut-off is

(a) $\dfrac{V_0}{\sqrt{(2V_0^2kt + 1)}}$ (b) $V_0\ e^{-kt}$

(c) $V_0/2$ (d) V_0

35. The position x of a particle varies with time, t, as $x = at^2 - bt^3$. The acceleration of the particle with be zero at time t equal to

(a) $\dfrac{2a}{3b}$ (b) $\dfrac{a}{b}$

(c) $\dfrac{a}{3b}$ (d) zero

36. Mark the correct statements
    (a) the magnitude of the velocity of a particle is equal to its speed
    (b) the magnitude of average velocity in an interval is equal to its average speed in that interval
    (c) it is possible to have a situation in which the speed of a particle is always zero but the average speed is not zero
    (d) it is possible to have a situation in which the speed of the particle is never zero but the average speed in an interval is zero
37. A particle moves along the x-axis as

$$x = u\,(t - 2s) + a\,(t - 2s)^2.$$

    (a) the initial velocity of the particle is u
    (b) the acceleration of the particle is a
    (c) the acceleration of the particle is 2a
    (d) at t = 2s particle is at the origin
38. Pick the correct statements
    (a) average speed of a particle in a given time is never less than the magnitude of the average velocity
    (b) the average velocity of a particle is zero in a time interval. It is possible that the instantaneous velocity is zero in the interval
    (c) the average velocity of a particle moving on a straight line is zero in a time interval. It is possible that the instantaneous velocity is never zero in the interval. (Infinite accelerations are not allowed)
    (d) None of these.
39. The acceleration of a particle as seen from two frames S_1 and S_2 have equal magnitude $4\ m/s^2$
    (a) the frames must be at rest with respect to each other
    (b) the frames may be moving with respect to each other but neither should be accelerated with respect to the other
    (c) the acceleration of S_2 with respect to S_1 may either be zero or $8\ m/s^2$

(d) the acceleration of S_2 with respect to S_1 may be anything between zero and 8 m/s^2

40. An object may have

(a) varying speed without having varying velocity

(b) varying velocity without having varying speed

(c) nonzero acceleration without having varying velocity

(d) nonzero acceleration without having varying speed

41. Mark the correct statements for a particle going on a straight line

(a) if the velocity and acceleration have opposite sign, the object is slowing down

(b) if the position and velocity have opposite sign, the particle is moving towards the origin

(c) if the velocity is zero at an instant, the acceleration should also be zero at that instant

(d) if the velocity is zero for a time interval, the acceleration is zero at any instant within the time interval

42. The velocity of a particle is zero at $t = 0$

(a) the acceleration at $t = 0$ must be zero

(b) the acceleration at $t = 0$ may be zero

(c) if the acceleration is zero from $t = 0$ to $t = 10$s, the speed is also zero in this interval

(d) if the speed is zero from $t = 0$ to $t = 10$s, the acceleration is also zero in this interval

43. What are *direction cosines* ? Show that is a unit be vector be resolved in terms of i, j and k the coefficients of i, j and k are the direction cosine of the vector.

44. What angle does the vector $i + \sqrt{2}\, j + 3k$ make with the axis of z and what is the value of its component in the y — z plane?

Ans. $30°, \sqrt{11}$.

45. If vectors A and B be respectively equal to $3i - 4j + 5k$ and $2i + 3j - 4k$, obtain the magnitudes of vectors A + B and A –

B. What will be the unit vectors parallel to A + B and A – B respectively ?

Ans. $\sqrt{27}$; $\sqrt{131}$; $\frac{1}{\sqrt{27}}(5i - j + k)$; $\frac{1}{\sqrt{131}}(i - 7j + 9k)$

46. (a) If A = 3i + 4j +4k and B = i + 4j + 4k, obtain the direction cosines of A + B and A – B.

Ans. 1/3, 2/3, 2/3, 1,0,0.

(b) What should be the vector which, on addition to vectors – 2i + 3j – 4k and – 2j + 4k, will give a unit vector along the axis of y? **Ans.** 2i.

47. If a and b are the vector $\overrightarrow{AB}, \overrightarrow{BC}$, determined by the adjacent sides of a regular hexagon, what are the vectors determined by the other sides, taken in order ?

Ans. (b – a). – a, – b, (a – b).

48. (a) The points A, B and C are respectively i + 2j – 3k, 3i – 4j + 5k and 5i – 10j + 13k. Show that vectors $\overrightarrow{AB}$ and $\overrightarrow{BC}$ are collinear.

(b) Deduce the condition for the vectors 2i + 3j – 4k and 3i – aj + bk to be parallel (or collinear).

Ans. a = – 9/2, b = – 6.

49. Deduce the condition for the coplanarity of three vectors A, B and C.

50. Show that the points A(4, 5, 1), B(0, – 1, – 1), C(3, 9, 4), and D(–4, 4, 4) are coplanar.

51. Two particles travel along the axes of z and y with velocities $v_1 = 3i$ and $v_2 = 4j$ respectively. If at the start their position be $x_1 = -4$ cm, $y_1 = 0$ and $x_2 = 0$, $y_2 = -5$ cm, obtain the position vector r giving the relative position of the second particle with respect to the first.

When will the two particles be the closest together ?

Ans. r = (4t – 5)j – (3t – 4)i; 1.28 s after start.

$\frac{d}{dt}|r|^3 = \frac{d}{dt}\left[(3t-4)^2 + (4t-5)^2\right] = 0$, whence : can be easily calculated.]

52. What is meant by the scalar product of two vectors ? show that it *is commutative as well as distributive.*

    Use the concept of scalar product to deduce the *cosine law.*

53. If a and b are two vectors given by $a = a_x i + a_y j + a_z k$ and b $= b_x i + b_y j + b_z k$, show that $a.b = a_x b_x + a_y b_y + a_z b_z$.

54. Obtain the scalar product of vectors (6i + 2j + 3k) and (2i – 9j + 6k) and also the cosine of the angle between them.

    Ans. 12; 12/77.

55. Define scalar and vector products of two vector. If i, j, k are unit vectors along x, y, and z axes respectively, show that ii = 1 and i.j = 0.

56. (a) Show that vectors A = 2i + 3j – 4k and B = 5i + 2j + 4k are perpendicular to each other.

    (b) Under what condition will the vectors A = 3i – 5j + 5k and B = 5i – j + bk be perpendicular to each other ?

    Ans. When b = – 4.

57. (a) If $|A + B| = |A - B|$, Show that A and B are perpendicular to each other.

    (b) What conclusions can you draw from the fact that scalar product A B = 0 ?

58. Forces acting on a particle have magnitudes 5, 3, 1 lb wt and act in the directions of the vectors 6i + 2j + 3k, 3i – 2j + 6k, 2i – 3j –6k respectively. These remain constant while the particle is displaced from the point A(2, – 1, – 3) to B (5, –1, –1). Find the work done by the forces, the unit of length being 1ft.

    Ans. 33 ft. lb.

    [**Hint:** Here, clearly, the three forces are 5(6i + 2j + 3k), 3(3i – 2j + 6k) and 1(2i – 3j – 6k) respectively. So that, resultant is (41i + j + 27k). For the rest, see worked example 11].

59. Show that if the scalar product of a vector r with each of three non-coplanar vectors be zero, then r must be the zero vector.

    [**Hint:** See worked example 6 (b)]

60. Show that if a and b are equal in magnitude and not-parallel, a + b and a – b are perpendicular.

61. Find the cosine of the angle between the vectors a = 3i + j + 2k and b = 2i – 2j + 4k. Find also the unit vector perpendicular to both a and b.

Ans. $\frac{3}{\sqrt{21}}; \frac{1}{\sqrt{3}}(i-j-k)$

62. If two vectors are 6i + 0.3j – 5k and i – 4.2j + 2.3k, find the vector product and by calculation prove that the new vector is perpendicular to the two.

63. What is the unit vector perpendicular to each of the vectors a = 2i – j + k and b = 3i + 4j – k ? Calculate the sine of the angle between these vectors.

Ans. $\frac{1}{\sqrt{155}}(-3i+5j+11k); \sqrt{\frac{155}{156}}$.

64. Show that the vector product is (i) *associative*, (ii) *not commutative.*

65. (a) Write down the values of (i) i × i, j × j, k × k (ii) i × j, j × k and k × i. **Ans.** (i) 0, 0, 0; (ii) k, i, j.

(b) Show that if b = c + ma, where m is any scalar, a × b = a × c, but that if a × b = a × c, b is not necessarily equal to c.

66. Express the magnitude of a × b in terms of the scalar product.

Ans. $(a \times b)^2 = a^2 b^2 - (a.b)^2$.

67. Show that if a, b, c be three vectors, we have a × (b + c) = a × b + a. × c.

68. Obtain the vector area of a triangle whose vertices are a, b and c. What is the condition for collinearity of the three points?

Ans. vector are = 1/2 (b × c + c × a + a × b);
a × b + b × c + c × a = 0.

69. Find the area of the parallelogram determined by vectors i + 2j 3k and – 3i – 2j +k.

What is the sine of the angle between the two vectors ?

Ans. $6\sqrt{5}$ sq.units; $3\sqrt{5/7}$.

70. Find the torque about the point i + 2j – k of a force represented by 3i + k acting through the point 2i – j + 3k.

71. Deduce the unit vector which is perpendicular to both the vectors P = 3i – 3j and Q = 6j + 5k.

Ans. $\frac{3}{\sqrt{774}}(5i+5j-6k)$.

72. A proton of velocity (3i + 2j) 10^8 cm/s enters a magnetic field 2000j + 3000k *gauss*. Calculate the acceleration. Also deduce the simultaneous electric field which will keep the particle undeflected.

Ans. 2.8 × 10^{15} (2i – 3j + 2k) cm/s^2; – 20i + 30j – 20k *esu*.

[**Hint:** $F = \frac{q}{c}(v \times B)$; if E be the required electric field $F = \frac{q}{c}(v \times B) + qE = 0$.]

73. The vectors i + 2j + 3k, – 3i + 2k and – 5k represent the edges of a parallelopiped; obtain its volume. **Ans.** 15 *units.*

74. (a) Deduce the condition for coplanarity of three vectors.

(b) Obtain the scalar triple product of (i) the vectors a, a, and b, (ii) the vectors a, b and c where a and b are parallel.

Ans. (i) 0; (ii) 0.

75. Vectors A, B and C are respectively given by 2i + j – 3k, i + 3j + 4k and 5j + mk. What should be the value of m so that (i) the volume of the parallelopiped, with its edges represented by the three vectors, be 15 units, (ii) the three vectors may be coplanar? **Ans.** (i) m = 15 or 8; (ii) m = 11.

76. (a) Write down the values of (i) k. k, (ii) [i j k], (iii) a × (b × c), (iv) b × (c × a) (v) c × (a × b).

Ans. (i) 1, (ii) 1, (iii) (a. c) b – (a. b)c, (iv) (a. b) c – (b. c)a, (v) (b. c) a – (c. a)b.

(b) Show that a × (b × c) + b × (c × a) + c × (a × b) = 0.

77. Given the three vectors A = 5i – 7j + 3k, B = – 4i + 7j – 8k, C = 2i – 3j, deduce the following: (i) A. B × C and A × B. C, (ii) A^2 C. **Ans.** (i) – 14. – 14; (ii) 83 (2i – 3j).

78. A body of mass 500 g, free to move, is subjected to a force of i + 2j + k newton. What velocity will it acquire after 4 seconds

and what distance will it have covered in that interval of time?

Ans. v = (8i + 16j + 8k) m/s;
S = 16i + 32j + 16k metre.

79. Show that in the case of a rigid body spinning about a fixed axis, the linear velocity (v) of a particle of the body is the cross product of the angular velocity ω and the position vector r of the particle relative to a fixed point on the axis.

80. A rigid body is spinning with an angular velocity of 4 radians per second about an axis parallel to (3j – k), passing through the point (1 + 3j –k). Find the velocity of the particle at the point 4i – 2j + k.

Ans. $v = \frac{4}{\sqrt{10}}(i - 3j - 9k)$, its magnitude $= 4\sqrt{\frac{91}{10}}$.

81. Obtain the values of (i) grad (1/r), (ii) [grad f (r)] × r.

Ans. (i) $-r/r^2$. (ii) 0.

82. Given that the electric field is the negative of the gradient of potential, *i.e.*, $E = -\nabla V$, obtain the value of the electric field if that of the potential be – Kxy, where K is a constant.

Ans. K (iy + jx)

83. If r = xi + yj + zk, show that (i) div r = 3, (ii) curl r = 0, (iii) div $(r^n r) = (n + 3)r^n$ and (iv) $\nabla^2 r^{-1} = 0$.

84. Show that (a) curl rϕ (r) = 0 and curl (a. r) a = 0.

85. Prove that curl grad ϕ = 0 and div curl f = 0.

86. The Laplacian Operator $\nabla^2 = \nabla . \nabla$ Why do we not have another operator $\nabla \times \nabla$?

87. What is meant by line and surface integrals ? Enunciate and prove Gauss's theorem.

88. Show that $\iint_3 F.dS = 6V$, where S is a closed surface, enclosing a volume V and F = xi + 2yj + 3zk.

89. Show that $\int_3 (x^2 i + y^2 j + z^2 k).\, ndS = 0$, where S denotes the surface of the ellipsoid $\frac{x^2}{a^2} + \frac{y^2}{b^2} + \frac{z^2}{c^2} = 1$.

90. Enunciate and prove *stokes' theorem.*

91. Show, using either Gauss's or Stokes' theorem, that $\int_3 \text{curl } f.ndS = 0$, where S is a closed surface.

92. Let $\vec{C} = \vec{A} + \vec{B}$

   (a) $|\vec{C}|$ is always greater than $|\vec{A}|$

   (b) it is possible to have $|\vec{C}| < |\vec{A}|$ and $|\vec{C}| < |\vec{B}|$

   (c) C is always equal to A + B

   (d) C is never equal to A + B

93. The displacement of a particle is proportional to the cube of time. Then magnitude of its acceleration

   (a) increases with time

   (b) decreases with time

   (c) constant but not zero

   (d) zero

94. Let the angle between two nonzero vectors $\vec{A}$ and $\vec{B}$ be 120° and its resultant be $\vec{C}$

   (a) C must be equal to |A – B|

   (b) C must be less than |A – B|

   (c) C must be greater than |A – B|

   (d) C may be equal to |A – B|

95. The magnitude of the vector product of two vectors $\vec{A}$ and $\vec{B}$ may be

   (a) greater than AB (b) equal to AB

   (c) less than AB (d) equal to zero

2

Reference Frames: Newtons Laws of Motion—Galilean Invariance

A PART FROM THE WELL UNDER

A part from the well understood mathematical terms like point, straight line etc., we shall repeatedly come across three other terms :

(i) Particle,

(ii) Event, and

(iii) Observer, as we proceed alongwith our subject. It is best, therefore, to understand at the very outset their exact connotations.

(i) *Particle :* A particle is ideally just a piece or a quantity of matter, having practically no linear dimensions but only a position, the measure of this quantity of matter being the mass of the particle.

Such idealised particles or point-bodies are, of coarse, convenient mathematical fictions and have no real existence. All that is implied is that, in our actual physical problems, if we happen to be dealing with a body whose size is very much smaller than the other quantities involved, we may to a first approximation, treat the body mathematically as a point-body or a particle. Thus, for instance, the bob of a simple pendulum, in relation to the length of the string carrying it or a planet, in relation to its distance from the sun and the stars, may be considered to be a particle. It is not always that simple, however For, the tiny electron, very close to a charged body is not regarded as a particle. It thus depends really on the circumstances obtaining in any given case whether or not a body is to be regarded as a particle. This will become-clearer as we proceed along.

(ii) *Event:* An event stands in oar mathematical model for anything that occurs suddenly or instantaneously at a point in space. It thus involves both, a position and a time of occurrence.

(iii) *Observer:* A person or an equipment which can locate record measure and interpret as event is called an observer.

REST AND MOTION

We are already generally familiar with these terms from our junior classes. Here, therefore, we may simply emphasize the fact that they have a meaning only in relation to a given frame of reference. A particle is taken to be at rest or in motion in relation to the given frame of reference, according as it continues, or does not continue, to occupy the same position in that particular frame of reference.

REFERENCE FRAME

A system of coordinate axes which defines the position of a particle or an event in two or three dimensional space is called a frame of reference. The essential requirement of a frame of reference is that it should be rigid

The simplest frame of reference is, of course the familiar Cartesian system of coordinates, in which the position of the particle is specified by its three coordinates x, y, along the three perpendicular axes, such that if r be the position vector of the particle with respect to the origin, we have

$$r = xi+yj+zk,$$

where i, j and k are the unit vectors along the three axes.

So that, velocity of the particle is given by

$$v = \frac{dr}{dt} = \frac{dx}{dt}i + \frac{dy}{dt}j + \frac{dz}{dt}k,$$

and its *acceleration* by $a = \frac{dv}{dt} = \frac{d^2r}{dt^2}$.

The origin of the frame of reference may not always be coincident with the position of the observer. The frame of reference must however, be at rest with respect to him.

Obviously, for complete identification of an exact in a reference frame, *i.e.*, for determination of its exact location as well as the exact time of its occurrence, we must, in addition to the usual three spatial coordinuum x, v and z, have yet another coordinate, viz., that of time t, perpendicular to the former three.

A reference frame, with such four coordinates x, y, a and t is referred to as a *space-time frame*—the four axes defining a four dimensional continuum, called space-time.

NEWTON'S LAWS OF MOTION AND THEIR LIMITATIONS

First Law

A body must continue in its state of rest or of uniform motion along a straight line unless acted upon by an external force.

Second Law

The rate of change of momentum is proportional to the impressed force and takes place in the direction of the force.

Third Law

To very action, there is an equal and opposite reaction. Thus, if F_{12} and F_{21} be the forces exerted on each other by two interacting bodies respectively, we have $F_{12} = F_{21}$.

Obviously, the first law defines farce as the *causal agent* of *motion change*, implying clearly that if there be no force applied, there will be no change in velocity and therefore, no acceleration or, that

$$\text{if } F = 0, \text{ we have } a = 0.$$

The second law gives s measure of force as the rate of change of momentum, implying that if p be the momentum of a particle = mv, where v is its velocity, we have

$$F = \frac{dp}{dt} = \frac{d}{dt}(mv).$$

If, therefore, m be constant and independent of the velocity of the particle, we nave

$$F = m\frac{dv}{dt} = ma$$

So that, again, if F = 0, a = 0, which is the first law.

Thus, *the first law is simply a special case of the second law when F = 0 and is thus contained in it*. Hence it is that the second law is taken to be the *real law of motion.*

The obvious and inherent *limitation* of the law is that the relation F = ma would not hold good in case m does not remain constant as, for example, in the case of (i) *a falling ruin drop*, rain drop, which gathers mass as it falls, duo to more water vapour condensing around it, (ii) a *rocket*, which sheds part of its mass as it moves forward, in the form of ejected burnt fuel, and (iii) *particles with relativistic velocities i.e.*, velocities which are an appreciable fraction of the velocity of light, (c) in free space, as we shall see law in the chapter following.

As to the third law of motion, it implies that the forces exerted by the two interacting bodies over each other are *equal and opposite provided they are both measured simultaneously*. A simultaneous measurement of The two forea is, of course, impossible but if the time taken by the interaction of the two bodies he sufficiently large compared with that taken by the light signal to travel from one to the other, this requirement may be said to be almost fulfilled for ordinary practical purposes. This means in other words, that *the law ceases to hold good for particles of atomic dimensions* for which obviously, this requirement cannot possibly be fulfilled. *This is an obvious limitation of the law.*

INERTIAL FRAMES OF REFERENCE

We have just discussed, Newton's laws of motion and their limitations. A very important point about the first two laws is that they do not hold good in each and every frame of reference. For a body at rest in one reference frame may not necessarily appear to be so in another. It may, for instance, appear to be moviaz in a circle in a frame of reference rotating with respect to the first. It is only in a very special frame of reference that the two laws of motion hold good. Such a frame of reference is called a Newtonian, inertial or Galilean frame of reference (because it was *Galileo* who first enunciated the law of inertia which we refer to as Newton's first law of motion). The fast law defines and the second identifies such a frame of reference to be *non-accelerating* and, therefore, also non-rotating. This means that in an inertial frame, the *acceleration* $a = d^2r/dt^2 = 0$ (because the force applied, $F = ma$ is zero), or that $d^2x/dt^2 = 0$, $d^2y/dt^2 = 0$, and $d^2z/dt^2 = 0$.

Now, suppose we have an inertial frame of reference S and another reference frame S' (with or without their respective coordinate axes parallel to each other), such that their orientations in space remain fixed and they are thus free to have only translatory motion with respect to one another. Suppose further that, with the origins O and O' of the two systems *initially coincident*, system S' has moved into the present portion, in time t, with a *constant velocity* v *relative* to S, so that $OO' = R = vt$.

The position vectors r and r' of a particle P, corresponding to the two frames of reference, are obviously related as r =R + r',

whence, r' = r – R = r – vt.

So that, we have d^2r/dt^2 = d^2r/dt^2 = [v being constant].

Clearly, d^2r/dt^2 is the acceleration of the particle in frame S' and d^2r/dt^2, its acceleration in the inertial frame S.

Thus, the acceleration of the particle is the *same* in the two frames of reference, indicating that if the partick be at rest in the inertial frame S, it would also appear to be at rest in frame S'. In other words, frame S' would also be an inertial frame.

We thus arrive at a general rule that *all frames of reference, moving with a constant velocity with respect to an inertial frame, are also inertial frames of reference.*

This means, as mentioned already, that inertial frames are *non-accelerating frames* or else a body at rest or in uniform motion in one frame may appear to have an accelerated motion in the other.

As will be readily seen, the rotating frame of reference, referred to above, in which a body, which is at rest in an inertial frame, appears to he moving in a circle (and thus having an acceleration) is *not* an inertial frame of reference. This again implies, as mentioned earlier, that *inertial frames are also non-rotating frames.*

Further, since Newton's laws of motion are obeyed only in inertial frames of reference, it follows as a natural consequence that, subject to the limitations mentioned under *mass remaining constant, the relation F = ma holds good only in inertial frames, not in now-inertial ones.*

Now, an inertial frame of reference being an unaccelerated one, the question arises as to -where to actually locate it. Obviously, the earth in not a suitable place for it since, as we know, apart from going round the awn, it also spins about its own axis. This imparts to a body in the frame of reference, located on the earth, a centripetal acceleration, whose value at the equator would be $\omega^2R = (2\pi/24 \times 60 \times 60)^2 \times 6.4 \times 10^8$ = 3.4 cm/s², which is by no means negligible. Nevertheless, being quite small for most of our everyday purposes, it is quite often neglected, and *the earth tafcen to be a satisfactory inertial* frame of reference. Hence, also the interior of a vehicle (train, car etc.), moving smoothly, with constant velocity, on the earth's surface, may be taken to be an equally satisfactory inertial frame of reference.

Strictly speaking, however, the earth being not quite satisfactory for the purpose, where else can we think of locating the frame of reference? Not even the sun is at rest but revolves about the centre of our galaxy, which itself moves through space. The nearest to something fixed in apace are the *fixed stars*, so called because they seem to preserve their arrangement. Not that we know them to be actually fixed in their positions in space, but their velocity from this fax off distance on the earth appears to be almost negligible. *Any rigid body, therefore, situated in remote space, far from attracting matter and fixed with respect to the fixed stars, may be taken to be an inertial Trowel of reference.*

Now, knowing full well that, in actual practice, motion is a always described in a relative frame of reference, *i.e.*, in a frame in which the position of a particle or a body is specified in relation to other material objects which may themselves be in motion relative to other objects, Newton insisted that there must exist a *fundamental or an absolute frame of reference*, which he called absolute space, with respect to which all motion must be measured. With the weight of his great authority behind the idea, the futile quest for thin elusive absolute reference frame continued for long.

Here, we may simply state that for most of our purposes, the reference frame, stationary with respect to the fixed stars, is good enough as our *fixed or absolute inertial frame of reference*. All other frames of reference in uniform motion relative to it, naturally, also serve as *equivalent inertial frames,* as also any reference frame whose origin coincides with that of an inertial frame even though its coordinate axes may be inclined to those of the latter.

The term *Galilean* frame of reference is used mostly in connection with *relativity*. Though here too, the earth may well be used as a Galilean frame in quite a number of oases in which its gravitational attraction is small compared with the other forces involved and its rotation is of no importance.

It only remains to be pointed out that the *concept of an inertial frame of reference is of real importance in physics for the following reasons.*

(i) *It is in respect of such a frame of reference that all our fundamental laws of physics have been formulated.*

(ii) *It is one of the basic assumptions of the special theory of relativity that the fundamental laws of physics can all be so expressed as to assume the same mathematical form in all inertial frames of reference.*

(iii) *It is endowed with the property of being isotropic with respect to mechanical and optical experiments ('isotropic' meaning same in all directions).*

GALILEAN TRANSFORMATION

An event or a physical phenomenon, observed simultaneously in two separate reference frames, will naturally have two separate sets of coordinates corresponding to the two frames of reference. Equations relating the two sets of coordinates of the event in the two systems are called transformation equations, because they enable the observations made in one system to be transformed, as it were, into those made in the other. And, if the two systems happen to be inertial or Galilean ones, the transformation is referred to as a Galilean transformation. Let us consider a few such transformations by way of illustration.

(a) *Transformation of position* : Let S and S' be two inertial frames of reference with observers stationed at their origins O and O' respectively, moving with a relative velocity v along the positive direction of x. Or, what is the same thing, let S be at rest and S' move with uniform velocity v along the +x direction. Needless to mention that the velocities we are considering here are all very much less than c, the velocity of light in free space *i.e.*, $v << c$.

Suppose some event occurs at A. Then, measuring tune from the instant the observers were just opposite each other, *i.e.*, setting the clocks in S and S' to zero when the origins O and O' of the two frames coincide with (or are opposite to) each other, the observer in frame S determines the position of the event by the coordinates x, y, z and the observer in S' by the coordinates x', y', z'. Since frame S' is in uniform motion relative to S, the x-measurements in S exceed those in S' by vt, the distance covered by S' in time t in the +x direction. We, therefore, have $x' = x - vt$. And, since there is no relative motion between S and S' alone the axes of y and z, we have $y = y'$ and $z = z'$.

Now, if we assume time to be independent of any frame of reference, we have the four transformation equations from S to S', viz.,

$$x' = x - vt \;...(i)\; y' = y...(ii)\; z, = z...\text{and } (iii)\, t'...t\, (iv)$$

Or, we could express the transformation in a general form in terms of the position coordinates of A m the two reference frames, as explained

$$x' = r - vt. \qquad ...(v)$$

The inverse transformation will naturally give

$x = x' + vt$, $y = y'$, $z = z'$, and $t = t'$. Also, $r = r' + vt$.

(b) *Transformation rf distance or length* : Suppose there are two events occurring simultaneously at A_1 and A_2, some distance apart, along the axis of x, such that their coordinates, as measured in the two frames of reference, are x_1, y, z; x_2, y, z and x'_1, y', z' x'_2, y', z' respectively. Then, dearly.

transformation equations for the event at A_1 (from S to S') are

$$x'_1 = x_1 + vt,\ y' = y,\ z' = z$$

And *transformation equations for the event at* A_2, (from S to S') are

$$x'_2 = x_2 + vt,\ y' = y,\ z' = z$$

Hence, distance between the two events is given by

$$x'_2 = x'_1 + \Delta x' = (x_2 - vt) - (x_2 - x_1) = \Delta x,$$

indicating that *the distance between the two events, measured by the observer in either frame of reference, is the same, irrespective of the values of v and t* (which need not be known). Only, the observations must be made simultaneously in the two frames of reference.

It will be easily seen that if Δx be the *length L of a rod, placed* along the axis of x in frame S and $\Delta x'$, its length L' along the axis of x' in frame S, we have $L'\ \Delta x' = \Delta x = L$. *i.e., it is the same in either frame of reference.*

Or, we could arrive at the same result more directly by considering the rod coordinates of the rod in the two frames of reference. Let these be x_1, y_1, z_1 and x_2, y_2, z, in frame S used x'_1, y'_1, z'_1 and x'_2, y'_2, z'_2 in frame S'. Then, *clearly length of the rod is n frame S, i.e.,*

$$L = \sqrt{(x_2 - x_1)^2 + (y_2 - y_1)^2 + (z_2 - z_1)^2}$$

and its *length in frame S', i.e.,*

$$L' = \sqrt{(x'_2 - x'_1)^2 + (y'_2 - y'_1)^2 + (z'_2 - z'_1)^2}$$

Since, as we have seen, $x'_2 - x'_1 = x_2 - x_1$, and similarly, $y'_2 - y'_2 = y_2 - y_1$ and $z'_2 - z'_2 = z_2 - x_1$, we have L' = L,

i.e., the length of the rod is the same in either frame of reference.

Now, *when a particular property is characterised by the same* number in the two frames of reference, it is said to be 'invariant to the *transformation*'. So that, we may say that a distance or a length is invariant to Galilean transformation.

(c) *Transformation of velocity :* Let us now consider the velocity measurements u And u' as made by the two observers in their respective frames of reference. For third, we coaster a displacement in the + x direction, say, and the time interval for it. Thus if x_1, t and x_2, t be the coordinates of the initial and final events characterising the displacement in frame S and x'_1, t'_1 and x'_2, t'_2, the corresponding coordinates in frame S', w have

velocity relative to frame S, given by $u = \frac{\Delta x}{\Delta t} = \frac{x_2 - x_1}{t_2 - t_1}$

or since $x_1 = x'_1 + vt'_1$, $x_2 = x'_2 + vt'_2$, $t_1 = t'_1$ we have

$$u = \frac{(x'_2 + vt'_2) - (x'_1 + vt'_1)}{t'_2 - t'_1} + \frac{x'_2 - x'_1}{t'_2 - t'_1} + u\frac{(t'_2 - t'_1)}{(t'_2 - t'_1)}$$

or $u = \frac{\Delta x}{\Delta t'} + v = u' + v.$

or $u' = u - v.$ ...(vi)

Again, we could arrive at the same result from relation (v) above, viz., r' = r – vt, differentiating which, we have

$$\frac{dr'}{dt} = \frac{dr}{dt} - v.$$

or $$u' = u - v \quad ...(vii)$$

indicating that *the velocities measured by the observers in the two frames of reference are not the same.* We, therefore, say that velocity is not invariant to Galilean transformation.

The inverse transformation (from frame S' to S) is, obviously, given by

$$u = u' + v. \quad ...(viii)$$

Here, relation (vii) or (viii) is often referred to as the Galilean law of addition of velocities.

(d) *Transformation of acceleration :* Considering acceleration a and a', as measured by the observers in their respective frames of reference, we have a = du/dt and a' = du'/dt.

Since, as just seen under (c), above, u' = n – v, we have

$$\frac{du'}{dt} = \frac{du}{dt} - 0 \qquad \text{(v being a constant.)}$$

or $$a' = a,$$

indicating that *the accelerations, as measured by the two observers in the two frames of reference respectively, are the same.* Acceleration is thus invariant to Galilean transformation.

FRAMES OF REFERENCE WITH LINEAR ACCELERATION

In Fig. 2.4 let frame S' be moving with a linear acceleration a_o with respect to the *inertial frame* S. Then, particle A or, in fact, any other particle at rest with respect to frame S will clearly appear to be moving with an acceleration —a_o with respect to frame S', And, therefore, a particle having an acceleration a with respect to the inertial frame S will appear to have an acceleration $a' = a - a_o$ in frame S'. So that, if m be the mass of the particle (assumed to remain unaltered in either frame), we have *observed force on the particle in frame S' given by*

$$F = ma' = m\,(a - a_o) = ma - ma_o,$$

where ma = F is clearly the *force on the particale in the inertial frame* S.

We, therefore, have $$F' = F - ma_o.$$

Or, putting $$ma_o = F_o,$$

we have $$F' = F - F_o$$

And if $$F = 0,$$

$$F' = -F_o.$$

Thus, even when no force is acting on the particale in frame S, a force $F_o = ma_o$ appears to be acting on the particle in frame S' which is, therefore, a *non-inertial frame.*

Incidentally, the relation $a' = a - a_o$ or $a = a' + a_o$ is called the *Galilean law of composition of accelerations.*

CLASSICAL RELATIVITY—GALILEAN INVARIANCE

In one of the corollaries to his laws of motion, *Newton* says : '*the motions of bodies included in a given space are the same among themselves whether that space is at rest or moves uniformly forward in a straight line'*. This clearly implies that if we are drifting along at a uniform speed in a closed space ship, *all the phenomena observed, and all the experiment performed inside the ship, will appear to be the same as if the ship were not in motion.* This is the *concept of Classical or Newtonian relativity, usually referred to as Galilean invariance.* It clearly means that *the fundamental physical laws and principles are all invariant to Galilean transformation.* Or, in plain and simple language, *the fundamental physical laws and principles are identical is all inertial frames of reference.* Let us see if this is so.

(i) *Newton's law of motion.* Considering the second law of motion which in the *real* law of motion and includes also the first law, let F be the force acting on a mass m in frame S (Fig. 2.4). Then, we have F = ma, where a is the acceleration produced in the mass. *Assuming the mass to be independent of velocity*, if F be the force acting on it in frame S', we have F' = ma', where a' is the acceleration produced in it in this frame. Since, as we know, a = a in the two inertial frames, we have

$$F = F', \qquad ...(i)$$

indicating that *the law is invariant to Galilean transformation.*

(ii) *The law of conservation of momentum.* The momentum of a body or a particle of mass m and velocity v is as we know, given by P = mv. And, the *law of conservation of momentum* merely states that *if a number of particles collide against each other, their total momentum after collision is the same as their total momentum before collision, irrespective of the collision being elastic or inelastic. Only, in the case of an elastic collision,* the entire kinetic energy of the particles before collision appears as their kinetic energy after collision and, in the case of an inelastic collision, part of the kinetic energy of the particles before collision appears in the form of heat (or some such sort of internal excitation energy) after the collision and only the balance as kinetic energy of the particles.

Thus, if we consider only two particles of masses m_1 and m_2 in frame S moving with velocities u_1 and u_2 respectively before collision and with velocities v_1 and v_2 after collision, the law of conservation of momentum demands that

$$m_1u_1 + m_2u_2 = m_1v_1 + m_2v_2. \quad ...(ii)$$

Taking the masses to remain unaltered in the two inertial frames, if the law of conservation of momentum is to be invariant to Galilean transformation, it should assume, in frame S' the form

$$m_1u_1' + m_2u'_2 = m_1v'_1 + m_2v'_2, \quad ...(iii)$$

where u'_1, u'_2 and v_1', v'_2 are the respective velocities of the particles before and after collision in frame S'.

This is actually found to be the case. For, as we know

$$u_1 = u'_1 + v, \; u_2 = u'_2 + v, \; v_1 = v_1' + v \text{ and } v_2 = v_2' + v,$$

where v is the relative velocity of frame S' with respect to frame S.

Substituting these values in relation (ii) above, we have

$$m_1 (u'_1 + v) + m_2 (u'_2 + v) = m_1 (v'_1 + v) + m_2 (v'_2 + v).$$

Or, $$m_1u_1' + m_2u_2' + v (m_1 + m_2) = m_1v_1' + m_2v_2' + v (m_1 + m_2).$$

Or, $$m_1u'_1 + m_2u_2' = m_1v_1 + m_2v'_2. \quad ...(iv)$$

showing that the *law of conservation of momentum is also invariant to Galilean transformation, i.e., holds good in all inertial frames of reference.*

It may be carefully noted that *although the values of momenta and changes of momentum of the individual particles are different in the two frames of reference, the total momentum of the two is conserved in either frame. Thus, whereas momentum, by itself, is not invariant to Galilean transformation, the principle of conservation is.*

(iii) *The law of conservation of energy.* Again, consider a collision between two particles in the inertial frame S. Taking their respective masses to be m_1 and m_2 and their respective velocities to be u_1, u_2 and v_1, v_2 before and after collision, we have, in accordance with the law of conservation of energy,

$$1/2\, m_1u_1^2 + 1/2\, m_2u_2^2 = 1/2\, m_1v_1^2 + 1/2\, m_2v_2^2 + E, \quad ...(v)$$

where E is part of the energy of the particles (before the collision) appearing in some other form, like heat etc.

Let the velocities of the two masses in frame S' before and after collision be u_1', u_2' and v_1', v_2' respectively. Then, with the values of the masses (m_1 and m_2) remaining unaltered and with E remaining the same (as established by actual experiment), the law of conservation of energy must assume the form

$$1/2\ m_1u_1'^2 + 1/2\ m_2u'_2{}^2 = 1/2\ m_1v_1'^2 + 1/2\ m_2v'_2{}^2 + E \qquad ...(vi)$$

if it is to be invariant to Galilean transformation.

That this is actually so may be easily seen by substituting in relation (v) above, the values of u_1, u_2, v_1 and v_2, obtained by Galilean transformation, as in case (ii) above, when we have

$$1/2\ m_1\ (u'_1{}^2 + v)^2 + 1/2\ m_2\ (u_2' + v)^2$$
$$= 1/2\ m_1\ (v_1' + v)^2 + 1/2\ m_2\ (v_2' + v)^2 + E,$$

where v, as we know, is the relative velocity of frame S' with respect to S.

Or, $1/2\ m_1\ (u'_1{}^2 + v^2 + 2u_1v) + 1/2\ m_2\ (u'_2{}^2 + v^2 + 2u'_2\ v)$

$= 1/2\ m_1\ (v'_1{}^2 + v^2 + 2v'_1v) + 1/2\ m_2\ (v_2'^2 + v^2 + 2v'_2v) + E.$

Or, $1/2\ m_1u_1'^2 + 1/2\ m_2u_2'^2 + v\ (m_1u_1' + m_2u_2')$

$= 1/2\ m_1v'_1{}^2 + 1/2\ m_2v_2'^2 + v(m_1v_1' + m_2v_2') + E.$

Since momentum remains conserved in frame S', we have

$$m_1\ u_1' + m_2u_3' = m_1v_1' + m_2v_2'.$$

And, therefore, $1/2\ ,m_1\ u_1'^2 + 1/2\ m_2u_2'^2 = 1/2\ m_1v_1'^2$
$+ 1/2\ m_2v_2'^2 + E,$

showing that energy is also conserved in frame S'.

In other words, *the law of conservation of energy* (*like that of conservation of momentum*) *is invariant to Galilean transformation.*

Now, simply because any two reference frames, moving with constant relative velocity, can be connected by means of a Galilean transformation, we should not rush to the conclusion that the fundamental laws of Physics are invariant of identical in all reference frames that can be so connected. There is a snag here, namely, we have assumed both *mass* and *time* to be invariant to Galilean transformation. There is little or no justification for it, as we shall see later when we are faced with invariance of the velocity of light (c) in free space. More correctly, the statement should be that the *fundamental physical laws are invariant or identical in all reference frames that can be connected, not by Galilean, but by Lorentz transformation*, which takes due account of the invariance of the value of c. For non-relativistic velocities (*i.e.*, for velocities very much less than c) Lorentz transformations too reduce to Galilean ones.

TRANSFORMATION EQUATIONS FOR A ROTATING FRAME OF REFERENCE

Consider an inertial frame of reference S and another reference frame S' whose origin and coordinate axes coincide with those of S.

Now, with the origins of the two frames still coinciding, let S' start rotating with a uniform angular velocity ω about the common axis of z, so that in time t, the axes OX' and OY' of S' have turned through angle ωt each with respect to axes OX and OY of S.

We have to determine the relation between the coordinates, x, z and x', y', z' of a particle P in the two frames of reference respectively, its position vector with respect to the origin being the same r n either frame.

Proceeding exactly, we have

$x' = xC_{xx}' + yC_{yx}' + zC_{zx}'$, $y' = xC_{xy}' + yC_{yy}' + zC_{zy}'$, and, of course,

$$z' = z.$$

Since $C_{xx}' = \cos XOX' = \cos \omega t$, $C_{yx}' = \cos YOX' = \cos (\pi/_2 - \omega t)$ $= \sin \omega t$ and $C_{zx}' = \cos XOZ' = \cos (\pi/2) = 0$, we have

$$x' = x \cos \omega t + y \sin \omega t, \qquad \text{...(v)}$$

Again, since $C_{xy}' = \cos XOY' = \cos (\pi/2 + \omega t) = -\sin \omega t$,

$C_{yy}' = \cos YOY' = \cos \omega t$ and $C_{zy}' = \cos ZOY' = \cos (\pi/2) = 0$,

we have $y' = -x \sin \omega t + y \cos \omega t,$ ...(vi)

and, as pointed out earlier $z' = z.$ ...(vii)

Relations (v), (vi) and (vii) are, therefore, the *transformation equations* in the case of the frame of reference S', rotating with a uniform angular velocity ω relative to the inertial frame S.

The inverse transformation equations (from S' to S) will obviously be

$x = x' \cos \omega t - y' \sin \omega t$, $y = x' \sin \omega t + y' \cos \omega t$ and $z = z'$.

Now, frame S is an *inertial frame*, and no force is applied to particle P. We, therefore, have

$$\frac{d^2x}{dt^2} = 0, = \frac{d^2y}{dt^2} = 0 \text{ and } \frac{d^2z}{dt^2} = 0.$$

On the other hand, differentiating expressions (v), (vi) and (vii). for x', y' and z' respectively with respect to t, we have

$$\frac{dx'}{dt} = -\omega x \sin \omega t + \omega y \cos \omega t + \frac{dx}{dt} \cos \omega t + \frac{dy}{dt} \sin \omega t.$$

Or, since $-x \sin \omega t + y \cos \omega t = y'$ [relation (vi), above], we have

$$\frac{dx'}{dt} = \omega y' + \frac{dx}{dt} \cos \omega t + \frac{dy}{dt} \sin \omega t. \qquad \text{...(viii)}$$

Similarly, $\frac{dy'}{dt} = -\omega x \cos \omega t - \omega y \sin \omega t - \frac{dx}{dt} \sin \omega t + \frac{dy}{dt} \cos \omega t.$

Or, since $x \cos \omega t + y \sin \omega t = x'$ [relation (v) above], we have

$$\frac{dy'}{dt} = -\omega x' - \frac{dx}{dt} \sin \omega t + \frac{dy}{dt} \cos \omega t. \qquad \text{...(ix)}$$

And, of course, $\frac{dz'}{dt} = \frac{dz}{dt}.$ ...(x)

Differentiating expressions (viii), (ix) and (x) once again, with respect to t, we have

$$\frac{d^2x'}{dt^2} = \omega \frac{dy'}{dt} - \omega \frac{dx}{dt} \sin \omega t + \omega \frac{dy}{dt} \cos \omega t.$$

Since from relation (ix) above,

$$-\frac{dx}{dt} \sin \omega t + \frac{dy}{dt} \cos \omega t = \left(\frac{dy'}{dt} + \omega x'\right), \text{ we have}$$

$$\frac{d^2x'}{dt^2} = \omega \frac{dy'}{dt} + \omega\left(\frac{dy'}{dt} + \omega x'\right) = 2\omega \frac{dy'}{dt} + \omega^2 x', \qquad \text{...(xi)}$$

Similarly, $\frac{d^2y'}{dt^2} = -\omega \frac{dx'}{dt} - \omega \frac{dx}{dt} \cos \omega t - \omega \frac{dy}{dt} \sin \omega t.$

Since from relation (viii) above,

$$\frac{dx}{dt} \cos \omega t + \frac{dy}{dt} \sin \omega t = \left(\frac{dx'}{dt} - \omega y'\right), \qquad \text{...(xii)}$$

we have $\frac{d^2y'}{dt^2} = -\omega \frac{dx'}{dt} - \omega\left(\frac{dx'}{dt} - \omega y'\right) = -2\omega \frac{dx'}{dt} + \omega^2 y'.$ ...(xii)

And $\frac{d^2z'}{dt^2} = \frac{d^2z}{dt^2}.$ ...(xiii)

It is thus clear from relations (xi) and (xii) above, that even though no force is acting on particle P in frame S, a force seems to be acting on it in frame S', producing an acceleration in it. *Frame S' is, therefore, a non-inertial frame of reference.*

NON-INERTIAL FRAMES—FICTITIOUS FORCES

We have seen that (i) a frame of reference in accelerated translational motion with respect to an inertial, and (ii) a reference frame in uniform rotation about an inertial are both *non-inertial* frames. Let us consider them a little more in detail.

(i) *The accelerated frame of reference.* If a force F = ma be acting on a particle in an inertial frame S, the force acting on it in a reference frame S', moving with an acceleration a_o with respect to S, is $F' = (ma - ma_o)$. Frame S' is, therefore, a non-inertial frame since Newton's second law of motion, obviously, does not hold good in its case. For, putting $-ma_o = F_o$, with ma, of course, equal to F, we have $-F' = F + F_o$. And, as we have seen already, if F = 0, $F' = F_o. = -ma_o$, *i.e.*, even when no force actually acts on the particle in the inertial frame S, a force equal to F_o *appears to be acting on it in frame* S'.

This force $F_o = -ma_o$ *does not actually exist but appears to come into being purely in consequence of the acceleration of frame* S' *with respect to* S. It is, therefore, aptly called a fictitious or a pseudo (false) force. And, clearly, *it is equal to mass times the acceleration of the non-inertial frame, with its sign reversed.* Or,

fictitious force F_o = mass (acceleration of non-inertial frame, with sign reversed).

Or, $F_o = m(-a_o) = -ma_o$.

As will be easily seen, *it is the force which, when added to the true force* (F) *in the inertial frame* S, *gives the observed force* F' *in the non-inertial or accelerated frame* S'.

This means, in other words, that Newton's second law of motion will also hold good in the non-inertial frame S' which will, therefore, also behave as an inertial frame of reference, *if we add to the true force* F *a fictitious force* $F_o = -ma_o$.

A familiar example of a fictitious force in an accelerated frame of reference is that on a particle at rest with respect to a lift which is descending with an acceleration equal to g, if m be the mass of the particale, the fictitious force acting on it is $F_o = -mg$. So that, the resultant force on it, as observed by an observer in the lift (the moving frame) is given by F' = true force acting on it, (*i.e.*, mg) + the fictitious force, – mg, Or, F' = mg – mg = 0, *i.e.*, the particle is weightless and thus remains suspended in the air, – a condition that obtains in artificial satellites.

Exactly similarly, a body on the palm of our hand appears to our palm to be weightless during the small fraction of time that we are jumping down from a table on to the ground. For, here too, the body is initially at rest with respect to our palm which moves down with an acceleration equal to g.

Another interesting point about a fictitious force is that *it enables us to determine whether or not a given frame of reference is accelerated.* For, if two frames be in uniform relative motion (zero acceleration), with respect to each other, they are obviously inertial frames, and it is impossible to find out which one is at rest and which one in motion. If, however, one of them be accelerated with respect to the other, fictitious forces come into play inside the accelerated frame. These forces are obviously different from the true or real forces. For, unlike the latter, which go on diminishing rapidly with distance and are almost zero at a large distance from a body, the former arise precisely due to the accelerated motion of the frame of reference and go on increasing with its acceleration. So that, whereas the real forces on particles or bodies at large distances from other bodies are negligible, the fictitious forces may have appreciable values. *If, therefore, we find a body or a particle, far away from other material bodies, to be acted upon by an appreciable force, we can safely declare that its frame of reference is an accelerated one.*

(ii) *The rotating frame of reference-Coriolis acceleration and force.* Let us now consider a reference frame S' (shown dotted) rotating with a *uniform angular velocity* ω with respect to an inertial frame S. To make matters simpler, let us assume the two frames of reference to have a common origin 0 and the axis of rotation of S' about S to be passing through 0.

Suppose now we have a moving particle P (like, for instance, a fly or an ant walking on a rotating piece of cardboard). The value of its position vector r at any given instant t will, obviously, be the same in either frame of reference, though its components along the three axes will naturally be different.

If the particle. (*i.e.*, the fly or the ant on the card board) be at rest with respect to the inertial frame S, it would appear to be moving with a relative linear velocity $- \omega \times r$ in the non-inertial; rotating frame S'. And, therefore, if its linear velocity in frame S be v = dr/dt, it will appear in frame S' to be v' = dr/dt – ω × r.

So that, we have

$$\left(\frac{dr}{dt}\right)_{S'} = \left(\frac{dr}{dt}\right)_{S} - \omega \times r. \text{ Or, } v' = v - \omega \times r.,$$

whence,
$$\left(\frac{dr}{dt}\right)_{S} = \left(\frac{dr}{dt}\right)_{S'} + \omega \times r. \qquad ...(i)$$

Or,
$$v = v' + \omega \times r.$$

This gives the linear velocity (v) of the particle in the inertial frame S at any given instant in terms of its velocity (v') in the rotating frame S' and its position vector r at that instant.

Now, equation (i) above, *relating the rate of change of a vector (r in this case) in two given frames of reference* (S *and* S') *is clearly a general one, applicable to any vector and may, therefore, be put in the* form of an *operator equation*

$$\left(\frac{d}{dt}\right)_{S} = \left(\frac{d}{dt}\right)_{S'} + \omega \times. \qquad ...(ii)$$

Applying this to the velocity vector v, in order to deduce a relationship between the accelerations of the particle in the two frames of reference, we have

$$\left(\frac{dv}{dt}\right)_{S} = \left(\frac{dv}{dt}\right)_{S'} + \omega \times r, \text{ i.e., } \frac{dv}{dt}\ \frac{dv'}{dt} + \bar{\omega} \times v.$$

Or, substituting the values of v, we have

$$\left(\frac{dv}{dt}\right)_{S} = \frac{d}{dt}[v' + \omega \times r']_{S'} + \omega \times (v' + \omega \times r).$$

Or

$$\frac{dv}{dt} = \frac{dv'}{dt} + \frac{d\omega}{dt} \times r + \omega\left(\frac{dr}{dt}\right)_{S'} + \omega \times v' + \omega \times (\omega \times r).$$

Clearly dv/dt is the acceleration of the particle, as observed in the inertial frame S. Denoting it by a, and denoting the acceleration dv/dt in the rotating frame S' by a, we have

$$a = a' + 2\omega \times v' + \frac{d\omega}{dt} \times r - \omega \times (\omega \times r)$$

$$[\ \therefore\ (dr/dt)\ S' = v.]$$

Since the angular velocity ω is uniform, $d\omega/dt = 0$. We, therefore have

$$a = a' + 2\omega \times v' + \omega \times (\omega \times r).$$

Here, $2\omega \times v'$ is what is called Coriolis acceleration, after G. Coriolis, who first discovered it; and $\omega \times (\omega \times r)$, the familiar centripetal acceleration. So that, we have

acceleration in observed acceleration Coriolis centripetal the inertial frame = in the rotating frame + acceleration + acceleration.

Hence, if m be the mass of the particle, we have *force acting on the particle in the rotating frame* = ma' = F' *and the true force acting onset in the inertial frame* = ma = F.

And, therefore, $ma = ma - 2m\omega \times v' - m\omega \times (\omega \times r)$.

Or, $$F' = F - 2m\omega \times v' - m\omega \times (\omega \times r).$$

But, as we know, the force acting in a non-inertial frame, *i.e.*,

$$F' = F + F_0,$$

where F_0 is the *fictitious force* acting in the non-inertial frame.

So that, here, *fictitious force* $F_0 = -2m\omega \times v' - m\omega \times (\omega \times r)$,

where $-2m\omega \times v'$ is the *Coriolis force* and $-m\omega \times (\omega \times r)$, the *centrifugal force.*

Thus, here, *fictitious force* F_0 = *Coriolis force* + *centrifugal force.*

It will thus be seen that the centrifugal force is a fictitious force *which acts on a particle at rest relative to a rotating frame of reference.* Being responsible for keeping the particle at rest in the rotating frame, *it is numerically equal to the centripetal force,* $\pi\omega^2 \times r_n$, where r_n is the distance of the particle from the centre of rotation (or the component of r normal to the axis of rotation) *but is oppositely directed, i.e*, outwards, away from the axis of rotation,.

It will also be seen that *Coriolis force too is a fictious force, which acts on a particle in motion relative to a rotating frame of reference.* It is proportional to the angular velocity (ω) of the rotating frame and to the velocity v' of the particle relative to it. Its direction is always perpendicular to that of v' and is obtained by a rotation though 90° in the opposite sense to that of ω.

It will obviously be zero if either v' *or* ω *is zero, i.e., if the particle is at rest relative to the rotating frame* (in which case the only fictious force acting on the particle will be the centrifugal force) or *if the reference frame be a non-rotating one.*

It will further be seen that particle P will move relative to the rotating frame S' in accordance with Newton's law of motion *if we add to the true force F, these two ficititions forces.*

EFFECTS OF CENTRIFUGAL AND CORIOLIS FORCES DUE TO EARTH'S ROTATION

Since the earth is rotating uniformly about its own axis, from west to east, a reference frame fixed on it is clearly a rotating frame of reference. As a result, the two fictitious forces, viz., the *centrifugal force* and the *Coriolis force* must act on a particle, *at rest and, in motion respectively*, relative to the rotating frame or reference. Let us see the effect of these two forces.

(i) *Effect of centrifugal force.* Let us consider a particle P at rest on the surface of the earth, in latitude ϕ. Then, because the particle is at rest in the rotating frame of reference of the earth, *there is no Coriolis force acting on it.* The only fictitious force effective is therefore, the centrifugal force $= m\omega \times (\omega \times R_n)$, where R_n is the component of the radius R of the earth, perpendicular to axis (or the distance of the particle P from the axis). Hence, if the *observed or apparent acceleration* of the particle, in latitude ϕ, be directed towards C, and its *true acceleration*, g, towards the centre (o) of the earth, we have

$$g\phi = g\ \omega \times (\omega \times R_n).$$

Now, if we take the axes, OY and OX, *along* and *perpendicular* to ω respectively, with i and j as the unit vectors along OX and OY as usual, we have

$$g = -g(\cos\phi\ i + \sin\phi\ j),\ w = \omega j \text{ and } R_n = R\cos\phi\ i.$$

Substituting in the expression for $g\ \phi$ above, therefore, we have

$$\begin{aligned} g\ \phi &= -g(\cos\phi\ i + \sin\phi\ j) - \omega j \times (\omega j \times R\cos\phi\ i) \\ &= -g(\cos\phi\ i + \sin\phi\ j) + \omega^2 R\cos\phi\ i, \end{aligned}$$

i.e., the *magnitude of the apparent acceleration* is

$$g\phi = \sqrt{(g\cos\phi - \omega^2 R\cos\phi)^2 + g^2\sin^2\phi}.$$

Since ω is small, we may neglect involving ω^4. So that, we have

$$g\phi = \sqrt{(g^2 \cos\phi - 2g\omega^2 R\cos^2\phi + g^2 \sin^2\phi}.$$

$$= \sqrt{g^2 - 2g\omega^2 R \cos^2\phi}.$$

Or, $$g\phi = g\left(1 - \frac{2\omega^2 R\cos^2\phi}{g}\right)^{\frac{1}{2}}$$

$$= g\left(1 - \frac{\omega^2 R\cos^2\phi}{g}\right)$$

very nearly,

i.e., $$g\phi = g - \omega^2 R \cos^2\phi.$$

This acceleration, as mentioned above, is directed towards C instead of O, the centre of the earth. So that, if the angle that this *apparent direction* PC makes with the true direction PO be ϕ, we have

$$\phi = \tan^{-1}\frac{g\cos\phi - \omega^2 R\cos\phi}{g\sin\phi} = \tan^{-1}\left[\left(1 - \frac{\omega^2 R}{g}\right)\cot\phi\right].$$

Thus, the effect of centrifugal force, due to rotation of the earth, is to reduce the effective value of g on its surface as also to slightly change its direction from the truly vertical,—towards the north and the south respectively in the two hemispheres.

Obviously, ϕ gives the deflection of a plumb line from the true direction of the gravitational force and will have its maximum value in latitude $\phi = 45°$.

(ii) *Effect of Coriolis force*. In case a body is in motion relative to the rotating frame of reference of the earth, the fictitious Coriolis force comes into play and two cases of interest arise: (a) *when the body is just dropped from rest so as to fall freely under the action of the gravitational force and* (b) *when it is given a large horizontal velocity* (case of a *projectile*).

In case (a), the horizontal component of the Coriolis force acting on the freely falling body deflects it a little from its truly vertical path. The vertical component, obviously, produces no such deflection but only affects the values of g.

To calculate this deflection or displacement of the falling body, let us take the axes of x, y and z along the *east, north* and *vertically upwards* respectively and i, j and k as the unit vectors along these axes. Then, if v be the velocity acquired by the body in time t taken by it to fall through a height h, we have

$\mathbf{v} = -\upsilon \mathbf{k}$ ($\because$ the velocity is directed *downwards*)

and $\omega = \omega \cos\phi\, \mathbf{j} + \omega \sin\phi\, \mathbf{k}$, where ϕ is the *latitude* at the place.

$\therefore$ *Coriolis acceleration*, say, $a_c = -2\omega \times v$

$$= -2\omega(\cos\phi\, \mathbf{j} + \sin\phi\, \mathbf{k}) \times (-\upsilon\mathbf{k})$$

$$= 2\omega\upsilon(\cos\phi\, \mathbf{i} - 0) = 2\omega\, \upsilon \cos\phi\, \mathbf{i},$$

i.e., the Coriolis acceleration on the body in latitude ϕ is $2\omega\upsilon \cos\phi$ along the positive direction of the axis of x or *is directed* towards *the east.*

The equation of motion of the body may, therefore, be put as

$$d^2x/dt^2 = 2\omega v \cos\phi.$$

Since v is the velocity acquired by the body in time t, we have $v = 0 + gt = gt$, since its initial velocity is *zero* and the acceleration due to gravity, g. So that,

$$d^2x/dt^2 = 2\omega gt \cos\phi.$$

$\therefore$ *x-component of the velocity of the body*, say, $v_x = dx/dt$

$$= \int 2\omega gt \cos\phi\, dt.$$

Or, $v_x = 2\omega g + \cos\phi\, t^2/2 + C$, where C is the *constant of integration.*

Since v_x is *zero* at the very start, when $t = 0$, we have $C = 0$.

And $\therefore$ $v_x = \omega g \cos\phi\, t^2$.

Integrating once again, we have

displacement along the axis of x, *i.e.*, $x = \int \omega g \cos\phi\, t^2\, dt$

$$= \omega g \cos\phi \left(\frac{t^3}{3}\right) + C'$$

where C' is another *constant of integration.*

Again, since at $t = 0$, there is no displacement, *i.e.*, $x = 0$, we have $C' = 0$.

Hence, $x = 1/2\ \omega g \cos\phi\, t^3$.

Now, t is the time taken by the body to fall through a height h. So that, its initial velocity being zero, we have $h = 1/2\ gt^2$, whence, $t = \sqrt{2h/g}$. Substituting this value of t in the expression for x above, we have

$$x = 1/3\ \omega g \cos\phi\ (2h/g)^{\frac{3}{2}} = \left(\frac{8}{9g}\right)^{\frac{1}{2}} h^{\frac{3}{2}} \omega \cos\phi.$$

Thus, the horizontal displacement of the body due to Coriolis force (i) *in latitude* ϕ is equal to $(8/9g)^{1/2}\ h^{3/2}\ \omega \cos\phi$ and (iii) *at the equator,* since $\phi = 0$, or $\cos\phi = 1$, it is equal to $(8/9g)^{1/2}\ h^{3/2}\ \omega$ (*i.e., the maximum*). And *it is always directed along the positive direction of the x-axis or towards the east.*

In case (b), if the horizontal velocity of the body, (or the projectile) be sufficiently large, so that it covers fairly large horizontal distances, the small *Coriolis force* gets sufficient time to act upon it, making the position vector turn at a constant rate of $-\omega \sin\phi$. Since *in the Northern hemisphere*, ϕ is positive, this rotation, as viewed from above, is *clockwise* (and hence the projectile gets deflected towards the right) and in the Southern hemisphere, it is *anticlockwise* (and the projectile thus gets deflected towards the left). This is known as Ferel's law.

FOUCAULT'S PENDULUM—DEMONSTRATION OF EARTH'S ROTATION

A Foucault's pendulum is just a simple pendulum, with a massive bob carried by a very long wire suspended from a rigid support, so as to have *equal freedom of oscillation in any direction.*

It *constitutes a simple device to demonstrate the rotation of the earth about its own axis* and was first used by Foucault for a public demonstration of the same in the year 1851, in Paris. The suspended mass in his case was 28 kg and the suspension wire about 70 metre long, so that the time-period of the pendulum was nearly 17 sec. Such a pendulum, once set oscillating, continues to oscillate for a long enough time.

The theory underlying the experiment is as follows:

If the pendulum were to be set up at the North pole of the earth, it may oscillate as a simple pendulum in a vertical plane which remains fixed in an inertial or Galilean frame of reference. Since the earth under it rotates from west to east with an angular velocity ω, the plane of oscillation of the pendulum would appear to an observer on the surface of the earth to be turning with an angular velocity $-\omega$, *i.e.*, in the opposite direction (from east to west) to that of the earth.

It is by no means necessary, however, that the pendulum should be situated right at one of the poles. An apparent rotation of its plane of oscillation, due to the earth's rotation may equally well be observed in *any latitude*, except, of course, at the equator. For, it can be shown that, *if the oscillations be small, the bob describes an ellipse with its centred the origin or the position of rest of the pendulum. The rotation of the earth (a rotating frame of reference) merely results in a Coriolis force coming into play due to the horizontal motion of the bob relative to it.* Acting, as usual, in a direction perpendicular to that of the velocity of the bob, *it cause precession of the plane of its oscillation, which thus rotates with an angular velocity* $-\omega \sin \phi$ *latitude* ϕ.

Actually, however, there are two superimposed rotations of the elliptical orbit of the pendulum, one depending upon the area of the elliptical orbit and hence called the *area effect* and the other, on the angular velocity, $-\omega \sin \phi$, due to rotation of the earth and called the *Foucault effect*. The former is a rotation in the *same* sense in which the ellipse is described (and will get reversed in sign if that sense is reversed) and the latter, a rotation *in a definite direction, viz., clockwise in the Northern, and anticlockwise in the Southern, hemisphere.*

Since we are interested only in the rotation of the ellipse in consequence of the earth's rotation or its Foucault rotation. We must eliminate the *are effect* which, in the absence of suitable precautions, may be so much greater than the *Foucault effect* as to completely mask it. This is usually done by setting the pendulum in motion *by first drawing it a little to one side with the help of a piece of thread tied to it and then burning the thread.* The angular velocity of the ellipse (*i.e.*, its rate of precession) due to the area effect is then negligible compared with its Foucault angular velocity, $-\omega \sin \phi$. that, in the case of a pendulum, set oscillating in this manner, the precession or the rotation of its orbit may be taken to be due entirely to Foucault effect *i.e.*, due to rotation of the earth alone.

Obviously enough, the time taken by one full rotation of the plane of oscillation of the pendulum is given by $T = 2\pi/($*Foucault angular velocity*) = $2\pi/\omega$ sin π. Or, since time taken by the earth to complete one rotation about its axis is 24 hours, we have $2\pi/\omega$ = 24 hours and, therefore, T = 24 sin ϕ hours.

As already mentioned, the rotation of the plane of oscillation of the pendulum will be clockwise in the Northern hemisphere and anticlockwise in the Southern hemisphere. There is thus clearly *a relative motion between the plane of oscillation of the pendulum bob and the earth in either hemisphere.* This relative angular shift of the place of oscillation is measured over a period of days to demonstrate the rotation of the earth about its axis.

Among other methods of demonstrating the rotation of the earth may be mentioned (i) the observation of the *fixed stars* which, in fact, gave us the very first indication regarding the earth's rotation about its axis, (ii) the observation of any fictitious forces, indicating the presence of an accelerated or a rotating frame of reference, as mentioned already under 2.11, and (iii) the familiar gyroscope.

TIME PERIOD AND ORBITAL SPEED OF A SATELLITE

A satellite is, in general, a smaller body revolving around a much larger body. Until recently, we had only *natural satellites,* like the earth revolving around the sun and the moon revolving around the earth in their respective orbits. The earth is thus a satellite of the sun and the moon, a satellite of the earth. We now have also *man-made or artificial* satellites, placed in orbit around the earth, with the help of powerful rockets. The principle underlying is the same in either case, viz., that the gravitational attraction on the satellite due to the larger body (or the primary) pulling it *inwards*, supplies the centripetal force and this is just balanced by the (fictitious) centrifugal force on it, pulling it *outwards*.

Thus, if m be the *mass of the satellite,* M, the *mass of the earth,* r, the distance between the earth and the satellite (*i.e.*, between their centres), we have *gravitational pull of the earth on the satellite* (*i.e., the centripetal force*), *inwards* $= \frac{mM}{r^2}G$, where G is the *gravitational constant.*

And *centrifugal force on the satellite, outwards* = $m\upsilon^2/r$, where υ is uniform speed of the satellite around the earth.

If ω be the *angular velocity* of the satellite, so that $\upsilon = r\omega$, we have

$$\frac{mMG}{r^2} = \frac{mr^2}{r} = \frac{m(r\omega)^2}{r} = mr\omega^2$$

whence, $\omega^2 = MG/r^3$. Or, $\omega = \sqrt{MG/r^3}$.

Since r = (R + h), where R is the radius of the earth and h, the distance of the satellite from the surface of the earth, we have

$$\omega = \sqrt{\frac{MG}{(R+h)^3}}.$$

Now, if g be the acceleration due to gravity on the surface of the earth, we have

$$g = MG/R^2 \text{ or } MG = gR^2.$$

Substituting this value of MG in the expression for w above, we have

$$\omega = \sqrt{\frac{gR^2}{(R+h)^3}}.$$

And, therefore, *time-period of the satellite,*

$$T = \frac{2\pi}{\omega} = 2\pi\sqrt{\frac{(R+h)^3}{gR^2}}$$

In case the orbit of the satellite lies close to the earth, *i.e.*, if h < < R, we have r = (R + h) = R. So that,

$$T = 2\pi\sqrt{\frac{R^3}{gR^2}} = 2\pi\sqrt{\frac{R}{g}}$$

Let us calculate the orbital speed of such a satellite.

Remembering that, in this case, r = (R + h) = R (the satellite being close to the earth), we have

$$\upsilon = r\omega = R\omega = R\sqrt{\frac{gR^2}{R^3}} = R\sqrt{\frac{g}{r}} = \sqrt{\frac{gR^2}{R}} = \sqrt{gR}.$$

So that, taking R = 6.4×10^8 cm and g = 980 cm/sec, we have

$$\upsilon = \sqrt{980 \times 6.4 \times 10^8}$$

$$= 7.92 \times 10^5 \text{ cm/s}^2 = 7.92 \text{ km/s}.$$

Thus, *the orbital speed of an earth satellite revolving close to the earth's surface is nearly* 8 km/s.

SOLVED EXAMPLE

Example 1:

The pilot of an aeroplane wishes to reach a point 200 miles east of his present position. A wind blows 30 miles per hour from the north west. Calculate b is vector velocity with respect to the moving air mass if his schedule requires him to arrive at his destination in 40 minutes.

(Take $\sqrt{2} = 1.4$).

Solution:

Let O be the position of the pilot at the given instant. Since he reaches the desired point 200 miles due east in 40 min., his velocity in this direction is $200 \times 60/40 = 300$ mi/hr.

The problem thus reduces to this, that 300 mi/hr is the reluctant of *the wind velocity v_w, and the pilot's own velocity v_p say*. By the Triangle law of vectors, therefore. OA represents the *wind velocity vector v_w*, AP, the *velocity vector v_p of the pilot* and OP, the *resultant velocity vector* R. So that, $R = v_w + v_p$, whence,

$$v_p = R - v_w$$

Now, $v_w = 30 \sin 45i - 30 \cos 45j = 15\sqrt{2}i - 15\sqrt{2}j$ and $R = 300i$, where 1 and j are the unit vectors along the east and the north directions respectively.

We, therefore, have $v_p = 300i - (15\sqrt{2}i - 15\sqrt{2}j) = 300i - 21(i - j)$.

The vector velocity of the polite with respect to the moving mass of air is thus $300\, i - 21(i - j)$.

Example 2:

A reference frame S' is moving with a constant velocity v with respect to an inertial frame S. If at t = 0, the position vector of O' (the origin of the moving frame S') with respect to O (the origin of the inertial frame S) be R, show that the position vectors r and r' of a point P with respect to frames S and S' respectively are related by the equation $r = R + vt + r'$.

Solution:

Since the reference frame S' is moving with a constant velocity with respect to the inertial frame S, it is also an inertial frame. The problem thus reduces to one of Galilean transformation of the position vector of a point in one Galilean frame of reference to that in the other.

Since, initially, at time t = 0, the position vector of the origin O' of the moving frame S' with respect to the origin O of the inertial frame S, is R, it would be (R + vt) in time t, when the former would have covered an additional distance vt. so that, vector OO' would be equal to R + vt.

The whole situation except that OO' now represents (R + vt) instead of R (because, in the case shown, O' coincided with O at t = 0), let us consider the position vectors r and r' of a point P with respect to the origins of the two frames of reference. It is clear at once that vector OP = vector OO' + vector O'P, *i.e.,*

$$r = R + vt + r'.$$

Example 3:

The earth, revolving around its own axis, is not, strictly speaking, an inertial frame of references. If, however, we take it to be so, what would be the error involved in 1 second, in the position of a particle close to its surface? (Radius of the earth = 6.4×10^8 cm).

Solution:

Obviously, due to rotation about its own axis, a particle on, or close to, the surface of the earth, experiences a centripetal acceleration equal to $\omega^2 R$, where ω is the *angular velocity* of the earth = $2\pi/24 \times 60 \times 60$ radians/s and R, its *radius* = 6.4×10^8 cm (given)

∴ *centripetal acceleration,*

$$a = \left(\frac{2\pi}{24 \times 60 \times 60}\right)^2 \times 6.4 \times 10^8 \text{ cm/s}^2.$$

$$= 3.385 \text{ or } 3.4 \text{ cm/s}^2.$$

Thus, the acceleration of a particle, measured in the earth's frame of reference, will differ by 3.4 cm/s^2 from that measured in a truly inertial frame of reference, and the position of a particle, as measured in the latter, will be altered by a distance $1/2\ at^2$ in t seconds or by 1/2 a = $1/2 \times 3.4$ or 1.7 cm in 1 second, when measured in the former.

The error involved in 1s in the position of the particle, as measured in the earth's frame of reference, would thus be 1-7 cm.

Example 4:

The initial positions of two particles are $x_1 = 10$ cm, $y = 0$ and $x_2 = 0$, $y = 10$ cm respectively. If the velocity of the first particle be -5×10^5 i m/s, calculate that of the second at the instant that the two just collide with each other. What is then the relative velocity of the first particle with respect to the second?

Solution:

Suppose the two particles collide after time t second. Then, clearly, *position vector of the first particle after time t, i.e.,* $r_1(t) = r_1 + v_1t = 10\,i + (-5 \times 10^5\,i)t = 10i + 5 \times 10^5$ *it.*

And, *position vector of the second particle after time t, i.e.,* $r_2(t) = r_2 + v_2t = 10j + v_2t$.

In order that the two particles may collide, we must, obviously, have

$$r_1(t) = r_2(t).$$

Or $\quad r_1(t) - r_2(t) = 0$, *i.e.*, $10i - 5 \times 10^5\,it - 10j - v_2t = 0$.

Or, $\quad (10 - 5 \times 10^5t)\,i - 10j - v_2t = 0$,

whence, $v_2t = (10 - 5 \times 10^5t)\,i - 10j$.

Since the second particle has only a velocity along the y-axis, the x-component of its velocity is *zero, i.e.*, $10 - 5 \times 10^5t = 0$, whence,

$$t = 10/5 \times 10^5 = 2 \times 10^{-5}\ \text{s},$$

So that, $v_2t = 10j$.

$$\text{Or, } v_2 = -\frac{10}{t}j = -\frac{10}{2 \times 10^{-5}}j = -5 \times 10^5\,j\ \text{cm/s}.$$

The *velocity of the second particle at the time of collision is thus* -5×10^5j cm/s.

And, *the relative velocity of the first particles with respect to the second* is, obviously, $v_r = v_1 - v_2 = -5 \times 10^5i + 5 \times 10^5j - i)$ cm/s.

Example 5:

Calculate the fictitious force and the observed (or total) force on a body of mass 5kg in a frame of reference moving (i) vertically upwards, (ii) vertically downwards, with an acceleration of 4 m/s^2.

(Take g = 9.8 m/s^2).

Solution:

Taking the *earth to be an inertial frame of references and the upward direction as positive, we have*

Case (I) *True weight of the body or the true force acting c it, i.e.,* F = mg = 5 (– 9.8) = – 49.0 *newton, i.e.,* 49.0 N *in the downward direction.*

And, the *fictitious force acting on it, i.e.,* F_o = *mass (reversed acceleration of the non-inertial frame)* = m (– a_o) = 5 (– 4) = – 20 *newton, i.e.,* 20 N *in the downward direction.*

∴ *observed or total force on the body,* F' = F + F_o = 49.0 + 20.0 = 69.0 N *in the downward direction, (i.e., the same as in a lift or deviator moving upwards with an acceleration of* 4 m/s²)

Thus, the fictitious force in the body is 20 N and the observed (or total) force on it, 69 N, *both acting downwards.*

Case (II) : *True force acting on the body*, as before,

F = mg = 5 (–9.8) = –40.0 *newton* = 49.0 N *downwards.*

And, *the fictitious force acting on the body,*

F_o = m (– a_o) = 5 [–(– 4)] = 20.0 N upwards.

∴ *observed or total force on the body,*

$$F' = F + F_o = -49.0 + 20.0$$
$$= -29.0 \text{ newton}$$
$$= 29.0 \text{ N downwards,}$$

(*i.e., the same as in a lift or elevator, moving downwards with an acceleration of* 4 m/s²).

Thus, in this case, *fictitious force on the body is 20N upwards and the observed* (*or total*) *force on it,* 29 N *downwards.*

Example 6:

Imagine two non-inertial frames S' and S" to be initially coincident with an inertial frame S, with S' at rest and S" having a velocity v along the axis of x and both subjected to the same acceleration a along this axi.

(i) The positions of the origins 0' and 0" of the two non-inertial frames with respect to the origin 0 of the inertial frame after time t;

(ii) the relations between the respective positions x' and x" of a particle, in frames S' and S", and its position x in frame S;

(iii) the equation of motion of a particle of mass m, subjected to a constant force F in frame S and hence also its equations of motion in frames S' and S", indicating any point of interest that emerges:

(iv) the relative velocity of frame S" with respect to frame S'.

What conclusion, if any, do you draw from the results obtained in regard to non-inertial frames, in general, moving with a constant relative velocity ?

Solution:

(i) Initially, the origins 0, 0' and 0" of the three reference frames are coincident. Therefore, the initial velocity of frames S' being zero and its acceleration along the x-axis, a, the position vector of its origin 0' after time t is given by

$$OO' = r_o' = 1/2\ ai^2 i,$$

where i is the unit vector along the axis of x.

i.e., displacement of 0' from O, along the x-axis, is $x' = 1/2\ at^2$, (Fig. 2.13).

And, since frames S" has an initial velocity v along the x-axis, as also an acceleration v along it, its position after time t is given by $OO = r_o' = (vt + 1/2at^2)i$

i.e., displacement of 0" from 0' along the x-axis is $x_o'' = vt + 1/2at^2$.

(ii) If the positions of a particle in frames S, S' and S" be respectively x, x' and x" we clearly have $x' = x - x_o' = x - 1/2at^2$ and $x'' = x - x_1 = x - \upsilon t - 1/2at^2$.

(iii) The equation of motion of a particle of mass m, subjected to a constant force F, in frame S; is , obviously, $F = m\dfrac{d^2r}{dt^2}$, where r is the position vector of the particle in frame S.

Similarly, if r' and r" be the position vectors of the particle in frames S' and S" respectively, the equation of motion of the particle in frame S' is

$$F' = m\frac{d^2r'}{dt^2}.$$

and its equation of motion in frame S'' , $F'' = m\frac{d^2r''}{dt^2}$.

In terms of components along the three Cartesian coordinate axes, we have

in frame S' $F_x' = m\frac{d^2x'}{dt^2}, F_y' = m\frac{d^2y'}{dt^2}$ and $F_z' = m\frac{d^2z'}{dt^2}$.

Since there is motion only along the x-axis (with y' = y'' and z' = z''), we need consider only the x-component. So that, we have

$$F'_{x0} = m\frac{d^2x'}{dt^2} = m\frac{d^2}{dt^2}\left(x - \frac{1}{2}at^2\right) = m\frac{d^2x}{dt^2} = ma.$$

Clearly, $m\frac{d^2x}{dt^2}$ is the x-component of force F on the particle in frame S. Denoting it by F_x, we, therefore, have

$$F'_x = F_x - ma. \quad \text{...(i)}$$

Proceeding exactly similarly, we obtain

in frame S'',

$$F''_x = m\frac{d^2x''}{dt^2} = m\frac{d^2}{dt^2}\left(x - u - \frac{1}{2}at^2\right) = m\frac{d^2x}{dt^2} - ma.$$

[$\because$ υ = *constant*]

$$\text{Or,} \quad F''_x = m\frac{d^2x}{dt^2} - ma = F_x - ma. \quad \text{...(ii)}$$

It will thus be seen that the same fictitous force ma appears in both the transformation equations I and II, in view, no doubt, of the *same acceleration* (a) *of the two frames S' and S'' with respect to* S.

As will be readily seen from the transformation equations I and II for frames S' and S'' respectively under (iii) above, *the applied and fictitious forces in frame S' are respectively equal to those in frame* S''.

(iv) The relative velocity of frame S'' with respect to frame S' is obviously given by

$$v''_r = \frac{dr'}{dt} - \frac{dr'}{dt} = (v + at)\,i - (at)\,i = vi.$$

What is true of the respective equality of the applied and fictitious forces in frames S' and S" may be said to be true, in general, in the case of all non-inertial frames moving with a constant relative velocity. This follows from the fact that all such frames, although non-inertial with respect to S, *are inertial with respect to each other in view of their constant relative velocity and, therefore, zero relative acceleration. The applied and fictitious forces in them are thus governed by Galiliean invariance.*

Example 7:

A mass of 1 kg is hurled horizontally due north with a velocity of 0.5 km per second in latitude 30° N. Obtain the magnitude and direction of the Coriolis force acting on the mass and its deflection in consequence there of in a time interval of half a second.

Solution:

Taking the axes of x, y and z along the east, the north and vertically upwards respectively, with i, j and k as the respective unit vectors along the three axes, as usual , we have

velocity of mass, v = 500 j m/s.

Now, the *angular velocity vector of the earth*, v, due to the spinning around of the earth from west to east, is directed along the axis. It, therefore, lies towards, the north at 30° with the horizontal in the given latitude ϕ (30° N). We therefore have

$$w = \omega \cos\phi\, j + \omega \sin\phi\, k = \omega\,(\cos 30^\circ\, j + \sin 30^\circ\, k).$$

where ω, as we know, is equal to $2\pi/24 \times 60 \times 60 = 0.727 \times 10^{-5}$ radian s.

$\therefore$ *Coriolis acceleration of the mass*

$$= -2\omega \times v = -2\omega\,(\cos 30^\circ\, j + \sin 30^\circ\, k) \times 500\, j$$

$$= 2\,(0.727 \times 10^{-5}) \times 1/2 \times 500\, i = 36.35 \times 10^{-5}\ \text{m/s}^2.$$

(Or, we could have obtained the Coriolis acceleration directly from the relation $a_c = 2\omega\upsilon \sin\phi$, since the *effective angular velocity* of the earth, here, is $\omega \cos(90 - \phi) = \omega \sin\phi$).

Because the mass is 1 kg, we have *Coriolis force acting on the mass*

$= 1 \times 36.35 \times 10^{-3} = 36.35 \times 10^{-3}$ N *towards the east.*

$\therefore$ *deflection of the mass due to this force in* $1/2$ s $= 1/2 \times 36.35 \times 10^{-3} \times (1/2)^2 = 4.544 \times 10^{-3}$ m.

Example 8:

Calculate the rate of rotation of the plane of oscillation of a pendulum in latitude 30° and hence obtain the time it will take to turn through a full right angle.

Solution:

We know that the time taken by the plane of oscillation of a pendulum to make one complete rotation in latitude ϕ is given by $T = 24/\sin\phi$ hours.

Also, if α be the angular velocity of the plane of oscillation in *this* latitude, its time-period (*i.e.*, time taken for one full rotation) must be $T = 2\pi/\alpha$.

We, therefore, have $\quad 2\pi/\alpha = 24/\sin\phi$,

whence $\quad \alpha = 2\pi \sin\phi/24$ radian/hr.

Or, *rate of rotation of the plane of oscillation of the pendulum* in latitude 30°

$$= 360^\circ \sin\phi/24 = 15^\circ \sin\phi \text{ per hour.}$$

Since, here, $\phi = 30^\circ$,

and, therefore, $\sin\phi = 1/2$, we have

rate of rotation of the plane of oscillation of the pendulum in latitude 30°

$$= 15 \times 1/2 = 7.5^\circ/\text{hour.}$$

$\therefore$ *time taken by the plane of oscillation to turn through one full right angle*

$$= 90/7.5 = 12 \text{ hours.}$$

[Or, we could have said that if T be the time-period of the plane of oscillation in hours, the time taken to turn through 90° must T/4 hours.

$$= \frac{24}{\sin\phi} \times \frac{1}{4} = \frac{24}{\sin 30^\circ} \times \frac{1}{4}$$

$$= \frac{24 \times 2}{4} = 12 \text{ hours}].$$

Example 9:

A pendulum is oscillating along the east-west direction at a place of latitude 30° N. After how much time will it start oscillating along the north west–south east direction?

Solution:

The problem here, obviously, is to calculate the time the plane of oscillation of the pendulum will take to turn through an angle of 135°. in latitude 30° N.

Since in latitude $\phi = 30°$, the rate of rotation of the plane of oscillation of the pendulum is (as we have seen, under: example 11 above) equal to 15° ϕ sin = 15° sin 30° = 7.5°/hr, we have, time taken by the plane of oscillation to turn through 135° = 135/7.5 = 18 hours.

Thus, *the pendulum will start oscillating along the north west—south east direction after* 18 *hours*

Example 10:

A satellite moves in a circular orbit around the earth at a height $R_c/2$ from the earth's surface, where R_e is the radius of the earth. Calculate its period of revolution.

Given $R_c = 6.38 \times 10^6$ *m;*

$g = 9.80$ *m/s²*

The time-period of a satellite revolving the earth at a height h from its surface is given by $T = 2\pi \sqrt{(R+h)^3/gR^2}$, where R is the radius of the earth

Here, $R = R_e$ and $h = R_e/2$. So that, time-period of revolution of the satellite,

$T = 2\pi \sqrt{(R_e + R_e/2)^3/g\,(R_e)^2} = 2\pi \sqrt{27\,R_e/8g}$.

Substituting the given values of R_e and g, therefore, we have time-period of the satellite,

$$T = 2\pi\sqrt{\frac{27\times 6.38\times 10^4}{8\times 9.80}} = 9.309\times 10^3 \text{ or } 9.3\times 10^3 \text{ s}$$

Example 11:

At what height from the surface of the earth must a satellite revolve in its orbit around the earth, concentric and coplanar with the equator, *such that it always appears to be at the same point over the earth's surface as viewed by an observer on the earth? (G = 6.67 × 10^{-8} c.g.s. units, R = 6.38 × 10^{8} cm and mass of the earth = 5.98 × 10^{27} g).*

Obviously, the satellite wil always apear to be at the same point over the earth's surface when its angular velocity is the same as that of the earch, viz, $\omega = 2\pi/24 \times 60 \times 60 = 7.27 \times 10^{5}$ *radians or 7.3 radians.*

Let r be the radius of the orbit of the satellite. The angular velocity of the satellite, i.e., $\omega = \sqrt{MG.r^3}$*. Or,* $\omega^2 = MG.r^3$*, whence,* $r^3 = MG/\omega^2$*. So that, substituting the values of M,G and ω, we have*

$$r^3 = \frac{5.98 \times 10^{27} \times 6.67 \times 10^{-8}}{\left(7.3 \times 10^{-5}\right)^2} = 74.85 \times 10^{27},$$

$$\text{whence, } r = (74.85 \times 10^{27})^{1/3}$$

$$= 4.2 \times 10^{9} \text{ cm.}$$

Now, r = (R + h), where R is the radius of the earth and h, the height of the satellite from the surface of the earth.

We therefore, have $h = r - R = 4.2 \times 10^{9} - 6.38 \times 10^{8} = 3.562 \times 10^{9}$ *cm.*

Thus, the satellite must revolve in its orbit around the earth at a height of 3.56 × 10^{9} *cm from the surface of the earth.*

Example 12:

Show that the motion of one projectile as seen from another projectile will always be a straight line motion.

Solution:

Since both projectiles are subject to the game acceleration—g, there is no relative acceleration between them. They must therefore, appear to be descending along a straight line path relative to each other, *i.e.*, the motion of one should appear to be a straight line motion as seen from the other.

Alternatively, we may proceed as follows;

Let the trajectories of the two projectiles, projected simultaneously, be described in the x-y plan, and let their initial velocities of projection and their angles of projection be u_1, u_2 and α_1, α_2, respectively.

Then, with the point of projection (O) as the origin if the position coordinates of points P_1 and P_2 at any given instant t be x_1, y_1 and x_2,

y_2 respectively, we have $x_1 = u_1 \cos \alpha_1 t$, $x_2 = u_2 \cos \alpha_2 t$, $y_1 = u_1 \sin \alpha_1 t - 1/2\, g\, t^2$ and $y_2 = u_2 \sin\alpha_2^t - 1/2\, t^2$. So that,

$$\frac{y_2 - y_1}{x_2 - x_1} = \frac{y}{x} = \frac{u_2 \sin\alpha_2 - u_1 \sin\alpha_1}{u_2 \cos\alpha_2 \cos\alpha_1}$$

$$= m, \text{ say.}$$

whence, y mx. which is the equation to a *straight line.*

Example 13:

To an observer, at re½t on the ground, a body, thrown vertically upwards in a uniformly moving frame S, appears to describe a parabolic path. What is the path of the body as it would appear to an observer in another reference frame S' in uniform motion parallel to S, when S' has (i) an identical velocity with S, (ii) a velocity equal and opposite to that of S, and (iii) a velocity twice that of S?

Solution:

Taking the parabolic path of the body in frame S to be described in the x – y plane (where the x-axis is along the horizontal), if the x and y-components of the velocity of the body in this frame be $v_{x'}$, and $v_{y'}$, clearly, v_x is also the velocity of frame S, along the x-axis.

In case (i), since frame S' moves with the *same* velocity as S, the x-component of the velocity of the body, as seen from S' will be, say, $v'x = v_x - v_{x'} = 0$ and its y-component v'_y will obviously remain the same as that in frame S, vis., $v_{y'}$ since neither frame has a vertical velocity.

The path of the body, therefore, will appear to an observer in frame S' to be a *vertical straight line.*

In case (ii), since frame S' moves with an equal and opposite velocity to that of S' the x-component of the velocity of the body, as seen from frame S', will obviously be $v'_x = v_x - (-v_x) = 2v_{x'}$, with the y-component, $v'_y = v_{y'}$, as in case (i).

So that, in time t, the *horizontal distance covered will be* $x = 2v_x t$ and the *vertical distance covered will be* $y = v_y t - 1/2gt^2$.. Or, eliminating t, we have

$$y = v_y \frac{x}{2v_x} - \frac{1}{2} g \frac{x^2}{4v_x^2},$$

or $$y = \frac{v_y}{2v_x}x - \frac{g}{8v_x^2}x^2,$$

which is an equation of the form $y = ax + bx^2$, the equation to a parabola, indicating that the path of the body in S will, in this case, appear to the observer in S' to be a *parabola, described in the same direction* as it appears to the observer at rest on the ground. Only, the x-component of the velocity of the body ($2v_x$) now appearing to be twice as much as to the stationary observer, the latus rectum of the *parabola* will be twice that of the one seen by the observer on the ground.

In case (iii), the frame S' having twice the velocity of S, we shall have x-component of the velocity of the body, as seen from S', *i.e.*, $v'_x = v_x - 2v_x = -v_x$ and the y-component, $v'_y = v_y$, as before.

This is again a case similar to that of the stationary observer on the ground, with only the x-component of the velocity of the body in S now reversed. To the observer in S', therefore, the path of the body in S will appear to be a *parabola,* similar in size and shows to that seen by the groundbased observer but *described in the opposite direction, i.e.*, a *parabola* which is a mirror-image of the one seen by the latter.

EXERCISES

1. The position vector s of two particles are respectively $r_1 = 4i + 6j + 4k/m$ and $r_2 = 2i + 3j + 2k/m$. The velocity of the first particle is $i - 2j + 2k$ m/s. How much greater should the velocity of the second the velocity of the second particle be so that the two may collide in 5 seconds?

   Ans. 0.2 (2i + 3j + 2k) m/s.

2. What is meant by a non-inertial frame of reference? And what is a fictitious force? Why is it so called? Under what condition will an accelerated frame of reference serve as an inertial frame ?

3. What will be the effective weight of a person carried vertically up in a rocket with an acceleration of 6g, if his actual weight on the earth is ϕ0 kg.?

4. What is meant by a frame of reference? Given examples to show that the state of motion of a body would appear to be different when viewed from different frames of reference. From which

particular frame of reference would you like to study this motion and why?

5. Enunciate Newton's laws of motion and discuss their limitations.
6. What is an *inertial frame of reference*? Show that all other framès or reference, with constant velocity relative to it, are also inertial frames. What are the characteristic properties and importance of such frames?
7. What is meant by Galilean transformation and Galilean invariance? Show that where as length (or distance) and acceleration are invariant to Galilean transformation, velocity is not.
8. A reference frame S' is moving with a constant velocity υ with respect to an inertial frame S and the position of its origin O' is given by R_0 at t = 0. If the position vector of a point in frame S be r, show that it will be i + vt + R_0 in frame S'.
9. Explain how the parabolic trajectory of a projectile would appear when observed from a frame of reference moving with (i) a velocity equal to the horizontal component of the velocity of the projectile, (ii) a velocity equal to twice the horizontal component of the velocity of the projectile, and (iii) a constant velocity inclined to the horizontal.

Ans. (i) a vertical straight line, (ii) original trajectory in the opposite direction (iii) a new parabolic trajectory.

10. Two inertial frames S and S' have their axes parallel and the position of the origin O' of frame S' relative to origin 0 of frame S is given by r_0 = i + 2j + 3k. Show that, if the position of a point P in frame S be (2,3,4) its position in frame S' would be (i + j + k).
11. If the respective axes of two frames S and S' be inclined to each other, with their origins coinciding, obtain the usual transformation equations and show that if frame S be inertial, frame S' too is inertial.
12. If in question 9 above, the position vector of the origin 0' of frame S' with respect to the origin 0 of frame S be $R_{0'}$, show that $x = x_0 + x'C'_{xx} + y'C'_{yx} + z'C'_{zx}$; $y = y_0 + x'C'_{xy} + y'C'_{yy} + z'C'_{zy}$ and $z = z_0 + x'C'_{xz} + y'C'_{yz} + z'C'_{zz}$.

13. Enunciate the laws of conservation of momentum and conservation of energy and show that they are both invariant to Galillean transformation.

14. A particle of mass m_1 moving with velocity υ collides head on against a particle of mass m_2 at rest. Taking the collision to be an *elastic* one obtain the velocities of the two particles and the fraction of the total kinetic energy acquired by the second particle after the collision. How will this fraction changes if $m_1 = m_2$?

Ans. $v_1 \dfrac{m_1 v - m_2 v_2}{m_1}$; $v_2 = \dfrac{2v}{1 + m_1 / m_2}$; $\dfrac{4 m_2 m_2}{(m_1 + m_2)}$;
the fraction will change to 1 *i.e.*, the entire energy of m_1 will be transferred to m_2.

15. If the collision be an inelastic one, so that the two particles coalesce together on collision, obtain their common velocity after the collision and the loss of kinetic energy suffered..

16. In a head on collision, a particle with an initial speed υ_0 strikes a stationary particle of the same mass. Find the velocities υ_1 and υ_2 of the two particles respectively after the collision if (a) half the original kinetic energy is lost, (b) the final kinetic energy is 50% greater thus the original kinetic energy.

Ans. (a) $\upsilon_1 = \upsilon_2 = 1/2\ \upsilon_0$, (b) $\upsilon_1 = 1/2 \upsilon_0 (1 + \sqrt{2})$, $\upsilon_2 = 1/2\ \upsilon_0 (1 - \sqrt{2})$.

3

Non Relative Particle Dynamics

GENERAL

Dynamics is the study of the motion of bodies and the relationship of this motion with the forces producing it. It is thus an important branch of mechanics and starts with the simplest of bodies, viz., particles, where a particle is just a '*quantity of matter having negligible dimensions.*

Now, ever since the time of *Newton*, the system of calculations used in dynamics, as elsewhere, has been based on his celebrated laws of motion and the assumed variance of space and time, as also that of mass, *i.e.*, of the assumption that the distance between two points, the interval between two events in space and the mass of a body at any point in space are the same for all observers, irrespective of their state of rest or motion. This system, referred to as the *Classical* or *Newtonian mechanics* enjoyed complete sway right up to the end of the last century and no doubts were residual questions asked as to its validity or accuracy.

The very turn of the century, Einstein put forward his Special theory of relativity (in the year 1905) based on the constancy of the speed of light (c) irrespective of any relative-motion between the source of light and the observer. Denying the invariance of both space and time, this theory leads to the conclusion that the mass (m) of a body increases with its velocity and is y times its rest mass (m_0) at velocity v, where $\gamma = 1/\sqrt{1-v^2/c^2}$ So that, the Newtonian mass or momentum of a fast moving body must be multiplied by the factor y to obtain its correct (relativistic) mass or momentum at its given speed. The system of calculations dealing with bodies moving with high velocities, approaching c, thus came to be known as the *Relativistic* or *Einsteinian mechanics*. At ordinary speeds, however, for which $v << c$, we have $Y \approx 1$ and the relativistic mechanics reduces to the *Classical* or *Newtonian mechanics*.

Again, in the year 1924, the French physicist, *Louis de Broglie*, put forward the view that just as light behaves like a wave on a macroscopic scale and as a particle (viz., a *photon*) on the microscopic scale (*i.e.*, in the absorption or emission of light), so also all small and moving particles (*electrons, protons, atoms, molecules* and *fundamental particles*, in general) as well as photons, exhibit a similar duality of nature in the reverse sense, *i.e.*, behave as corpuscles or particles on a macroscopic scale and as waves on a microscopic scale, with a frequency and a wavelength, called *de Broglie wavelength*, associated with them—this wavelength being given by $\lambda = h/mc$ for a photon (its velocity being c) and by $\lambda = h/mv$ for a particle, in general, where h is the well known *Plancks' constant*, equal to 6.62×10^{-27} erg-sec.

Thus, with the wave-like nature of tiny particles, their behaviour cannot possibly be predicted by means of either the Classical or the Relativistic mechanics. Indeed, *Werner Heinsberg* enunciated in the year 1927 his famous *principle of uncertainty*, according to which *the position and the momentum of a particle cannot both be measured simultaneously with an equal degree of accuracy*, meaning thereby that if the position of the particle be measured accurately, the inaccuracy in the measurement of its momentum increases correspondingly and *vice versa*, such that the *product of the uncertainty* Δr *in the position of the particle at any given instant and the uncertainty* Δp *in its momentum at the same instant is equal to, or greater than, the Planck's constant h, i.e.*,

$$\Delta r \, \Delta p \geq h, \qquad [\text{where } h = 6.62 \times 10^{-27} \text{ erg-sec.}$$

This uncertainty is by no means due to any want of perfection in the measuring instruments or the experimenter but is *inherent* in the very nature of things,—the measurement of one (position or momentum) disturbing the other by an unknown amount.

This strange behaviour of matter on the atomic scale lies in the domain of *Quantum mechanics* (or wave *mechanics*), due to the painstaking mathematical investigations of *Schrodinger, Heinsberg, Bohr* and *Dirac*, the contribution of *Dirac* beings in main, the incorporation into it of the relativistic effects in conformity with the Special Theory of Relativity.

On account of the very small value of h, however, the uncertainty in the measurement of position and momentum of a particle is appreciable only if it be of atomic size and is quite negligible for larger particles. So that, in the case of relatively larger bodies, quantum mechanics too gets reduced, for all intents and purposes, to classical mechanics,

Thus, except in the case of bodies moving very fast ($v \to c$) or having extremely small or atomic size, the classical mechanics gives results quite in conformity with experimental facts. And, since in everyday life we mostly come across bodies of relatively large size, moving with speeds far less than that of light (c), we shall concern ourselves, in general, with only the classical mechanics except in cases where it happens to be patently inapplicable.

EQUATIONS OF MOTION OF AN UNCHANGED PARTICLE

As mentioned already, the classical mechanics is based on Newton's laws of motion and the invariance of space, time and mass. So that, starting with the 2nd law, which is *the* law of motion including, as it does, both the first and the third laws, we have

force equal to rate of change of momentum, i.e.

$$F = \frac{dp}{dt} = \frac{d}{dt}(mr). \qquad [\because p = mv.$$

Since mass remains constant, we have $F = m\frac{dv}{dt} = m\frac{d^2r}{dt^2} = ma$ ...(I)

where $dv/dt = d^2r/dt = a$, the acceleration of the particle.

(i) Now, putting relation I above as $d^2r/dt^2 = F/m = a$, we have, on integrating with respect to t,

velocity of the particle a install t, i.e., dt/dt = at+ C_1,

where C_1 is the constant of integration.

Since at $t = 0$, $dr/dt = C_1$, we have

initial velocity of the particle (*i.e.*, its velocity just before the force is applied to it) equal to C_1.

Representing the *initial velocity* by the usual notation u and the final velocity at instant r by v, we, therefore, have

$$\frac{dr}{dt} = v = u + al \qquad ...(i)$$

(ii) Integrating equation (i) again with respect to t. we have *position of the particle at instant t* given by $r = ut + 1/2\ at^2 + C_2$, where C_2 is another constant of integration.

Again, at $t = 0$, $r = C_2$, where C_2 represents the *initial position vector* r, of the particle.

So that, $r = ut + 1/2at^2 + r_0$, whence. $(r - r_0) = ut + 1/2\ at^2$.

Clearly, $(r - r_0) = S$, the *displacement* vector of the panicle at instant t.

We, therefore, have $S = ut + 1/2\ at^2$. ...(ii)

(iii) We have *acceleration of the particle*, a = dv/at.

Taking its dot *product* with the velocity v of the particle at instant t, we have, v.a = v.dv/dt.

Or, since v = dr/dt, we have dr/dt.a = v.dv/dt,

integrating which with respect to t, we have

$$\int \frac{dr}{dt}.adt = \int v.\frac{dv}{dt}dt.$$

or $r.a = 1/2\ v.v + C_3$.

or $r.a = 1/2v^2 + C_3$,

where C_3 is yet another constant of integration.

At $t = 0$, clearly, $v = u$

and $r = r_0$.

So that, $r_0.a = 1/2u^2 + C_2$.

or $C_3 = a.r_0 - 1/2u^2$.

Substituting this value of C_3 in the expression above, we therefore have $r.a = 1/2v^2 + a.r_0 - 1/2u^2$.

or $1/2(v^2 - u^2) = a\ (r - r_0)$

or $v^2 - u^2 = 2aS$.

or $v^2 = u^2 + 2aS$. ...(iii)

These relations, (i), (ii) and (iii), are the three ***general equations of motion of a particle.***

Let us now consider *two particular cases.*

(a) *Case of a body falling freely under the action of gravity* : If we consider a body acted upon only by the gravitational force of attraction due to the earth, the acceleration on it (called the *acceleration* due to gravity) acts vertically downwards, so that, taking the origin at the surface of the earth and the x-axis vertically upwards the have *equation of motion* of the body or the particle, $md^2x/dt^2 = mg$.

or *acceleration of the particle*, $d^2x/dt^2 = -g$.

This expression, on integration with respect to f, gives the velocity r of the panicle at a given instant t, *i.e.*, $dx/dt = v = -gt + C_1$, where C_2 is the constant of integration.

Since at $t = 0$, $v = u$, the initial velocity of the panicle, we have $C_2 = u$, and therefore,

$$dx/dt = v = -gt + u$$

or $$v = u - gt. \quad ...(iv)$$

Again, integrating the above expression with respect to t, ve have *position of the body or the particle at instant t given by*

$x = ut - 1/2gt^2 + C_2$, where C_2 is another constant of integration.

Again, at, $t = 0$, $x = x$, so that, $C_2 = x_0$.

And, therefore, $x = ut - 1/2gt^2 + x$,

whence, $(x_0 - x) = -ut + 1/2\ gt^2$.

Since $(x, -x) = h$, the *height through which the body has fallen*,

we have, $h = ut + 1/2gt^2$. ...(v)

Finally, we have $g = -dv/dt$.

Taking the dot product of this expression with v, we have

$$v.g = -v.dv\ dt.$$

or, $(dx/dt).g = -v.dv/dt$, $[\because v = dx/dt$.

Integrating which with respect to t, we have

$$\int \frac{dx}{dt}.g.dt = \int -v.\frac{dv}{dt}dt.$$

or $x.g = -v^2/2 = C_3$, where C_3 is a constant of integration.

Since at $t = 0$, $x = x_0$ and $v = u$,

we have $x_0g = -u^2/2 + C_3$,

whence $C_3 = u^2/2 + x,g$.

$$\therefore \quad xg = -v^2/2 + u^2/2 + x_0g.$$

or, since $(x_0 - x) = h$, we have

$$v_2 = u^2 + 2gh, \quad ...(vi)$$

whence, the *velocity acquired by a body starting from rest and falling freely through height h is, clearly,* $v = \sqrt{2gh}$.

For a body thrown vertically upwards, we simply put $-h$ in place of h in the expressions deduced above, when we obtain the following corresponding relations:

$$v = u - gt \quad \text{...(vii)}$$

$$h = ut - 1/2gt^2 \quad \text{...(viii)}$$

and $$v_2 = u^2 - gh \quad \text{...(ix)}$$

So that, from relation (ix), *the maximum height attained by the body* (when $y = 0$) is $h = u^2/2g$ and, from relation (vii), *the time taken* to do so is $t = u/g$.

(b) *The projectile* : In the cases of motion considered above, the motion of the body or the particle takes place *along the direction of the applied* force and the problem is thus *one-dimensional.* Let us now consider a *two-dimensional* problem, viz., the motion of a *projectile.*

A projectile is a body or a particle, projected at an angle to the horizontal, this angle being referred to as its *angle of elevation or projection* and the path taken by it as its *trajectory.*

Thus, suppose a particle is projected with velocity u from a point O at an angle a with the horizontal OX, (Fig. 3.1). Let OY be an up ward vertical through 0 and let the position of the particle at time t after projection be P (x, y), as shown.

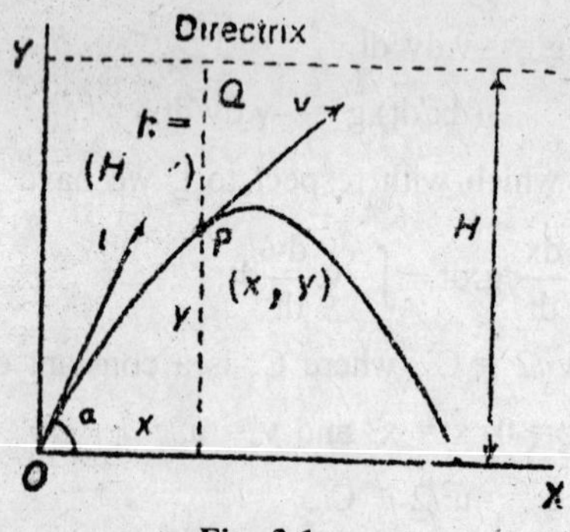

Fig. 3.1

Then, assuming no resistance due to air, the particle is throughout its motion, under the influence of the uniform acceleration due to gravity alone, acting vertically downwards. Its equations of motion along the horizontal and the vertical are, therefore, $md^2/dt^2 = 0$ and $md^2y/dt^2 = -mg$ respectively, *i.e.*, $d^2x/dt^2 = 0$ and $d^2y/dt^2 = -g$.

Since u is the velocity of projection of the particle at an angle α with the horizontal, its components *along the horizontal and the vertical* are respectively

$$u_x = u \cos \alpha$$

and $$u_y = u \sin \alpha.$$

Then, since the particle has no acceleration along the horizontal, its position after time t along this direction is given by

$$x = u_x t = u \cos \alpha t \qquad \text{...(x)}$$

And since there is an acceleration –g acting on it along the vertical, its position after time t along this direction is, in accordance with relation (viii) above, given by

$$y = u_y t^2 - 1/2gt^2 = \sin \alpha t - 1/2gt^2 \qquad \text{...(xi)}$$

From relation (x), t = x/u cos α. Substituting this value of t in relation (xi), we, therefore, have

$$y = u \sin\alpha \frac{x}{u \operatorname{cod}\alpha} \frac{1}{2} g \left(\frac{x}{u \cos\alpha} \right)^2$$

or $$y = x \tan \alpha = \frac{gx^2}{2u^2 \cos^2 \alpha} \qquad \text{...(xii)}$$

indicating *that the trajectory of the projectile is parabolic in shape.*

Now, relation (xii) may be put as

$$\frac{g - x^2}{2u^2 \cos\alpha} - x \frac{\sin\alpha}{\cos\alpha} + y = 0$$

$$\text{or } x^2 - x \frac{\sin\alpha}{\cos\alpha} . \frac{2u^2 \cos^2 \alpha}{g} + y \frac{2u^2 \cos^2 \alpha}{g} = 0$$

$$\text{or } x^2 - \frac{2u^2 x \sin\alpha \cos\alpha}{g} + \frac{2u^2 \cos^2 \alpha . y}{g} = 0$$

$$\text{or } x^2 - \frac{2u^2 x \sin\alpha \cos\alpha}{g} + \frac{u^4 \sin^2 \alpha \cos^2 \alpha}{g^2}$$

$$= \frac{u^4 \sin^2 \alpha \cos^2 \alpha}{g^2} - \frac{2u^{42} \cos^2 \alpha y}{g}.$$

$$\text{or} \left(x - \frac{u^2 \sin \alpha \cos \alpha}{g} \right)^2 = \frac{2u^2 \cos^2 \alpha}{g} \left(y - \frac{u^2 \sin^2 a}{2g} \right) \quad ...(xiii)$$

showing that *the vertex of the parabola is at the point $u^2 \sin \alpha \cos \alpha/g$, $u^2 \sin^2 \alpha/2g$, and that the length of the latus rectum is $2u^2 \cos^2 \alpha/g$. The directrix is thus horizontal and at a height $u_2 \cos^2 \alpha/2g$ above the vertex or u^2/g above the horizontal line OX.*

The *velocity of the projectile* at any instant t is given by $v^2 = v_x^2 + v_y^2$, where v_x and v_y are the velocities, of the particle at time t along the axes of x and y respectively.

There being no acceleration along the x-axis, $v_x = u_y = u \cos x$. And, the acceleration along the axis of y being $-g$, $v_y = u_y - gt = u \sin \alpha - gt$, in accordance with relation (vii) above. So that,

$$v^2 = v_x^2 = (u \cos x)^2 + (u \sin \alpha - gt)^2.$$

$$= u^2 \cos^2 \alpha + u^2 \sin^2 x - 2 u \sin \alpha gt + g^2t^2.$$

$$= u^2 - 2 u g \sin \alpha t + g^2 + t^2 = u^2 - 2g (u \sin \alpha t - 1/2gt^2).$$

or $v^2 = u^2 - 2gy$,

whence, $v = \sqrt{u^2 - 2gy}$

This *velocity v is the same as that the particle would acquire in falling freely to P from a point on the directrix vertically above P.*

Range of the projectile : Putting $y = 0$ in relation (xii) above,

we have $\tan \alpha = gx/2u^2 \cos^2 x$.

$$\text{or} \qquad x = \frac{\sin\alpha}{\cos\alpha} \left(\frac{2u^2 \cos^2 \alpha}{g} \right) = \frac{\sin\alpha \, 2u^2 \operatorname{cis}\alpha}{g}$$

$$\text{or} \qquad x = \frac{u^2 \sin 2\alpha}{g}$$

This gives the *maximum horizontal distance covered by the projectile* or the *range of the projectile on the horizontal plane.* Denoting it by R, we therefore, have

$$R = u^2 \sin 2\alpha/g.$$

For a given range R and for a given velocity of projection, u, there are two possible angles of elevation or projection, viz., α and $(\pi/2 - x)$ and, clearly, the range (R) has its maximum value u^2/g, when sin $2\alpha = 1$ *i.e.*, when $2\alpha = 90°$ or $\alpha = 45°$ or $\alpha/4$.

SOME DEFINITIONS AND PRELIMINARY RELATIONS

Before proceeding with the study of the behaviour of a charged particle in an electric or a magnetic field, let us refresh our memory as to the meanings of some of the terms we shall have occasion to use so frequently in our discussion- as also recapitulate some simple preliminary relations.

(i) *Electric field and intensity af the field* : The area round about an electric charge q, within which its influence is perceptible, is referred to as its *electric field*, and the force experienced by a unit positive charge placed at any point in the field is called the *strength* or the *intensity* of the field at that point, or, usually, simply the *field* at that point and is denoted by the symbol E. Having both magnitude and direction, it is a vector quantity and hence, the intensity at a point due 10 a number of charges is equal to the vector sum of the intensities there due to the individual charges.

Further, if the intensity of the field be the same at all points in it, it is said to be a *uniform electric field.*

If follows, therefore, that the force experienced by a charge q, when placed at a point where the intensity of the field is E, is equal to qE.

It may as well be mentioned here that in the *electrostatic system of units*, q is measured in *esu* or *statcoulombs*. The *practical unit* of *charge* in this system is the *coulomb* = 3×10^3 *esu* or statcoulombs. In the *Rationalised* MKS (or the SI) *system*, which is now increasingly coming into vogue, the unit of Charge is the *coulomb* the same as the practical unit in the electrostatic system of units.

Since, the intensity of the electric field E is the force on unit charge its unit in the *esu* system is *dyne/esu*. And since *field intensity a also* defined as rase of change of *potential with distance*, and the unit of potential in the electrostatic system is the statvolt, *the unit of intensity in this system is also 1 statvolt/cm*. The practical unit of E a this system is volt/cm, where 1 volt is 1/300th of the *esu*, *i.e.*, = 1/300 *statvolt.*

In the RMKS (or the SI) *system*, the unit of electric intensity is *Newton/coulomb* (*i.e.*, N/C) or 1 volt/meter, where 1 newton = 10^5 dynes. We thus have 1 N/1C = 1 volt/1 meters = 10^5 dynes/3 × 10^5 stat coulombs = 1/3 × 10^4 statvolts/cm.

(ii) *Kinetic energy of a charged particle in an electric field* : Since the potential (V) between two points in an electric field is defined as the work done in moving a unit positive charge from one posit to the other (quite independently of the path taken), it is clear that the work done in taking a charge q from one point to the other (along any path) will be qV.

This work appears in the form of increased A.E. of the particle. So that if m be the *mass* of the particle carrying the charge q sad v, the velocity acquired by it, we have

$$1/2\ mv^2 = qV = \text{charge} \times \text{potential difference}.$$

Thus a particle carrying a positive charge *gains* or *loses* energy according as it moves from a higher to a lower potential or from a lower to a higher one. Naturally, the reverse will be the case if the charge carried by the particle be negative.

In the electrostatic system, this K.E. is measured in *ergs* and in the RMKS (or the SI) system, in joules. More often than not, however, it is measured in *electron volts* (eV), where 1 eV = 1.60 × 10^{-12} ergs = 1.60 × 10^{-19} joules.

(iii) *Magnetic field and strength of a magnetic field—Flux density:* The region in which magnetic effects can be detected is called a *magnetic* field and the strength of a magnetic field at any point in it is measured by the *magnetic flux density* B at that point, where flux density is defined as *the magnetic flux per unit area perpendicular to the direction of the flux*. In the emu or the *Gaussian system*, its unit is the gauss and in the RMKS (or the SI) system, it is *weber/metre*^{2} (Wb/m^2) or *testa* (T), where 1 Wb/m^2 (or T) = 10^4 gauss.

The flux density (B) is a *vector quantity*, having magnitude as well as *direction* (which is the same as that of the lines of magnetic flux).

(iv) *Force on a moving charge in a magnetic field* : We know that the force acting on a conductor of length δl carrying a current I in a magnetic field of strength B is given by $F_{mag} = BI\ \delta l \sin\theta$, where 9 is the angle that the conductor makes with the direction of B. This force is perpendicular to both the field (B) and the current (I) and its direction is given by the familiar *Fleming's left hand rule.* Putting it in vector notation, we have

$$F_{mag} = I\,\delta l \times B = \delta l \times B,$$

where δl and I are obviously *parallel vectors.*

If the quantities δl, I and B be measured in *emu*, F_{mas} is in *dynes*, but if they be measured in RMKS (or SI) units, F_{mag} is in newtons (where 1 N = 10^5 dynes).

Now, as we know, *a moving charge constitutes a current and must, therefore, experience a similar force in a magnetic field.*

Thus, if v be the velocity of a charge q moving in a direction making an angle θ with the field (B), as shown in Fig. 3.2, the distance covered by it in time $\delta t = v.\delta t$. This is equivalent to a conductor of length $\delta l = v\delta t$, inclined to the direction of B at an angle θ and carrying a current = $q/\delta t$ (∵ current is the rate of flow of charge).

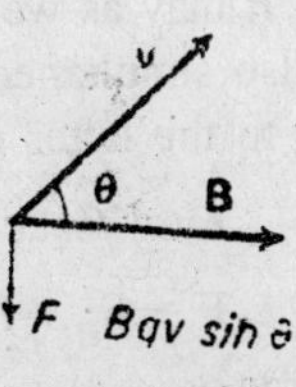

Fig. 3.2

∴ force acting on the charge, *i.e.*,

$$F_{mag} = \frac{q}{\delta t}\, v\delta t \times B \sin\theta = qv\,B \sin\delta.$$

or, in vector notation, $F_{mag} = qv \times B$

This force is called *Lorentz force* and its direction, as given by the *right-handed screw rule*, is perpendicular to both v and B.

If the quantities q, v and B be measured in emu, F_{mag} is in dynes, but if measured in RMKS (or SI) units, it is in *newtons.*

Quite often, however, q and B are measured in the Gaussian system of *units, i.e.*, q in *esu* and B in emu (or gauss). We have, therefore, to convert q into emu by dividing it by c (the speed of light in free space), because 1 esu of charge = l/c of 1 emu of charge. In such a case, then, the expression for F_{mag} becomes

$$F_{mag} = \frac{q}{c}\, v \times B$$

It will be noted that *for a force to act on a charged particle in a magnetic field,*

(i) *the particle must be in motion, i.e.*, v ≠ 0, because a stationary charge does not constitute a current;

(ii) *the velocity* v *of the particle must not be in the same direction as* B, or else, under the rule of vector product, $v \times B$ would be equal to zero.

Further, since the force (F_{mag}) acts in a direction perpendicular to v, the velocity of the particle, *there is no change in the magnitude of the velocity but only a continuous change in its direction.*

This means, in other words, that *there is no change in the kinetic energy of a charged particle during its motion in a magnetic field.*

It may as well be mentioned here that in case the charged particle is also simultaneously subjected to an electric field E, the force on it due to the latter is $F_{et} = qE$. And therefore,

total force acting on the charged particle, i.e.,

$$F = F_{el} + F_{mag} = qE + q/c\ v \times B.$$

This force F, due to the two fields together, is also called *Lorentz* force.

CHARGED PARTICLE IN A UNIFORM AND CONSTANT ELECTRIC FIELD

We have seen under accordance with Newton's second law of motion, the force acting on a particle is given by $F = d^2r/dt^2$, where m is the mass of the particle (remaining constant at non-relativistic speeds) and d^2r/dt, its acceleration.

Now, imagine a particle of mass m and carrying a charge q to be placed in a uniform and constant electric field E. The force acting on [it will obviously be qE. If the acceleration acquired by the particle under the action of this force be $a = u^2r/dt^2$, we clearly have

$$ma = md^2r/dt^2 = qE,$$

whence, $d^2r/dt^2 = q/m\ E$. ...(i)

Integrating expression (i) with respect to t, we have

velocity of the particle, $v = \frac{dr}{dt} = \frac{qE}{m} t + C_1$

where C_1 is the constant of integration.

Clearly, at $t = 0$, $v = u$, the *initial velocity* of the particle. So that $C_1 = u$. We therefore, have

$$\frac{dr}{dt} = v = \frac{qE}{m} t + u \qquad \text{...(ii)}$$

As will be easily seen, this is a relation similar to v = u + at for an uncharged particle, since qE/m here is equal to a.

Again integrating relation (ii), we have

$$r = \frac{1}{2} \frac{qE}{m} t^2 + ut + C_2$$

where C_2 is another constant of integration.

Again, at t = 0, r = r_0, the initial position vector of the particle, and we have $r_0 = C_2$.

So that, $$r = \frac{1}{2} \frac{qE}{m} t^2 + ut + r_0 = \frac{1}{2} at^2 + ut + r_0. \qquad \text{...(iii)}$$

The vector diagram corresponding to it is as shown in Fig. 3.3, where Oa represents the *initial position vectors* r_0 at t = 0, AB *represents* ut, *the distance covered in time t due to the initial velocity* u and BC represents 1/2 at² or $\frac{1}{2} \frac{qE}{m} t^2$, *the distance covered on account of the acceleration acquired due to field E,* directed along BC. The resultant displacement r of the particle is the *vector sum* of all the three and is represented by OC, in accordance with the polygon law of addition of vectors.

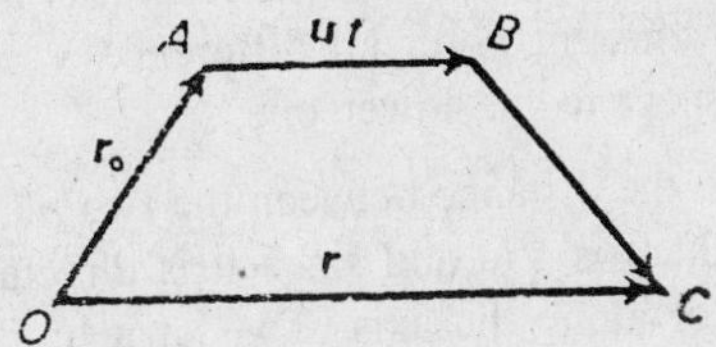

Fig. 3.3

Two particular cases : Having discussed above the general case of the motion or displacement of a charged particle when placed in an electric field, let us consider two particular cases, viz., (i) *when the electric field acts along the direction of motion of the particle* and (ii) *when it acts in a direction perpendicular to it.*

(i) It will be easily seen that in the case of both p and r_0 being *directed along the electric field* E, the acceleration produced in the particle will also be directed along E. The electric field is, in this case called a longitudinal field and the acceleration acquired by the particle, a longitudinal acceleration. The resultant position vector r will now also be directed along E, thus making the whole problem a simple *one-dimensional* one.

Also, if $r_0 = 0$ and $u = 0$, *i.e., if the particle is initially at the origin and starts from rest*, we shall again have a one-dimensional problem with

$$r = \frac{1}{2}at^2 \quad \frac{1}{2}\frac{q}{m}Et^2 \text{ in the direction of E,}$$

exactly in the manner that the distance covered by an uncharged particle, starting from rest, is given by $S = 1/2\ at^2$.

So that, in either case, we may replace, the vectors r_0, r, u, T v and E by their respective scalar magnitudes.

(ii) Let us now discuss the case when the electric field is a transverse one, *i.e., in a direction perpendicular to the initial velocity of the particle*. In this case, the acceleration acquired by the panicle is also referred as the transverse acceleration.

Thus, imagine a beam of charged particles, each of *mass* m and carrying a *charge* q, to be travelling along the axis of x with velocity v and passing in between two metallic plates P_1 and P_2 (Fig. 3.4), maintained at a constant potential difference V, with the upper plate positive with respect to the lower one.

Then, if d be the distance between the two plates, the electric field set up between them is V/d and since it is directed from the upper to me lower plate, or along the axis of y, let it be denoted by E,

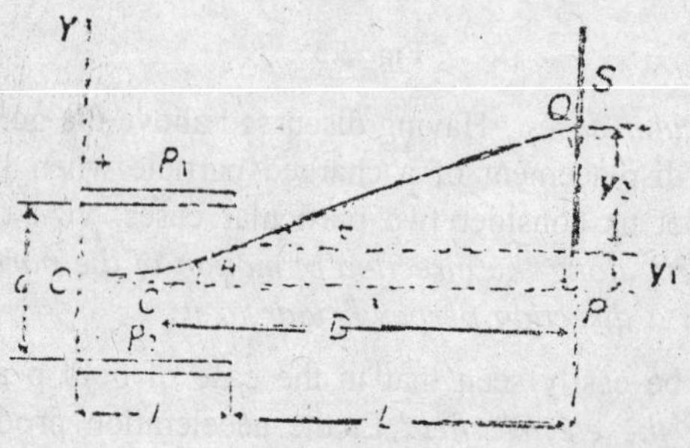

Fig. 3.4

As soon as the beam of charged particles enters in between the plates at the origin O, it is subjected to this transverse electric field E_y. Since this field is perpendicular to the velocity v_r of beam the latter remains unaffected and the beam, therefore, continues to cover the horizontal distance between the plates with the same velocity v_0. If l be the length of the plates or that of the electric field set up between them, the time taken by the beam to cover this distance is l/V_x.

During this time, obviously, each particle of the beam acquires a *transverse acceleration* $a = q.E_y/m$ and hence a transverse velocity $v_x = 0 + at$.

or $$v_y = 0 + q.\frac{E_y}{m}t = \frac{qE_y}{m}.\frac{l}{v_p},$$

[∵ the initial transverse velocity is *zero.*

∴ *transverse displacement of the beam* (upwards or downwards, according as the charge on the particles is negative or positive) during the time $t = l/v_x$ that it takes to cover the entire electric field of length l is say, $y_1 = 0 + 1/2at^2$,

i.e., $$y_1 = \frac{1}{2}\frac{qE_y}{m}.l^2 \frac{qE_y}{2mv_x^2}.l^2.$$

This is clearly the equation to a parabola ($qE_y/2mv_x{}^2$ being a constant quantity). So that, the-beam of charged particles takes a parabolic path in between the two plates, as shown.

It will be seen that the problem here is a *two-dimensional* one, similar to that of a particle thrown horizontally and subjected to the downward acceleration due to gravity (*i.e.*, similar to that of a projectile). The important difference, however, is that, unlike the gravitational force acting on the uncharged particle, we can here alter the electric field and hence the force acting on the particle, as desired. Also, the electric field is effective only within a distance l, the length of the two plates, beyond which it suddenly drops to zero and the beam, therefore, moves straight along the direction in which it emerges from the electric field between the plates.

If there be a screen S at a distance L from the ends of the plates, the beam will strike it at, say, Q, making an angle θ with the x-axis, instead of at P, along it, as it would if it were undeflected. The *total vertical displacement* of the beam is thus PQ = y, which is clearly made up of two parts: y_1 within the plates (or the electric field) and y_2, on emergence from the plates (or the electric field).

As we have seen above, $y_1 = \frac{q.E_y}{2mv_x^2}.l^2.$

And $y_2 = L \tan \theta,$

where $\tan\theta = \frac{v_y}{v_x} = \frac{qE_y/mv_x}{v_x} = \frac{qE_y l}{mv_x^2}$

So that, $y_2 = \frac{LqE_y l}{mv_x^2}.$

And, therefore, *total displacement* of the beam,

$$y = y_1 + y_2 = \frac{qE_y l^2}{2mv_x^2} + \frac{LqE_v l}{mv_x^2}$$

or $$y = \frac{qE_y l^2}{mv_x^2}\left(L + \frac{l}{2}\right)$$

Since $qE_y l/mv_x^2 = \tan\theta$ and $(L+l/2)$ = distance D of the screen from the centre C of the electric field between the plates, we have

$$y = D \tan \theta.$$

This is the *principle* underlying the *cathode ray oscillograph, television picture* tubes and a host of other experiments, like determination of e/m (*i.e.*, *charge to mass ratio*) for an electron etc.

In all these cases, an electron beam is obtained from a hot cathode or filament, surrounded by a hollow metallic cylinder, kept at a negative potential with respect to the cathode and called the *modulator*. This helps to make the electrons emitted by the cathode into a compact beam. The electron beam is then subjected to a high accelerating potential V', say, between the cathode and a pair of anodes, so that the energy acquired by each electron is eV', where e is the *charge* on an electron. If m be the mass of an electron and v_x, the velocity acquired by it along the axis of x, clearly,

$$1/2mv_x^2 = ev',$$

whence, $v_x = \sqrt{2eV'/m}$

In the case of the *cathode ray oscillograph*, (Fig. 3.5), this high velocity electron beam is passed through two pairs of plates which, when a potential difference is applied to them, have a vertical and a horizontal electric field set up in between them and thus deflect the electron beam passing through them vertically and horizontally respectively. Thus, activising any one pair of plates at a time, the beam may be deflected vertically or horizontally, as desired. The deflection of the beam is noted on the fluorescent screen from the position of the bright spot produced first by the undeflected beam at P and then by the deflected beam at Q, say, in the case of vertical deflection.

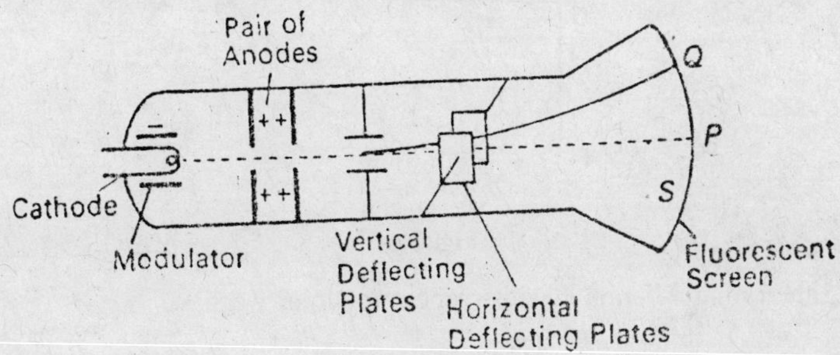

Fig. 3.5

The cathode ray oscillograph is put to a number of uses, one of which is to demonstrate in a spectacular manner the formation of Lissajous' figures when a particle is subjected simultaneously to two simple harmonic motions at right angle to each other.

CHARGED PARTICLE IN AN ALTERNATING ELECTRIC FIELD

Suppose we have a particle of *mass m* and carrying a charge q placed in an alternating electric field whose intensity E at any instant t is given by $E = E_0 \sin \omega t$, where E_0 is the *maximum or the peak value* of E, *i.e.*, the amplitude of the electric field vector and $2\pi/\omega$, its time-period.

If, therefore, the acceleration produced in the particle under the action of this field be d^2r/dt^2, we have

$$m\frac{d^2r}{dt^2} = qE = qE_0 \sin \omega t,$$

whence, $$\frac{d^2r}{dt^2} = \frac{qF_v}{m}.\sin \omega t \qquad ...(I)$$

indicating that *the acceleration of the particle too, like the applied electric field, varies sinusoidally with time,* as shown in Fig. 3.6, *having an amplitude (or peak value)* qE_0/m, and *the same time-period* $2\pi/\omega$ *as that of the applied field.*

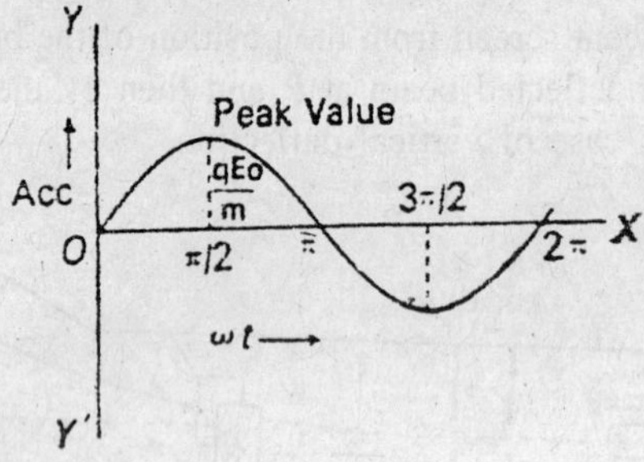

Fig. 3.6

Integrating relation I with respect to time, we have

$$\frac{dr}{dt} = v = -\frac{qF_0}{m\omega}\cos\omega t + C_1$$

where C_1 is the constant of integration to be determined from the initial or boundary conditions.

Initially, i.e., at $t = 0$, when the electric field (E) is zero, the initial velocity of the particle too is zero and we have

$$0 = -\frac{qF_0}{m\omega} + C_1.$$

$$\text{or } \frac{dr}{dt} = v = \frac{-qF_0}{m\omega}\cos\omega t + \frac{qE_0}{m\omega}$$

$$\text{or } \frac{dr}{dt} = v = \frac{-qF_0}{m\omega}(1 - \cos\omega t) = \frac{q2E_0}{m\omega}\sin^2\omega t \qquad \text{...(i)}$$

indicating that the *velocity of the particle at any given instant is always positive, in the direction of* E_0 and hence unidirectional, *though it varies after the manner of a sin*^{2}*—curve*, as shown in Fig. 3.7, and has an amplitude or peak value $2gE_0/m\omega$.

Again, integrating relation ii, with respect to t, we have

$$r\left(\frac{qE_0}{m\omega}\right)t - \left(\frac{qE_0}{m\omega^2}\right)\sin\omega t + C_2,$$

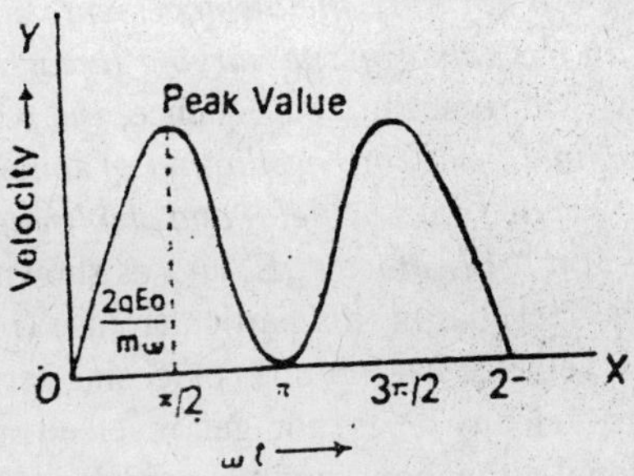

Fig. 3.7

where C_2 is another constant of integration.

Since at t = 0, r = r_0, the initial position vector, we have

$$r = \left(\frac{qE_0}{m\omega}\right) t - \left(\frac{qE_0}{m\omega^2}\right) \sin \omega t + r_0, \qquad \text{...(iii)}$$

where r, may be zero or have a finite value, depending upon the choice of the origin.

However, if r = 0 at t = 0. We have r_0 = 0. In that case, therefore,

$$r = \left(\frac{qE_0}{m\omega}\right) t - \left(\frac{qE_0}{m\omega^2}\right) \sin \omega t \qquad \text{...(iv)}$$

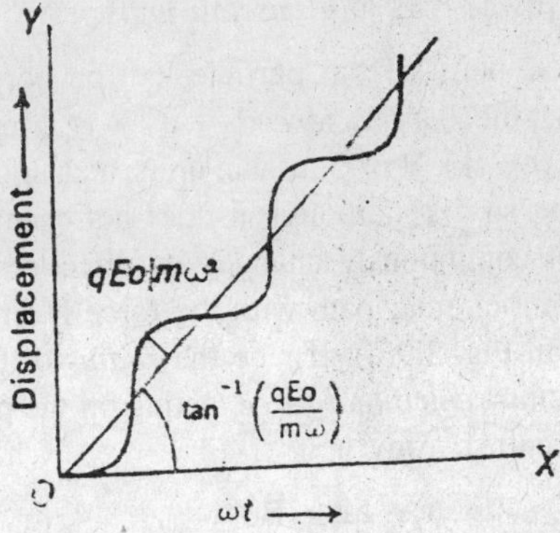

Fig. 3.8

It will thus be seen that under the initial or boundary condition chosen here (viz., r = 0 at t = 0), *the displacement vector* (t) is the sum *of two vectors, with the value of one varying linearly, and that of* the *other, sinusoidally*, with time. In consequence, *the resulting motion of the particle is a simple harmonic oscillation of lime-period* $2\pi/\omega$ (*the same as that of the applied electric field) and amplitude* $qE_0/m\omega^2$, *super-imposed on a constant drift velocity* $qE_0/m\omega$, as shown in Fig. 3.8. This is obvious from the fact that in this particular case (r = 0 at t = 0) the expression for the velocity has no sine or cosine term attached to it, indicating that the velocity does not get reversed at any time (*i.e.*, remains unidirectional throughout) and the particle continues to advance in one and the same direction.

CHARGED PARTICLE IN A UNIFORM AND CONSTANT MAGNETIC FIELD

We have seen under when a particle carrying a charge q moves with velocity v in a magnetic field B, it is acted upon by a *Lorentz force* V = qv × B in emu, RMKS or SI units, or equal to q/c v × B in *Gaussian units* (*i.e.*, with q in esu and B in emu).

In the absence of any other force, therefore, (the force due to gravity being negligible compared with the Lorentz force), if m be the mass of the particle, and d^2r/dt^2, its acceleration we have

$$m\frac{d^2r}{dt^2} = m\frac{dv}{dt} = qv \times B \text{ in emu, RMKS or SI units} \quad ...(i)$$

or $$= (qv/c) \times B \text{ in } \textit{Gaussian units}.$$

Now, if the velocity of the particle, (v) be perpendicular to the magnetic field (B), the *Lorentz force* F = qv × B in emu, RMKS or SI units, or, equal to (qv/c) x B in Gaussian units, acts upon it in a direction perpendicular to v; so that, although it does not change the magnitude of the velocity, it continuously changes its direction, resulting in the particle moving in a circular path with the force F directed towards its centre, as shown in Fig. 3.9(a). If r be the *radius* of this circular path, clearly, the balancing *centrifugal force,* acting on the particle *outwards*, (away from the centre) = mv^2/r.

So that, $$mv^2/r = qv \times B,$$

whence, r = mv/qB, if emu, RMKS or SI units be used or r = mvc/qB if Gaussian units be used. ...(ii)

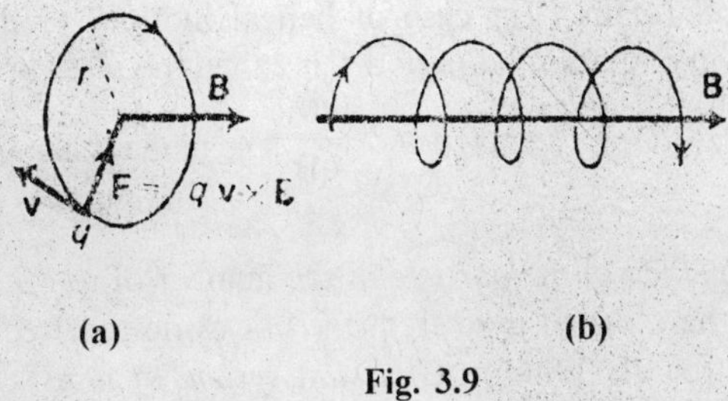

(a) (b)

Fig. 3.9

If, however, the velocity v of the particle be inclined to the magnetic field at an angle other than 90° (*i.e.*, be not normal to the field), it may be resolved into a component v_1, *parallel to the field*, and a component v_2, at right angles to it. The *Lorentz force acting on the particle* will then be $gv_2 \times B$ in emu, RMKS or SI units, or $(qv_2/C) \times B$ in Gaussian units, making the particle move in a circular path of radius $r = mv_2/qB$, if emu, RMKS or SI units be used or, $r = mv_2c/qB$ if Gaussian units be used....(iii)

And, since the component v_1 of the velocity, parallel to the field B makes the particle move in the direction of the field, the combined effect of the circular and the linear horizontal motion is to produce a *helical motion* (or a *moving* circle), as shown in Fig. 3.9 (b), with the direction of the magnetic field (B) as axis.

The radius r of the circular, or the helical path, described by a charged particle in a uniform magnetic field is sometimes called gyro radius or cyclotron radius.

Note : It will be noted that in expressions (ii) and (iii) above, for r, the terms mv and mv_2 denote the momentum of the particle in the plane perpendicular to B. The relations would, therefore, hold good even in the case of relativistic velocities if we use the appropriate value of the momentum (p) in place of mv or mv_2. These relations may, therefore, be used to obtain the value of p for high energy particles.

If ω be the angular velocity of the particle in circular or helical motion in the magnetic field, we have $\omega = v/r$ (or v_2/r) $= qB/m$ (or qB/mc if Gaussian units be used). And, therefore,

time-period of the particles (or the time taken by it to describe one *full circle*) is given by $r = 2\pi/\omega = 2\pi/qB$ (or $2\pi mc/qB$ in *Gaussian system of units*).

The distance through which the particle advances forward as it completes one full rotation (in case of helical motion) or the linear distance it covers in one full time-period T is called the *pitch of the helix* and is clearly equal to $v_1 T \frac{2\pi m}{qB} v_1$ [or, $\frac{2\pi Mc}{qB} v_1$ in Gaussian system of units]

And, the *frequency of the particle* or the number of rotations made by it per second, here called gyro frequency or cyclotron frequency, n = 1/T = qB/2πm [or qB/2πmc in *Gaussian system of units*].

As will be readily seen, *the gyro frequency* or cyclotron frequency, *of a particle is quite independent of the velocity, and hence also of the energy, of the particle so long as <he velocity ties in the non-relativistic region, i.e., so long as* v << c. This, as shall see in the next article, is the principle underlying the cyclotron.

This expression for frequency no longer holds good as v → c, for then the mass of the particle varies in accordance with the relation $m = m_0\sqrt{1 - v^2/c^2}$, where m_o is the rest mass of the particle.

Alternatively, we could tackle the problem analytically as follows:

Let the magnetic field be directed along the axis of x, so that B = Bi. Then, expressing v in terms of its components along the three coordinate axes, relation I above may be put in the form

$$m\frac{d}{dt}(v_x i + v_y j + v_z k) = q^2(v_z i + v_v j + v_x k) \times (Bi)$$

$$\text{or } \frac{dv_x}{dt}i + \frac{dv_y}{dt}j + \frac{dv_z}{dt}k = \frac{qB}{m}(0 + v_x j - v_y k). \quad \text{...(iv)}$$

Equating the coefficients of i on either side of the equation, we have

$$\frac{dv_x}{dt} 0 \quad \text{...(v)}$$

indicating that the *particle does not accelerate in the direction of* the field or that v_x, the velocity of tie panicle in this direction (*i.e.*, along the field) remains constant.

This velocity, in our earlier discussion above, we have denoted by v_1. So that, $v_x = v_1$.

Similarly, equating the coefficients of j and t on either side, we have

$$\frac{dv_x}{dt} = \frac{dB}{m} v_x \quad \text{...(vi)}$$

and $$\frac{dv_x}{dt} = \frac{dB}{m} v_y \qquad ...(vii)$$

Differentiating relation (vi) with respect to t, we have

$$\frac{d^2 v_y}{dt^2} = \frac{dB}{m}\frac{dv_x}{dt}$$

∴ substituting the value of dv_x/dt from relation (vii) above, we have

$$\frac{d^2 v_y}{dt^2} = \left(\frac{dB}{m}\right)^2 v_y.$$

Or, putting $qB/M = \omega$, we have $$\frac{d^2 v_y}{dt^2} = -\omega^2 v_y \qquad ...(viii)$$

As will be readily seen, this is an equation similar to that for the *simple harmonic motion* of a particle except that *in place of the displacement y of the particle, we have here its velocity* v_y. On that analogy, therefore, the solution of this equation is

$$v_y = A \sin(\omega t + f)^+, \qquad ...(ix)$$

where the values of the constant A and ϕ can be obtained directly from the initial or boundary conditions.

From relation (vi), therefore, we have

$$\frac{d}{dt}\left[A \sin(\omega t + \phi)\right] = \frac{dB}{m} v_x = \omega v_z,$$

i.e., $$A\,\omega \cos(\omega t + \phi) = \omega v_z.$$

or $$v_z = A \cos(\omega t + \phi). \qquad ...(x)$$

Squaring and adding relations (ix) and (x), we have

$$v_y^2 + v_x^2 = A^2.$$

But we have, in our earlier discussion, taken the velocity of the particle in the Y – Z plane to be v_2. So that, $A^2 = v_2^2$. or, $A = v_2$.

Since $v_y = d_y/dt$ and $A = v_y$ we have from relation (ix),

$$\frac{d_y}{dt} = v_2 \text{ sn } (\omega t + \phi) \qquad ...(xi)$$

And since $v_x = d_x/dt$, we have from relation (x) above,

$$d_x/dt = v_2 \cos(\omega t + \phi) \qquad ...(xii)$$

Equations (xi) and (xii) on integration, respectively yield

$$y = -\frac{v_2}{\omega}\cos(\omega t + \phi)$$

and $$z = \frac{v_2}{\omega}(\omega t + \phi)$$

squaring and adding which, we have

$$y^2 = z^2 = \frac{v_2^2}{\omega^2} r^2$$

This is the equation to a circle of radius r in the Y – Z plane,

where $r = v_2/\omega = v_2 \frac{dB}{m} = mv_2/qB,$

which is thus the value of the *gyro radius* or the *cyclotron radius.*

Since the particle has also a constant horizontal velocity v_1 along the axis of x (or the direction of the field), the resultant motion of the particle is helical, with the axis of the helix lying along the direction of the field (B).

Note : It may be mentioned again that if we use *Gaussian units, i.e.*, if we take q in esu and B in emu (or gauss), we must use q/c instead of q in all the expressions obtained above, since q esu of charge = q/c emu of charge.

THE CYCLOTRON

This is a device to obtain high energy charged particles (like *protons*, deutrons, α-particles) and was originally developed by *E.O. Lawrence and U.S. Livingston* in the year 1931.

Construction: As shown in Fig. 3.10(a), it essentially consists of *two short,, semi-circular hollow cylindrical boxes* D_1 and D_2, usually of copper, and called the *dees* on account of their shape like the letter D. These are placed in an evacuated cylindrical steel tank (Fig 3.10(b)), with their edges parallel but with a small diametrical gap G between them.

The whole assembly is arranged to lie in between the pole-pieces of a large and powerful electromagnet NS, of about the same diameter as that of the two dees; so that, the magnetic field is parallel to the axis of the tank and perpendicular to its base and hence also to the plane of the dees, with a flux of the order of 1.5 *weber-metre*^{2} *or 15000 gauss.*

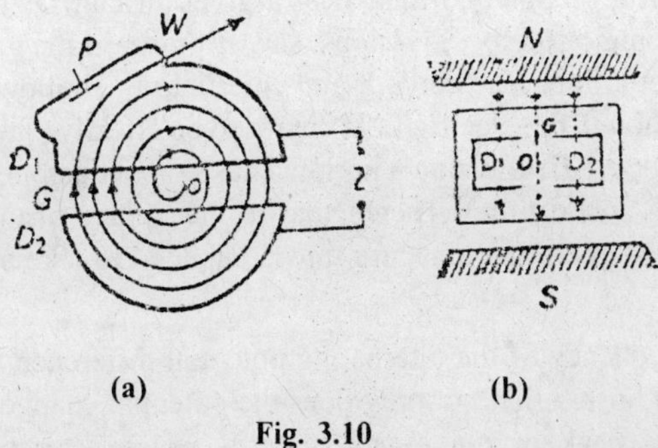

(a) (b)

Fig. 3.10

The dees, also referred to as the *electrodes,* are connected to a radio frequency oscillator in order to produce a high alternating potential difference across them, of the order of 10^4 *hertz* (*i.e.*, 10^7 *cycles per second*).

The *positive ions* to be accelerated may be obtained in one of the following ways:

(i) The required gas, (for example, *hydrogen* for obtaining *protons, heavy hydrogen* for obtaining *deutrons* and *helium* for obtaining *α-particles*) is filled into the steel tank at a low pressure of about 10^{-3} mm of mercury and a filament, at or near the centre of the tank, *just outside the dees,* is heated. With a small potential difference applied between the filament and the tank, the electrons emitted by the hot filament acquire sufficient energy to be able to ionise some of ions. Some of these positive ions then find their way into the gap between the two dees.

(ii) Alternatively, what is known as a *capillary ion source* is used. The ions here too are produced in the same manner as indicated under (i) above, but *in a separate or an auxiliary chamber* and led from there into the steel tank through narrow opening or a capillary tube which is so directed that there is again a narrow vertical column of positive ions near the centre O of the gap G between the two dees.

Both the methods have their advantages and disadvantages into which, however, we need not go here.

Principle underlying and working of the cyclotron. On account of the high frequency alternating potential difference applied to the dees

D_1 and D_1 they alternately acquire positive and negative potentials about 10^7 times in a second. So that, if at a given instant D_1 happens to be negative, some of the positive ions, say, *protons*, in the gap G between the dees, get attracted towards it and enter into the hollow space within it. Now, although the electric field ceases to be effective inside D_1, (there being no electric field inside a hollow charged conductor), the magnetic field acts in a direction perpendicular to their plane of motion and the positive ions, or the protons, are thus compelled to take a circular path inside D_1.

The frequency of the alternating potential difference applied to the dees is so adjusted that not only does the potential change direction but has also its peak or maximum value V when, after covering their semicircular path through D_1, the positive ions, or the protons, emerge out into the diametrical gap G; so that, they now find D_2 to be negative and D_1 positive, and therefore cross over from D_1 to D_2, *each acquiring an amount of energy qV in the process* (where q is the charge on it). Hence, as under the action of the .magnetic field perpendicular to them, they again describe a semi-circular path inside D_2, they emerge out into the gap G with a higher velocity.

If this frequency of the alternating potential difference be further equal to the gyro frequency or the cyclotron, frequency $qB/2\pi m$ (by suitable adjustment of the magnetic field B), the frequency of rotation of the ions or the protons then being quite independent of their velocity inside the dees, they always emerge out into the gap from either dee to find that the dee have, in the meanwhile, changed their signs of potential. The motion of the ions through the dees is thus in resonance with the alternating potential applied. Every time, therefore, the ions gain an amount of energy qV each as they cross over from one dees to the other, with the result that they continue to move in circular paths of continually increasing radii inside the two dees, thus describing an ever-expanding spiral, as shown in Fig. 3.10(a), until the radius of their path becomes nearly the same as the outermost radius of either dee.

By this time, the *ion beam* acquires a sufficiently high velocity and is brought out of dee D_1 through a thinly covered window W by means of a deflecting plate P which is maintained at a negative potential. To make sure that this negative potential is with respect to the earth, *it is so arranged that the window W ties at right angles to the diametrical gap G between the dees, i.e.*, OW is perpendicular to the gap G, because, then, the instantaneous potential difference across the gap G is zero at the time that the ions or the protons emerge out at W. (This is so because the time taken by the ions to cover the distance from the gap to W, inside D_1, is one-quarter of their time-period and since the potential difference across the gap had its peak or maximum value when the ions crossed the gap from D_2 to D_1, it must, after a quarter of a time-period, have an instantaneous value zero). Thus, deflected by the negative plate P, the high energy beam of positive ions (or protons) is allowed to fall on the target, placed in an evacuated side-tube attached to the steel tank.

As we know, the maximum radius of the circular path of the ions is given by $r_{max} = mV_{max}/qB$, where v_{max} is the maximum velocity acquired by the ions and B, the flux density of the field. So that,

$V_{max} = qB, r_{max}/m$. And, therefore,

kinetic energy acquired

$$= \frac{1}{2} mv^2_{max} = \frac{1}{2} m\left(\frac{qBr_{max}}{m}\right)^2 = \frac{1}{2}\frac{q^2B^2r^2{}_{max}}{m}$$

For a high energy ion beam, therefore, we require a high magnetic flux B and a large value of the maximum radius (r_{max}) of the circular path of the ions, or, a large cyclotron radius.

It may appear from the expression above that the energy of an ion beam is quite independent of the potential difference across the dees, since at a lower potential across them the ions may just have to make a larger number of revolutions to acquire the same amount of energy as they would acquire in a smaller number of revolutions if the potential difference across the dees were high. Theoretically, this seems to/be all right but, in actual practice, for a variety of reasons it is always desirable not to make the number of revolutions of the ions inside the dees very large and *this naturally necessitates a high potential difference across the dees.*

We cannot, however, indiscriminately go on increasing the potential difference across the dees to increase the energy of the ion beam because at higher velocities thus acquired by the ions, relativistic effects manifest themselves and their mass increases in accordance with the relation $m = m_0/\sqrt{1 - v^2/c^2}$, , (where m^ is their rest mass), their time-period, *i.e.*, the time taken to make one complete revolution inside the dees thus increases or their frequency ($qB/2\pi m$) decreases and no longer agrees with the frequency of the alternating potential difference applied, *i.e.*, the resonance between the two is destroyed. They do not, therefore, acquire energy on crossing over from one dee to the other and hence cease to be accelerated. *There is thus a limit up to which the ions can be accelerated in a cyclotron, the maximum energy limit being roughly* 25 MeV. A cyclotron, it will be readily seen, is thus *wholly unsuitable for accelerating electrons to form a high energy electron* beam, since on account of their small mass, even a comparatively small increase in their energy results in their acquiring a very high velocity, giving rise to relativistic effects.

The remedy for this lies in either decreasing the frequency of the applied alternating potential difference so as to be in step with the decreasing frequency of rotation of the ions, keeping the magnetic field (B) unaltered or, increasing the magnetic field with increase in the mass of the ions, keeping the frequency of the alternating potential difference unaltered, so that $qB/2\pi m$ remains constant. A device of the former type is called a *syncho-cyclotron* and that of the latter type, a *synchrotron.* We are not, however, concerned with these here.

MAGNETIC FOCUSSING

Particles carrying the *same charge* (q) and having the *same momentum* (p), even if moving in different directions, can all be brought to focus at very nearly the same point on a screen by means of a suitably applied magnetic field. This is called *magnetic focussing*, and since the particles come to a common focus after describing an angle of 180° from their point of entry into the magnetic field, we qualify it as 180° magnetic focussing.

If, on the other hand, the particles carry the *same charge* but have *different momenta*, they are brought to focus at different points on the screen.

To take the second case first, if a beam of particles, each carrying a charge q, *enters normally*, through a slit S in a screen, [Fig. 3.11 (a)], into a magnetic field of flux density B perpendicular to the beam (*i.e.*, perpendicular to the plane of the paper in the case shown), a particle of the beam is acted upon by a force equal to qvB, where v is the velocity of the particle. Under the action of this force, the particle takes a circular path, such that if m be the mass of the particle and r, the radius of its circular path, we have

$$mv^2/r = qvB, \text{ whence, } r = mv^2/qvB = mv/qB. \qquad \text{...(i)}$$

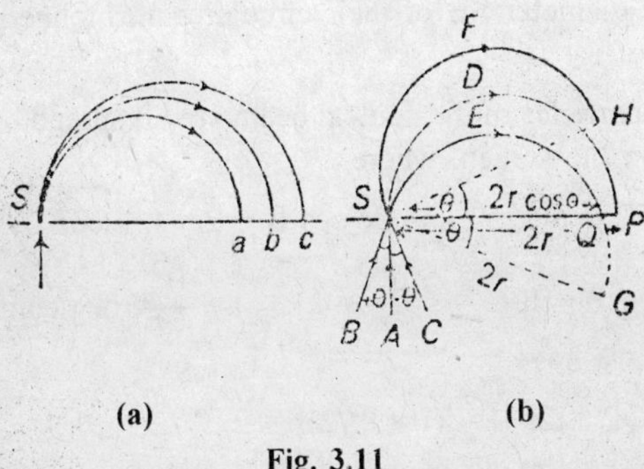

(a) (b)

Fig. 3.11

Now, q and B being constants, *i.e.*, the same for each particle, r varies with the momentum p = mv of the panicle. So that, *panicles having different momenta come 10 focus at different points a, b, c on the screen.* The device thus acts as a momentum selector.

Obviously, even if the velocities of the panicles in the beam be arranged to be the same, their momenta will still be different on account of their *different masses*. So that, particles of different masses will then come to focus at different points. This is the principle underlying the *mass spectrographs*, like the *Bainbridge mass spectra-graph*, for the separation of *isotopes, i.e., elements having the same atomic number and chemical properties but different atomic weights.*

Let us now consider the first case, viz; the *case of particles carrying the same charge and having the sane momentum.*

Thus, suppose we have a conical beam of particles, all carrying the same charge q and having the same momentum p = mv, entering the magnetic field through a slit S as shown in Fig. 3.11(b). Then, the particles entering the slit normally along AS describe, under the action of the magnetic field, the trajectory SDP, thus covering an angle of 180° from S and coming to focus oa the screen at P, such that the diameter of the circular path taken a SP = 2r.

The particles entering the slit Along AS at an angle +θ, with AS describe the trajectory SEQ and come to focus at Q, such that the diameter SG of their circular pith is also inclined to SP at an angle θ, and, therefore, the chord SQ = 2r cos Q.

Similarly, the particles entering the slit along CS at an angle –θ with AS describe the trajectory SFQ and also come to focus on the screen at Q, with the diameter SB of their circular path inclined to SP at an angle θ.

Thus, the particles of the conical beam are all spread over the small distance QP on the screen, where

$$QP = SP - SQ = 2\ r - 2r \cos \theta = 2r\ (1 - \cos \theta).$$

Or, since $\cos\theta = 1 - \frac{\theta^2}{2} + \frac{q^4}{4} - .... = 1 - \frac{\theta^2}{2}$, neglecting the higher powers of θ), we have

$$QP = 2r\ [l - (I - \theta^2/2)] = 2r.\theta^2/2 = r\theta^2.$$

θ^2 being negligible for a small value of θ, points Q and P will be practically coincident or the same.

In other words, *all particles having the same charge and momentum came to focus at the same point P, after cowing an angle of 180° from their point of entry (S) into the magnetic field.*

If the screen be a photographic plate and the slit (S) parallel to the magnetic field, we shall have line images of the slit corresponding to particles of different momenta at points like a, b, c in the first case [Fig. 3.11(a)] and a single thin line image of the slit at P in the second case, [Fig. 3.11(b)], the line images corresponding to all particles being coincident there.

Determination of relative abundance of isotopes in an element : Accelerating the ions of a given element through slit S. [Fig. 3.11(a)] by means of in adjustable potential difference and using another slit at P, we can estimate the relative abundance of the different isotopes in the element. For, obviously only those ions will pass through the slit at P for which the ratio momentum to charge, *i.e.*, mv/q is *constant* = Br (see relation /above). These ions are allowed to send a current through a sensitive galvanometer and the deflection noted. The accelerating potential is now altered so that another set of ions, with a slightly different value of m/q now enter the slit S and emerge through the slit of P, producing a slightly different deflection in the galvanometer. These deflections in the galvanometer being proportional to the number of ions present in the beam sod responsible for the current through it, we can easily estimate from them the relative abundance of the different ions present in the beam and hence that of the different isotopes present in the given element.

CHARGED PARTICLE IN A COMBINED ELECTRIC AND MAGNETIC FIELD

Having studied the behaviour of a charged particle in an electric and a magnetic field separately, let us now see how it will behave in a combined electric and magnetic field, (i) *when the two fields are parallel (or antiparallel)* and (ii) *when they are crossed, i.e.*, at right *angles to one another.*

(i) *When the magnetic and electric fields are parallel* : Thomson parabolas. Let a magnetic field B and an electric field E be both along the axis of y [Fig. 3.12(a)], and let a particle of *mass m* and carrying a charge q be travelling along the negative direction of the z-axis with velocity v, *i.e.*, let

$$\mathbf{B} = B\mathbf{j},\ \mathbf{E} = E\mathbf{j} \text{ and}$$

$$\mathbf{v} = -v\mathbf{k}.$$

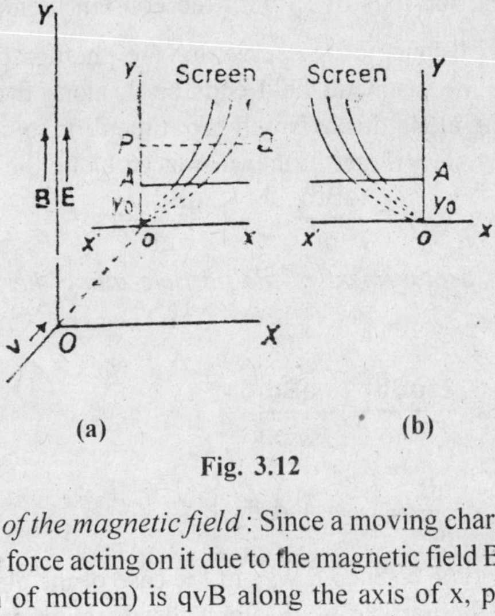

(a) (b)

Fig. 3.12

(a) *Action of the magnetic field* : Since a moving charge is equivalent to a current, the force acting on it due to the magnetic field B (perpendicular to its direction of motion) is qvB along the axis of x, perpendicular to both B and v, in accordance with Fleming's left hand rule. So that, its acceleration along this axis is $a = d^2/di^2 = qvB/m$, assumed to be constant (in view of any change in velocity of the particle on account of the two fields being small or negligible).

If the length of the region occupied by the magnetic field be d_1, clearly enough, the time taken by the particle to cross the field, say, t

$= d_1/v$, because the force due to the magnetic field being along the axis of x and that due to the electric field, along the axis of y, there is no force acting along the axis of z (at right angles to both the axes of x and y) and the velocity of the particle along this axis therefore remains constant at v.

During this time t, the particle will, therefore, move or get displaced along the axis of x through a distance $x_1 = ut + 1/2\,at^2 = 0 + \frac{1}{2}\frac{qvB}{m}\left(\frac{d_1}{v}\right)^2$ since its initial velocity along the axis of x is zero, *i.e.*,

$$x_1 = \frac{1}{2}\frac{qvBd_1^2}{mv}$$

And the velocity acquired by the particle along the axis of x in time t, when it *just emerges from the magnetic field,* is given by $v_x = 0 + at = \frac{qvB}{m}.\frac{d_1}{v} = \frac{qBd_1}{m}$. From this point onwards, therefore, it will move (along the axis of x) with the constant velocity v.

Now, if the distance of the screen or the photographic plate from the point where the magnetic field ends be d_2 along the axis of z, the particle travelling along this axis will take time d_2/v, to cover it. During this time the particle will get further displaced along the axis x through a distance $x_2 = v_x.d_2/v = \frac{qBd_1}{m}.\frac{d_2}{v} = \frac{qBd_1d_2}{mv}$

Thus, *total displacement of the particle along the x-axis,*

i.e., $$x = x_1 + x_2.$$

or $$x = \frac{1}{2}\frac{qBd_1^2}{mv} + \frac{qBd_1d_2}{mv}$$

or $$x = \frac{qB}{mv} + \left(\frac{d_1^2}{2} + d_1d_2\right) \qquad ...(i)$$

(b) *Action of the electric field :* In the case of the electric field, the force acting on the particle is qE along the direction of the field, *i.e.*, along the axis of y and hence its acceleration along this axis is qE/m. So that, proceeding exactly as in the case of the magnetic field above, we have

total displacement of the particle along the y-axis,

i.e., $$y = y_1 + y_2.$$

or $$y = \frac{1}{2}\frac{qEd'^2_1}{mv^2}\frac{qEd'_1 d'_2}{mv^2}$$

or $$y = \frac{qE}{mv^2}\cdot\left(\frac{d'^2_1}{2} + d'_1 d'_2\right), \qquad ...(ii)$$

where d'_1 is the length of the region occupied by the electric field and d'_2, the distance of the screen or the photographic plate from the point where the electric field ends.

Eliminating v between relations (i) and (ii), we therefore have

$$\frac{x^2}{y} = \frac{q}{m}\cdot\frac{B^2}{E}\left(\frac{d_1^2/2 + d_1 d_2}{d_1^2/2 + d'_1 d'_2}\right) = \frac{q}{m}\frac{B^2}{E}.k, \qquad ...(iii)$$

where k is a constant, equal to $\left(\frac{d_1^2/2 + d_1 d_2}{d'^2_1/2 + d'_1 d'_2}\right)$, depending upon the geometry of the apparatus.

This equation represents a *parabola about the y-axis, if q/m be* constant, indicating that *all particles having the same charge to mass ratio q/m, but different velocities, fall at different points on the same parabola*, those with a different charge to mass ratio falling on a different parabola.

In Fig. 3.12(a) only one branch of the parabolas is shown. The other branch is obtained, as shown in Fig. 3.12(b), by reversing the magnetic field, so that the two fields become antiparallel.

It will be seen from relation II above that the displacement of the particle along the axis of y (*i.e.*, due to the electric field) is inversely proportional to the square of the velocity (v) of the particle, so that particles moving with the *maximum velocity* (v_{max}) suffer the *minimum displacement*, given by

$$y_0 = \frac{qE}{mv^2_{max}}\left(\frac{d'^2_1}{2} + d'_1 d'_2\right).$$

If a particle with the maximum velocity has been accelerated

through a potential difference V. its *kinetic energy* is given by $1/2mv_{2max} = qV$, whence, $mv^2_{max} = 2qV$. Substituting this value of mv^2_{max} in the relation for y_0 above, we have

$$y_0 = \frac{qE}{2v}\left(\frac{d'^2_1}{2} + d'_1 d'_2\right),$$

which is obviously the same for all particles moving with the maximum velocity V_{max} irrespective of their charge to mass ratio q/m. *No particles can thus suffer a displacement less than* y_0 *along the axis of y and there can, therefore, be no trace of the parabolas along the y-axis* below OA, where OA = y_0 (Fig. 3.12). That is why these parts of the parabolas have been shown dotted in the figure.

The point where the two branches of the parabolas meet, when produced backwards, gives the position of the origin O.

Since these parabolas were first obtained by J.J. *Thomson* in his experiment to determine the charge to mass ratio (e/m) for positive rays (sec next article), they are referred to as *Thomson parabolas.*

It will be seen that dividing relation (ii) by (i), we have

$$\frac{y}{x} = \frac{E}{vB}\left(\frac{d'^2_1/2 + d'_1 d'_2}{d^2_1/2 + d_1 d_2}\right),$$ the equation to a straight line, indicating that *all particles having the same* velocity (y) fall on *some such straight line as PQ, parallel to the x-axis, irrespective of their charge to mass ratio (a/m).*

Note: It may be pointed out once again that if Gaussian units be used, *i.e.*, if q be taken in esu and B in ems (or gauss), θ must be replaced by q/c in all the expressions above.

Determination of e/m for positive rays : It was first shown by Goldstein in the year 1886, that if a perforated cathode be used in a discharge tube in which the pressure is a little higher than in the usual ones for the production of cathode rays (so that there is an appreciable amount of residual gas in the tube), luminous streamers arc seen travelling into the back region of the cathode (Fig. 3.13), away from the anode, *i.e.*, in the opposite direction to that of the cathode rays (which travel towards the anode), the colour of the streamers depending upon the nature of the residual gas inside the tube.

These came to be known as *canal rays*, since they appeared to issue from the cylindrical channels or canals in the cathode, or *positive rays*, since they carried a positive charge on them, as evidenced by their travelling away from the anode as also by the direction of their deflection by a magnetic or an electric field.

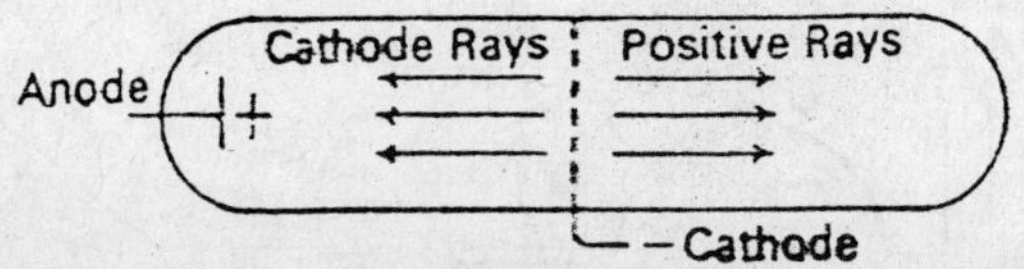

Fig. 3.13

These rays, as we now know, are just *streams of positive ions* produced because of the fist moving electrons, rushing from the cathode towards the anode, *ionising the gas*, *i.e.*, knocking out electrons from some of the gas atoms they collide with. These positive ions carry a charge equal in magnitude to e (the charge on an electron) or a multiple of e, according as one or more electrons have been knocked out of an atom.

J.J. Thomson was the first to have devised a method, in the year 1913, to determine their *charge to mass ratio e/m* by the application of a combination of a magnetic and an electric field, parallel to each other, when he obtained parabolas on a fluorescent screen, such as the ones we have discussed above. The essential features of the apparatus used by him are shown in Fig. 3.14.

The gas in question is allowed to enter *bulb* B at a slow rate rough an *inlet i* and allowed to pass out through an outlet O, so that the conical part of the apparatus to the right remains highly evacuated. When a potential difference of 30 to 50 thousand volts is applied across an *aluminium anode* A and a cathode C, consisting of an aluminium rod carrying along its axis a copper tube of a fine bore (1 mm or less), positive rays are formed in bulb B. Most of these rays strike against the cathode which thus gets heated. In order to cool it, therefore, it is surrounded by a water jacket J, as shown. Some of them, however, passing through the narrow bore of the cathode, form a compact beam.

This beam is subjected to a combination of magnetic and electric fields by means of an electromagnet NS whose pole-pieces P_1 and P_2 are iron plates which are electrically insulated from the rest of the electromagnet by means of mica plates M_1 and M_2 and which also serve as the two plates of a capacitor to which a desired potential difference may be applied to produce an electric field in the same reason *as, and parallel with*, the magnetic field due to the electromagnet. As iron shield I is arranged, as shown, to screen off the discharge in bulb B from any stray external fields.

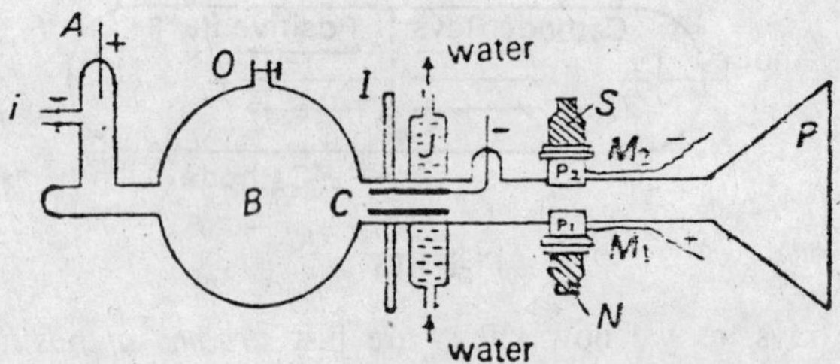

Fig. 3.14

The positive-ray beam, thus subjected to parallel magnetic and electric fields, suffers deflection in the manner already discussed and describes arcs of parabolas on the fluorescent, screen or plate P, corresponding to the positive ions constituting the beam,

From the coordinates x, y of any point on a parabola, the charge to mass ratio e/m for the positive ion corresponding to that particular parabola can then be obtained from relation (iii) above, which, on substituting e for q, gives

$$\frac{e}{m} = \frac{x^2}{y} \cdot \frac{e}{B^2}\left(\frac{d'^2_1/2 + d'_1 d'_2}{d^2_1/2 + d_1 d_2}\right)$$

(ii) When the magnetic and electric fields are crossed or mutually perpendicular. Let us now consider the case of a charged particle in a combined magnetic and electric field perpendicular to each other.

Let the electric field E be along the axis of y and the magnetic field B, along the axis of z (Fig. 3.15). So that, E = Ej and B = Bk.

If at any given instant, the velocity of the particle, subjected to the two fields simultaneously, be v, we have $v = v_y i + v_y j + v_z k$.

If the mass of the particle be m and the charge carried by it, q, its *equation of motion* is

$$m\frac{dv}{dt} = qE + q(v \times B)$$

$$= q[E + (v \times B)]$$

or

$$m\frac{dv}{dt}(v_x j + v_y j + v_y k)$$

$$= q[Ej + (v_x i + v_y j + v_z k) \times Bk]$$

$$= qEj + qB\,(v_y i - v_x j),$$

i.e., $$\frac{dv_x}{dt}i + \frac{dv_y}{dt}j + \frac{dv_z}{dt}k$$

$$= 1/m\ [qEj + qB\,(v_y i = v_x j)]$$

$$= \left(\frac{qE}{m} - \frac{qBv_x}{m}\right)j\ \frac{qBv_y}{m}i$$

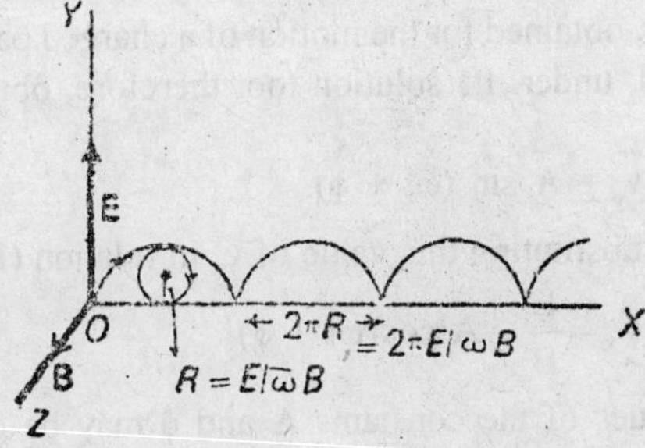

Fig. 3.15

Thus, the component accelerations along the three coordinate axes are

$$\frac{dv_x}{dt} = \frac{qB}{m}v_y \quad ...(i)$$

$$\frac{dv_y}{dt} = \frac{qB}{m} - \frac{qB}{m}v_x \quad ...(ii)$$

and $$\frac{dv_z}{dt} = 0 \quad ...(iii)$$

From relation (iii) we have v_z = a *constant*,

indicating that *the velocity component of the particle along the direction of the magnetic field remains unaltered in the crossed field.* If, therefore, this component of the velocity of the particle be initially *zero*, its motion will be restricted entirely to the X- Y plane.

Let us, therefore, deduce the values of the components v_x and v_y.

Differentiating relation (ii) with respect to (we have

$$\frac{d^2v}{dt^2} = -\frac{qB}{m}.\frac{dv_x}{dt}.$$

Or, substituting the value of dv_x/dt from relation (i), we have

$$\frac{d^2 v_y}{dt^2} = -\frac{q^2 B^2}{m^2} v_y$$

Or, putting $qB/m = \omega$, the angular frequency of the particle, we have

$$\frac{d^2 v_y}{dt^2} = -\omega^1 v_y,$$

which is the equation of a *simple harmonic motion*, in which the displacement of the particle is replaced by its velocity v_y, the same as equation (viii), obtained for the motion of a charged particle in a uniform magnetic field, under. Its solution too, therefore, obtained in the same manner, is

$$v_y = A \sin(\omega t + \phi). \qquad \text{...(iv)}$$

And, on substituting this value of v_y in relation (ii) above, we have

$$v_x = \frac{E}{B} - A\cos(\omega t + \phi) \qquad \text{...(v)}$$

where the values of the constants A and ϕ may be obtained from the initial conditions

Thus, if initially the particle be at rest, we have $v_x = v_y = v_z = 0$ at $t = 0$. So that, from relations (iv) and (v) we obtain $\phi = 0$ and $A=E/B$.

And, thus,
$$v_y = \frac{E}{B}\sin \omega t \qquad \text{...(vi)}$$

$$v_x = \frac{E}{B} - \frac{E}{B}\cos \omega t = \frac{E}{B}(1 - \cos \omega t) \qquad \text{...(vii)}$$

and, of course, $v_z = 0$. ...(viii)

To obtain the values of the coordinates, we integrate these equations, when we have

$$x = \int v_x \, dt = \int \frac{E}{B}(1 - \cos\omega t)dt$$

$$= \frac{E}{B}\left(t - \frac{\sin\omega t}{\omega}\right) + C_1$$

where C_1 is a constant of integration.

Since at $t = 0$, $x = 0$, we have $C_1 = 0$. And, therefore,

$$x = \frac{E}{B}(\omega t - \sin\omega t) \qquad \text{...(i)}$$

$$y = \int v_y \, dt = \int \frac{E}{B} \sin \omega t \, dt$$

$$= -\frac{E}{B\omega} \cos \omega t + C_3,$$

where C_2 is another constant of integration.

Since at t = 0, y = 0, we have $C_2 = E/B_\omega$. And, therefore,

$$y = -\frac{E}{B\omega} \cos \omega t + \frac{E}{B\omega} = \frac{E}{B\omega}(1 - \cos \omega t) \quad \text{...(i)}$$

and $\qquad z = 0. \qquad$ ...(iii)

Equations I, 11 and 111 indicate that the particle has (a) *a constant velocity E/B along the x-axis*, (b) *a clockwise circular motion with angular velocity* ω in the X-T plane and (c) zero velocity along the *z-axis i.e., the particle executes a cycloidal motion in the X- Y plane, with the head of the cycloid having a velocity E/B along the x-axis,* the same as executed by a particle on the ran of a wheel of radius R = E/ωB roiling along the x-axis, as shows in the figure, with the maximum displacement along the y-axis equal to the diameter of the circle, *i.e.*, 2R = 2E/ωS and the distance between two consecutive cusps of the cycloidal path equal to 2πR = 2πE/ωB.

Now, two points may be noted with interest:

(i) In a-frame of reference, moving along the -c-axis with the same velocity as the head of the cycloid. viz., E/B (instead of remaining stationary), the x-component of the velocity of the particle, relative to the observer in the moving frame, will obviously become $v_x = -A \cos(\omega t + \phi)$, the y and z-components remaining unaltered. As a result, the particle will execute a circular or a helical motion, with angular velocity ω according as $v_x = 0$ or $\neq 0$. This, it will be recalled is exactly the type of motion executed by a charged particle under the influence of a uniform magnetic field alone.

This means, in other words, that *in this particular moving frame of reference, the electric field becomes ineffective.*

(ii) As we have seen, the velocity of the charged particle along the z-axis remains unaffected in the crossed magnetic and electric fields, obviously because neither the magnetic nor the electric field exerts any force on it. And if $v_y = 0$, *i.e.*, if the panicle has no velocity along the y-axis, the force on it due to the magnetic field, in a direction perpendicular to that of the y-axis, will also the zero. So that, the only force acting on it due to the magnetic field will be because of its velocity r, along the axis of x and this, as we know, will be equal to qv_xB *along the negative direction of the y--axis* (in accordance with Fleming's left hand rule).

And since the force acting on the particle due to the electric field *E will be qE along the positive direction of the y-axis, the two will just balance each other and there will be no motion of the particle along the y-axis if* $q\,v_xB = qE$. *Or,* $v_x = E/B$, *i.e., if us velocity component along the x-axis be equal to the ratio between the electric and the magnetic fields.*

Now, in relation (v) above, if $v_x = E/B$, we have $A = 0$. It follows, therefore, that under these conditions (viz., $v_y = 0$ and $qv_xB = qE$ or $v_x = E/B$), the circular part of the motion of the particle will vanish altogether and it will execute a linear motion along the axis of x with a constant velocity $r_x = E/B$.

Since this expression for v_x involves-neither m nor q, it follows that a applies equally well to all charged particles irrespective of their mass and the sign and magnitude of the charge they any.

Such an arrangement of crossed fields is known as a velocity selector or a velocity filter, since it is used for *selector* a beam of electrons or ions having a desired constant velocity $v = E/B$ (and *filtering* out others) by suitably adjusting the value of either field, as, for example, in Thomson's experiment for determining the *charge to mess ratio e/m* for electrons and in Bainbridge's *mass spectrograph* etc.

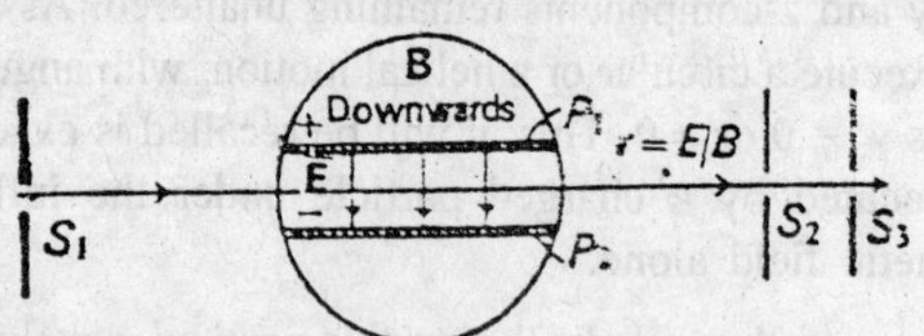

Fig. 3.16

Thus, if we apply a potential difference to plaits P_1 and P_2, (Fig. 3.16), an electric field E is produced *in the plane of the paper* in the direction P_1 to P_2. If we also simultaneously produce a magnetic field B *within the same region,* (as indicated by the circle), *perpendicular to the plane of the paper* and, therefore, also to the electric field, and directed downwards, by means of a powerful electromagnet, such that E/B remains constant, we have the desired crossed i fields to serve as velocity selector.

For, if a stream of ions or electrons, having different velocities, be alleged to pass in between the plates, any particles having a y-component will move up or down (according to the sign of the charge on them and will be stopped by one or the other plate, whereas those having zero y-component will continue to move straight along the x-axis without suffering any deflection one way or toe other. And, be effect of the component of the velocity of the particle will merely be to spread the beam along the axis of z, *i.e.*, perpendicular to the plane of the paper, so that we shall have a wide beam of velocity V_x issuing out of the plates P_1 and P_2. If, however, we arrange a slit S_1 along the x-axis, as shown, it stops the electrons or ions with a z-component of velocity, so that only those having the x-component of velocity V_x pass in between the plates, *i.e.*, through the crossed fields, and a narrow beam of electrons or ions of the required velocity emerges from them which may be further narrowed down, if so desired by passing it through a couple of slits such as S_2 and S_3.

The velocity V_x of the electron or the ion beam may be adjusted as desired by suitably changing E, and hence the ratio E/B by altering the potential difference across the two plates.

SOLVED EXAMPLES

Example 1:

The adjoining figure shows a particular charge distribution, 1/2 q, –q, 1/2q. Show that for r >> a, the electric field at P is given by 3Q/ r^4, where Q = qa^2.

Solution:

Here, clearly, *electric field at* P *due to the charge* 1/2 q *to the left of the point* O *is*

$$= \frac{\frac{1}{2}q}{(r+a)^2}r = \frac{q}{2(r+a)^2}r \text{ directed outwards,}$$

where r is the *unit vector* along OP.

Electric field at P *due to the charge* 1/2 q *to the right of the point O*

$$= \frac{q}{2(r-a)^2} r \text{ also directed outwards,}$$

and *electric field at p due to the charge* – q at O $= \frac{-q}{r^2}\hat{r}$ *directed inwards.*

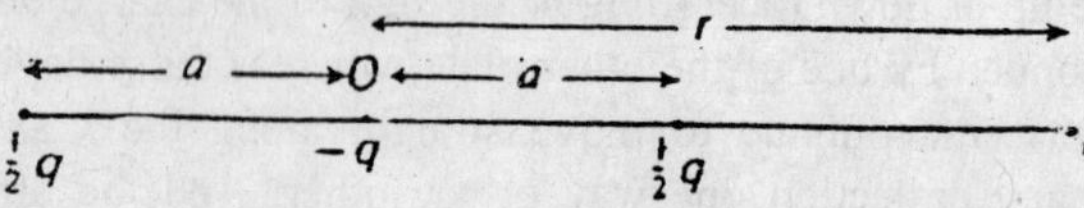

$\frac{1}{2}q$, $-q$, $\frac{1}{2}q$, a , O , a , r , P

Fig. 3.17

∴ *resultant electric field at* P *due to the three charges,* say, E

$$= \left(\frac{q}{2(r+a)^2} + \frac{q}{2(r-a)^2} - \frac{q}{r^2}\right) r$$

$$= \frac{q}{2r^2}\left[\left(1+\frac{a}{2}\right)^{-2} + \left(1-\frac{a}{2}\right)^{-2} - 2\right]\hat{r}.$$

Or, expanding by Binomial theorem and remembering that (a/r)< < 1 or a < <r, or, what is the same thing. r >>a,. we have

$$E = \frac{q}{2r^2}\left[\left(1 - \frac{2a}{r} + \frac{(-2)(-3)}{2!}\cdot\frac{a^2}{r^2}\right)\right.$$

$$\left. + \left(1 + \frac{2a}{r} + \frac{(-2)(-3)}{2!}\cdot\frac{a^2}{r^2}\right) - 2\right]\hat{r}$$

$$= \frac{q}{2r^2}\left(\frac{3a^2}{r^2} + \frac{3a^2}{r^2}\right) r = \frac{3qa^2}{r^4} r, \text{ directed outwards.}$$

Or, *magnitude of the electric field at P due to the three charges* $= 3qa^2/r^4 = 3Q/r^4$ *directed outwards.* [∵ $qa^2 = Q$

Example 2:

A cathode ray oscillograph has deflecting plates of length 2 cm and separation 0.5 cm. Calculate the potential difference (in volts) between the plates which will cause angular deviation of 0.04 radius in an electron beam of speed 8.0 × 10³ cm/sec. [e/m = 5 × 10 esu/gm.].

Solution:

Let the potential difference required to the applied to the plates be V volts = V/300 statvolts or esu.

$\therefore$ electric field in between the plates

$$= \frac{\text{p.d.}}{\text{distance between the plates}} = \frac{V}{300 \times 0.5} = \frac{V}{150} \textit{ statvolts/cm.}$$

Hence *acceleration produced in an electron in between the plates*

$$= \frac{\text{force acting on the electron}}{\text{mass of the electron}} = \frac{Ve}{150 \times m} \text{cm / sec,}$$

where e is the charge on the electron and m, its *mass*.

$$= \frac{V}{150} \times 5 \times 10^{17} = \frac{1 \times 10^{17}}{30} \text{cm / sec}^2$$

Now, the electron remains in between the plates for an interval of time

$$t = \frac{\text{length of the plates}}{\text{horizontal velocity of the electron } (V_z)}.$$

Here, *length of the plates* = 2 cm and *horizontal velocity of the electron* = 8 × 10^8 cm/sec.

So that, $t = \frac{2}{8 \times 10^8} = 0.25 \times 10^{-8} = 2.5 \times 10^8$ sec.

$\therefore$ vertical velocity acquired by the electron in time t, say, v_2

$$= 0 + at = 0 + \frac{V \times 10^{17}}{30} \times 2.5 \times 10' = \frac{5V}{6} \times 10^7 \text{ cm / sec.}$$

$\therefore$ *angular deviation of the electron,*

$$\theta = \frac{V_y}{V_z} \text{radian} = 0.04 \text{ radian} \textit{ (given).}$$

Or, $\frac{5V \times 10^7}{6 \times 8 \times 10^3} = 0.04$, whence, $V = \frac{6 \times 8 \times 10^8 \times 0.04}{5 \times 10^7}$

$$= \frac{19.2}{5} = 3.84 \text{ volts.}$$

Thus, the *potential difference that must be applied to the plates* is 3.84 *volts.*

Example 3:

(a) Calculate the intensity of the electric field which produce in an electron an acceleration equal to that due to gravity. (mass of an

electron = 0.91 × 10^{-27} gm., charge on it, e = 4.80 × 10^{-10} esu and g = 980 cm/sec²).

(b) Obtain the ratio of the electrostatic force between two electrons distant r apart and the gravitational force between them. (G = 6.67 × 10^{-8} dynes-cm²/gm²).

Solution:

(a) Let E be the required intensity of the field. Then, clearly, force acting on an electron placed in this field = Ee and, therefore, *acceleration produced in the electron,* a = force/mass = Ee/m cm/sec², *i.e.,*

$$a = \frac{E(4.8\times10^{-10})}{0.91\times10^{-27}}\,\text{cm}/\text{sec}^2.$$

This must be equal to the acceleration due to gravity, *i.e.,* 980 cm/sec².

We, therefore, have $\dfrac{E(4.8\times10^{-10})}{0.91\times10^{-27}} = 980,$

whence, $E = \dfrac{980\times0.91\times10^{-27}}{4.8\times10^{-10}}$

$$= \frac{980\times0.91}{4.8}\times10^{-17} \textit{ statvolts/cm.}$$

Or, $E = \dfrac{980\times0.91}{4.8}\times10^{-17}\times300 = \dfrac{98\times0.91\times10^{-14}}{1.6}$

$$= 5.57 \times 10^{-27} \text{ volts/cm.}$$

(b) The *electrostatic force between the two electrons,* say,

$$F_E = \frac{e.e}{r^2} = \frac{e^2}{r^{2'}}$$

and *gravitational force between them, say,* $F_G \; \dfrac{m\times m}{r^2}G = \dfrac{m^2}{r^2}G.$

∴ *ratio of dectrostastic force and gravitational force, i.e., FE/*F_G

$$= \frac{(4.8\times10^{-10})^2}{(0.91\times10^{-27})^2\times6.67\times10^{-8}} = 4.17\times10^{42}.$$

Example 4:

A particle of charge q and mass m enters in electric field—Ej between two plates with an initial velocity v_o *i. (i) Calculate the forces* *acting on it in the and y directions. (ii) will a force in the y-direction influence the x-component of the velocity ? (iii) Solve for v_z an v_y as functions of time and write the complete vector equation for the velocity v acquired after time t. (iv) Taking the origin at the point of entry, write the complete vector equation for the position of the particle as a function of time while the particle is between the plates.*

Solution:

(i) The x-direction being perpendicular to the direction of the field which is along the y-direction, *there will be no force acting on the particle along the x-direction, i.e.,* $F_x = 0$. *In the y-direction,* naturally, *the force acting on the particle = charge on the particle × intensity of the field, So that,* $F_y = -qEj$.

(ii) A force in the y-direction will have no component in the x-direction and can, therefore, have *no effect* on the velocity of the particle in this direction.

(iii) The initial velocity of the particle is given to be $v_0 i$, along the x-direction. Since there is no force acting on the particle is this direction , its velocity in this direction remains unaltered at $v_0 i$. And, since a force $-qEj$ acts on the particle along the y-direction, its acceleration along this direction is $a = -qE/m$. The velocity acquired by the particle along this direction in time t is, therefore,

v = 0 + at (the initial velocity in this direction being zero), *i.e.*,

$$v = = -\frac{qE}{m}t.$$

The *complete vector equation for the velocity of the particle* is thus $v = v_0 i - \frac{qE}{m}t$.

(iv) The distance covered by the particle along the x-direction in time t is $v_0 t$ and the distance covered by it along the y-direction is $ut + 1/2\ at^2 = 0 + 1/2\ \frac{qE}{m}t^2$.

If therefore, r be the position vector of the particle after time t, we have

$$r = r_0 + v_0 ti - \frac{qE}{m}t^2 j.$$

Since the origin is taken at the point of entry, $r_0 = 0$. We, therefore, have *vector equation for the position of the particle,*

$$r = v_o t i - \frac{qE}{m} t^2 j.$$

Example 5:

In example 7 above, if the charged particle be an electron, having an initial energy 10 [10] ergs, if the electric field strength be 0.01 statvolt/cm and if the length of the plates be 2 cm, find (a) the velocity, vector as it leaves the region between the plates, (b) the angle (v, i) for the particle as it leaves the plates, (c) the displacement of the electron and (d) the point of intersection of the x-axis with its direction as it leaves the field.

Solution:

(a) Here, if v_o be the initial velocity of the electron, we have 1/3 m $v_o^2 = 10^{-10}$ ergs, where m is the mass of electron, so that

$$v_o = \sqrt{\frac{2\times10^{-10}}{m}} = \sqrt{\frac{2\times10^{-10}}{9.1\times10^{-28}}}$$

cm/sec along the axis of x, *i.e.*, equal to v_x.

Since this velocity remains unaffected by the electric field, the time taken by the electron to cross the field-length l between the plates is given by $t = l/v_o = 2/v_o$ sec.

∴ *velocity acquired by the electron in time t sec, i.e.,*

$$v = v_o i + \frac{eE}{m}\left(\frac{1}{v_o}\right) j = v_o i + \frac{eE}{m}\left(\frac{1^2}{v_o^2}\right)^{\frac{1}{2}} j,$$

where eE/m is the acceleration acquired by the electron.
Or,

$$v = \left(\frac{2\times10^{-10}}{9.1\times10^{28}}\right)^{\frac{1}{2}} i + \frac{4.8\times10^{-10}(10^{-2})}{9.1\times10^{-28}}\left(\frac{4\times9.1\times10^{-28}}{2\times10^{-10}}\right)^{\frac{1}{2}} j.$$

$$= \left(\frac{2\times10^{18}}{9.1}\right)^{\frac{1}{2}} i + \frac{4.8\times10^{16}}{9.1}(2\times9.1\times10^{-18})^{\frac{1}{2}} j.$$

Or, $v = 4.7 \times 10^3 i + 2.25 \times 10^7 j$ cm/sec.

(b) The angle θ that the electron makes with the x-axis as it leaves the plates is given by the relation $\tan\theta = v_y/v_x = 2.25 \times 10^7/4.7 \times 10^8 = 0.0475$, whence, $0 = 2°44'$.

(c) The displacement of the electron (see Fig. 4.4) is the distance covered by it along the y-direction in time t, *i.e.*, equal to y = 0 + 1/2 at^2 , because the initial velocity along the y-direction is zero.

Or *displacement,*

$$y = \frac{1}{2}\frac{E}{m}t^2 = \frac{1}{2}\frac{eE}{m}\frac{l^2}{v^2} = \frac{1}{2}\frac{4.8\times10^{-10}\times10^{-}}{9.1\times10^{-28}}\left(\frac{4\times9.1\times10^{-28}}{2\times10^{-10}}\right)$$

$$= \frac{1}{2}\left(\frac{4.8\times10^{16}}{9.1}\right)(2\times9.1\times10^{-18}) = 4.8\times10^{2}\text{ cm.}$$

(d) If x be the distance of the point of intersection of the direction of the electron, as it emerges from the field, with the axis of x, we have

$$\tan\theta = \frac{y}{l-x}, \text{ whence, } l-x = \frac{y}{\tan\theta}.$$

Or,
$$x = l - \frac{y}{\tan\theta} = 2 - \frac{4.8\times10^{-2}}{4.78\times10^{-2}} = 2 - 1 = 1\text{ cm.}$$

Thus, the point of intersection of the direction in which the electron emerges from the field and the axis of x lies *at a distance of* 1 cm *from the origin or the point of entry of the electron into the field.*

Example 6:

Alpha particles of speed v_x = 2.0 × 10^9 cm/sec pass through 10 cm length of a field where the alternating electric field E_x = 5 × 10^3 sin (ωt + θ) volts/cm exists. Calculate the difference of speed between the fastest and the slowest particle emerging from the field.

Solution:

Here, clearly, the *amplitude* or the *peak value* of the field, *i.e.*, E_0 = 5 × 10^3 *volts per* cm or 5 × 10^3/300 *statvolt*/cm.

With the high speed of the particles, the values of the field may be assumed to remain the same during the small fraction of time they take to cross just 10 cm length of it. So that, an α-particle (with its positive charge) gains energy when the field is positive and loses an equal amount of energy when the field is negative, the maximum gain or loss obviously occurring when the field has its peak value.

∴ assuming the field to be at its peak positive value, we have

force acting on the α-particle = q × E_0 and hence

energy gained by the particle in crossing 10 cm *of the field*

$= q \times E_o \times 10.$

The particle will, therefore, have the maximum speed.

Similarly, when the field is at its peak negative value, the *energy lost by the particle in traversing 10 cm of the field* $= q \times E_o \times 10$. So that, the particle will now have the minimum speed.

The difference in the energies of the fastest and the slowest particles will thus be $2q \times E_o \times 10$.

Now, an α-particle carries a positive charge of magnitude 2e, where e is the charge on a proton (equal to 4.8×10^{-10} esu). So that, *difference in the energies of the fastest and the slowest particles, i.e.,*

$\Delta E = 2 \times 2e \times E_o \times 10 = 2 \times 2 \times 4.8 \times 10^{-10} \times (5\times 10^3/300) \times 10 = 3.2 \times 10^7$ ergs.

Since energy $E = 1/2\, mv^2$, where m is the mass and v, the velocity of the α-particle, we have

$$\Delta E = mv\Delta v, \text{ whence, } \Delta v = \Delta E/mv.$$

∴ substituting the values of ΔE, m (equal to 4 times the mass of a proton) and v (given), we have

difference in the velocities of the fastest and the slowest particles, i.e.,

$$\Delta v = \frac{3.2\times10^7}{4\times1.6725\times10^{-24}\times2.0\times10^8} = \frac{3.2\times10^8}{4\times1.6725\times2}$$

$$= \frac{0.4\times10^8}{1.6725} = 2.39\times10^7 \text{ cm/sec.}$$

Example 7:

A stream of protons in first accelerated through a potential difference V until it acquires a constant velocity V and then passed through two parallel plates to which an alternating potential $V_o \sin \omega t$ is applied. If the frequency of the alternating potential be not very large and the time of transit of the stream through the plates comparatively small, show that the stream emerges from the plates with a velocity equal to $\left[\frac{2e}{m}(V + V_o \sin \omega t)\right]^{\frac{1}{2}}$, *where e is the charge on a proton and m, its mass.*

Solution:

Since v is the velocity acquired by a proton on being accelerated through a potential difference V, we have $1/2\ mv^2 = eV$.

On being subjected to the alternating potential $V_0 \sin \omega t$, the proton will acquire an additional energy $eV_0 \sin \omega t$. If, therefore, v' be its velocity *now,* clearly, *in crease in its energy* $= 1/2mv'^2 - 1/2mv^2 = 1/2m\ (v'^2 - v^2)$.

We, therefore, have $1/2m\ (v'^2 - v^2) = eV_0 \sin \omega t$. Or, $mv'^2 = 2eV_0 \sin \omega t + mv^2 = 2eV_0 \sin \omega t + 2eV$,

whence, $v'^2 = \dfrac{2eV_0 \sin \omega t}{m} + \dfrac{2eV}{m} = \dfrac{2e}{m}(V + V_0 \sin \omega t)$.

Or, *velocity of the proton stream on emerging through the plates, i.e.,*

$$v' = \left[\frac{2e}{m}(V + V_0 \sin \omega t)\right]^{\frac{1}{2}}$$

Example 8:

For a proton initially travelling with velocity $v_x = 5 \times 10^8$ cm/sec, calculate the following: (i) Force experienced in a magnetic field $B = 2000j + 4000k$ gauss. (ii) Acceleration acquired in an electric field $E = 200i + 100j$ volts/cm. (iii) Transverse deflection in travelling a length $L_x = 10$cm in as electric field $E_y = 200$ volts/cm.

Solution:

(i) We know that force acting on a charge q moving perpendicular to a magnetic field B, with velocity v is given by $F = \dfrac{q}{c}(v \times B)$, using *Gaussian units.* So that,

$$F = \frac{q}{c}(5 \times 10^8 i)\ (2000i + 4000k).$$

Here, $q = e = 4.8 \times 10^{-10}$ *esu* and $v = v_x = 5 \times 10^{-8}$ cm/sec. We, therefore,

have $F = \dfrac{4.8 \times 10^{-10}}{3 \times 10^{10}} \times 5 \times 10^8 \times 2000\ (k - 2j) = 1.6 \times 10^8\ (k - 2j)$ *dynes.*

Or, *magnitude of the force,* $F = 1.6 \times 10^{-8} \sqrt{5} = 3.578 \times 10^{-8}$ dynes and its *direction cosines,* clearly, 0, $-2\sqrt{5}$ and $1\sqrt{5}$.

(ii) We know that the force experienced by a particle carrying a charge q in an electric field E is equal to qE. Therefore, if π be the *mass* of the particle and a, the *acceleration* produced in it, we have

$$m\,a = q\,E, \quad \text{whence, } a = qE/m.$$

So that *acceleration of the proton in the electric field. i.e.,*

$$a = \frac{4.8\times10^{-10}}{1.67\times10^{-24}}\left(\frac{200i+100j}{300}\right) = \frac{4.8\times10^{-10}}{1.67\times10^{-24}}\left(\frac{2i+j}{3}\right)$$

∴ *magnitude of the acceleration,*

$$a = \frac{4.8\times10^{-10}}{1.67\times10^{-24}\times3}\sqrt{5} = \frac{1.6\times\sqrt{5}}{1.67}\times10^{14}$$

$$= 2.142 \times 10^{14} \text{ cm/sec}^2,$$

and its *direction cosines,* 2√5 1√5 and 0

(iii) The *transverse displacement of the particle* will clearly be $y = 1/2\ at^2$, where acceleration a = (*charge* × *electric field*/ *mass* = qE_y/m and $t = L_x/v_x$.

So that,

$$y = \frac{1}{2}\frac{4.8\times10^{-10}}{1.67\times10^{-24}}\times\frac{200}{300}\times\left(\frac{10}{5\times10^8}\right)^2 = \frac{1.6}{1.67\times25} = 0.0383 \text{ cm.}$$

Example 9:

(a) A stream of protons and deutrons in a vacuum chamber enters a uniform magnetic field. Both protons an deutrons have been subjected to the same accelerating potential; hence kinetic energies of the particles are the same. If the ion stream is perpendicular to them magnetic field and the protons move in a circular path of radius 15 cm., find the radius of the path traversed by the deutrons. Given that the mass of a deutron is twice that of a proton.

(b) Calculate the Lorentz force on a proton moving with velocity (2i + 3j) 10^8 cm sec in an electric field of intensity (i + 2j + 3k) statvolts/ cm and a magnetic field of (300j + 300k) gauss.

Solution:

(a) We know that the radius of the path traversed by a charged particle in a magnetic field is given by r = mvc/qB, using *Gaussian units.*

So that, *radius of the path of the proton,* $r_p = m_p\ v_p\ c/qB$ and *radius of the path of the deutron,* $r_D = m_D v_D c/qB$.

$$\therefore \quad \frac{r_D}{r_p} = \frac{m_D v_D c}{qB} \times \frac{qB}{m_p v_p c} = \frac{m_D v_D}{m_p v_p}.$$

Now, we are given that $m_D = 2m_p$. And, since the kinetic energies of the two particles are the same, their velocities must be inversely proportional to the square roots of their masses. For,

$$im_D v_D^2 = 1/2 m_p v_p^2 \text{ and } \because \frac{v_D}{v_p} = \sqrt{\frac{m_p}{2m_p}}\ \frac{1}{\sqrt{2}} \quad \text{Or,} \quad v_D = \frac{v_p}{\sqrt{2}}.$$

$\therefore\ r_D/r_p = 2m_p \times v_p \sqrt{2}/m_p \times v_p = \sqrt{2}$.

Since $r_p = 15$ cm (given), we have $r_D = r_p \sqrt{2} = 15\sqrt{2} = 21.21$ cm.

Thus, *the radius of the circular path of the deutron = 21.21 cm.*

(b) As we know, *Lorentz force* $F = qE + \frac{q}{c} v \times B$. Substituting the given values, therefore, we have

$$F = 4.8 \times 10^{-10}(i + 2j + 3k) + \frac{4.8 \times 10^{-10}}{3 \times 10^{10}}(2i + 3j)10^3 \times (300j + 300k)$$

$$= 4.8 \times 10^{-10}[(i + 2j + 3k) + 2k - 2j + 3k)] = 4.8 \times 10^{-10}(4i + 5k)$$

$\therefore$ *magnitude of the force,* $F = 4.8 \times 10^{-10}(\sqrt{16} + 25)$

$$= 4.8 \times 10^{-10}\sqrt{41} = 3.07 \times 10\text{--}9 \text{ dyne},$$

and the direction cosines $= 4/\sqrt{41}.0.\ 5/\sqrt{41}$.

Example 10:

An electron of energy 20 electron volts moving in a direction perpendicular to a magnetic field $B = 10^2$ Wb/m^2 describes a circle of radius r calculate the value of r. (1 eV = 1.6×10^{10} joules; mass of electron may be taken to be 9×10^{31}kg).

Solution:

Let v be the velocity of the electron perpendicular to the magnetic field. Then, if m be its mass, we have

$$1/2\, mv^2 = 20 \text{ eV} = 20 \times 1.6 \times 10^{-10} \text{ joules}.$$

$$\text{whence, } v = \left(\frac{2 \times 20 \times 1.6 \times 10^{-19}}{m}\right)^{\frac{1}{2}} = \left(\frac{64 \times 10^{-19}}{m}\right)^{\frac{1}{2}} \text{ m/sec.}$$

Now, as we know, the radius of the circle described by a charged particle in a uniform magnetic field or the *gyro* (*or cyclotron*) *radius*, as it is called, is given by r = mv/qB. So that, we have

$$r = \frac{m(64\times10^{-19})^{\frac{1}{2}}}{\sqrt{m\times q\times B}} = \frac{\sqrt{m(64\times10^{-19})^{\frac{1}{2}}}}{q\times B}$$

$$= \frac{(9\times10^{-31}\times64\times10^{-19})^{\frac{1}{2}}}{1.6\times10^{-19}\times10^{2}}$$

$$= \frac{3\times8}{1.6}\times10^{-8} = 15.0\times10^{-8} \text{ metre.}$$

Thus, the *radius of the circular path described by the electron* = 15.0×10^{-8} *metre.*

Example 11:

An electron of velocity v = (2i + 3j) 10^8 cm/sec enters a region of uniform magnetic field B = 500i gauss, so that its path becomes helical.

(i) In what direction does the axis of the helix lie? (ii) Calculate the radius of the helix. (iii) Calculate the number of rotations as the electron advances 10 cm along the axis of the helix.

Solution:

Here, the velocity v of the electron may be resolved into two rectangular components, $v_1 = 2 \times 10^8$i cm/sec along the direction of the magnetic field B = 500i gauss and $v_2 = 3 \times 10^8$j cm/sec, perpendicular to it, the former component being responsible for the forward motion of the electron along the field and the latter for its circular motion.

(i) *the axis of the helix lies along the direction of the magnetic field, i.e., along the axis of x;*

(ii) *the radius of the helix* is given by $r = mv_2c/qB$. Taking $m = 9 \times 10^{-28}$ gm. $q = 4.80 \times 10^{-10}$ esu, $v_2 = 3 \times 10^8$ cm/sec and B = 500 gauss (given).

we have
$$r = \frac{9\times10^{-28}\times3\times10^{8}\times3\times10^{10}}{4.80\times10^{-10}\times500} = \frac{81}{4.80\times500}$$

$$= 0.03375 \approx 0.034 \text{ cm.}$$

(iii) The *forward distance covered by the electron in one full rotation, i.e., pitch of the helical path*

$$\frac{2\pi mc}{qB}v_1 = \frac{2\pi \times 9 \times 10^{-28} \times 3 \times 10^{10}}{4.80 \times 10^{-10} \times 500} \times 2 \times 10^8.$$

$\therefore$ *number of rotations made by the electron as it advances through a distance of*

$$\text{cm} = \frac{10}{\text{pitch of the helix}} = \frac{10 \times 4.80 \times 10^{-10} \times 500}{2\pi \times 9 \times 10^{-28} \times 3 \times 10^{10} \times 2 \times 10^8}$$

$$= \frac{4.80 \times 5}{2\pi \times 54} \times 10^3 = 70.74.$$

Example 12:

An electron describes a helix of radius 10 cm and pitch 3 cm in a magnetic field of 50 gauss. Calculate the components of its velocity along and perpendicular to the field. (Take mass of electron = 9 × 10^{28} gm).

Solution:

Let the components of the velocity of the electron along and perpendicular to the magnetic field be v_1 and v_2 respectively, the former responsible for its for r = mvc/qB if *Gaussian units* be used; *i.e.*, if q be in *esu* and B *in emu or gauss.*

Ward motion along the direction of the field and the latter for its rotation, making its resultant path a helix.

The *radius of the helix*, therefore, is given by the relation $r = mv_2c/qB$ (see § 4.6),

$$\text{whence, } v_2 = \left(\frac{qB}{mc}\right)r = \left(\frac{4.80 \times 10^{-10} \times 50}{9 \times 10^{-28} \times 3 \times 10^{10}}\right)10$$

$$= (8.9 \times 10^8)10 = 8.9 \times 10^9 \text{ cm/sec.}$$

And, since *pitch of the helix* $= \frac{2\pi mc}{qB}v_1 = 3$, we have

$$v_1 = \frac{3}{2\pi}\left(\frac{qB}{mc}\right) = \frac{3}{2\pi}(8.9 \times 10^8) = 4.25 \times 10^8 \text{ cm/sec.}$$

Thus, the *components of the velocity of the electron along and perpendicular to the magnetic field* are respectively 4.25 × 10^9 cm/sec.

Example 13:

An electron enters a transverse magnetic field B at a point P with velocity v along the axis of x and, describing a curved path of radius of curvature r under the action of the field, emerges from it at a point Q where the distance between the two points is x, Show that (i) its displacement along the y-axis as it emerges from the field is y $= r\left[1-\sqrt{1-\left(\frac{x}{2}\right)^2}\right]$ *and (ii) its angular deflection* $\theta = qBx/mvc$.

Solution:

In Fig. 3.18, let the dotted circle represent the region of the magnetic field B perpendicular to the plane of the paper and directed downwards, and let P be the point of entry of the electron into the field (in a direction perpendicular to it), PQ, the curved path of radius of curvature r described by it in the field and Q, its points of emergence from the field in a direction tangential to the curved path at Q, such that the horizontal distance between P and Q is x.

Then, clearly, *displacement suffered by* the electron in the magnetic field is y and the *angular displacement* suffered by it, θ where θ is the angle that the curved path subtends at its centre of curvature 0.

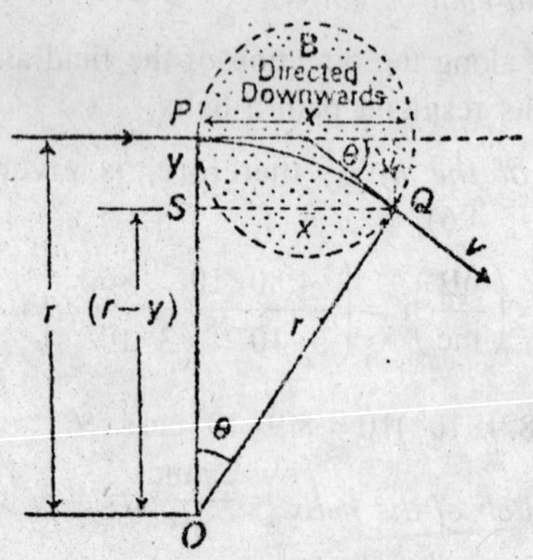

Fig. 3.18

(i) Drop QS perpendicular on to OP. Then, OP = r, SP = y and OS = (r – y).

Now, in the right angled triangle OSQ, we have $OQ^2 = OS^2+SQ^2$.

Or, $r^2 = (r - y)^2 + x^2$,

whence, $r - y = \sqrt{r^2 - x^2}$.

Or, $y = r - \sqrt{r^2 - x^2}$.

Or, *displacement* $y = r\left(1-\sqrt{1-\frac{x^2}{r^2}}\right) = r\left[1-\sqrt{1-\left(\frac{x}{r}\right)^2}\right]$.

(ii) Again, from Δ OSQ, we have $x = r \sin^{\theta}$. Or, since θ is always small, we have $x = r\theta$, whence, $\theta = x/r$.

But r, as we know, is given by the relation $r = mvc/qB$. So that,

angular displacement suffered by the electron, $\theta = qBx/mvc$.

Alternatively, we could obtain θ as follows: The angular velocity of the electron inside the magnetic field being $\omega = qB/mc$, we have $\theta = \omega t$ there t is the time taken by the electron to cover the distance x in the magnetic field and, therefore, equal to x/v. So that, $\theta = \omega x/v = qBx/mvc$.

Example 14:

The dees of a cyclotron, used for accelerating protons, have radii of 50 cm. each and an alternating potential difference of frequency 10 megacycles per second and maximum value 10^4 volts is applied to them. Obtain. (i) the strength of the magnetic field of the cyclotron, (ii) the velocity and the kinetic energy acquired by a proton, (iii) the maximum number of revolutions made by the proton and (iv) the time spent by the proton inside the dees.

Solution:

(i) We have *cyclotron frequency,* $n = qB/2\pi mc$, where q is in esu and B in emu or gauss. So that,

magnetic field

$$B = \frac{mc}{q}2\pi n = \frac{1.67\times10^{-24}\times3\times10^{10}\times2\pi\times10\times10^{6}}{4.8\times10^{-10}}$$

$$= \frac{1.67\times3\times2\pi}{4.8}\times10^{3} = 6557 \text{ gauss.}$$

(ii) We know that the maximum velocity of the proton (corresponding to the radius of its orbit, *i.e.*, the outer radius of the dees) is given by v = qBr/mc (using *Gaussian units*, as here). So that,

velocity acquired by the proton,

$$v = \frac{4.8\times10^{-10}\times6557\times50}{1.67\times10^{-24}\times3\times10^{10}} = \frac{8\times6557}{1.67}\times10^{2}$$

$$= 3.142 \times 10^9 \approx 3 \times 10^9 \text{ cm/sec.}$$

And ∴ *energy acquired by the proton,* $E = 1/2\ mv^2 = 1/2\ (1.67 \times 10^{-24})\ (3 \times 10^9)^2$

$$= \frac{15.03}{2}\times10^{-6} = 7.5\times10^{-6} \text{ ergs.}$$

$$= \frac{7.5\times10^{-6}}{1.6\times10^{-12}} = 4.7\times10^{6} \text{ electron volts.}$$

(iii) If N be the number of rotations made by the proton inside the dees before emerging out, we have *energy gained by it* = 2NeV ergs, because in one rotation it crosses the gap between the dees two times and each time acquires an energy eV ergs.

Since 2 NeV ergs = 2NV electron volts, we have $2NV = 4.7\times 10^6$,

whence, $$N = \frac{4.7\times10^6}{2V} = \frac{4.7\times10^6}{2\times10^6} = 2.35\times10^2 = 235.$$

Thus, the *number of revolutions made by the proton inside the dees = 235.*

(iv) The frequency of the alternating potential difference applied to the two dees being 10 megacycles per second, the time spent by the proton inside the dees during one rotation = $1/10 \times 10^6$ sec.

Hence, *time spent by the proton inside the dees in* 235 *rotations*

$$= 235/10 \times 10^6 = 2.35 \times 10^{-5} \text{ sec}$$

Example 15:

Find the distance separating singly charged ions, of potassium isotopes of atomic weights 39 and 40 and the same kinetic energy, at the focus of 180° mass spectrometer if the former ions follow a path of 50 cm. radius of curvature. What is the maximum permissible spread θ on each side of the initial ion beam so that the two focal lines may not overlap?

Solution:

Let v_1 and v_2 be the velocities of the two types of ions and m_1 and m_2 their atomic weights respectively. Then, since their kinetic energy is the same, we have $v_2/v_1 = \sqrt{m_1/m_2}$,

$$\text{Now,} \quad r = \frac{cmv}{qB} \therefore r^1 = \frac{cm_1 v_1}{qB} \text{ and } r_2 = \frac{cm_2 v_2}{qB}.$$

$$\text{Or,} \quad \frac{r_2}{r_1} = \frac{r_2}{50} = \frac{m_2}{m_1}\frac{v_2}{v_1} = \frac{m_2}{m_1}\sqrt{\frac{m_1}{m_2}} = \sqrt{\frac{m_2}{m_1}} = \sqrt{\frac{40}{39}},$$

$$\text{whence, } r_1 = 50\left(1+\frac{1}{39}\right)^{\frac{1}{2}} = 50\left(1+\frac{1}{2}\frac{1}{39}+\ldots\right) = \left(50+\frac{50}{78}\right)\text{cm.}$$

$\therefore$ *separation of the two focal lines* $= 2(r_2 - r_1) = 2\left(50+\frac{50}{78}\right)$

$$= 2 \times \frac{50}{78} = 1.28 \text{ cm.}$$

Now, in order that the two lines may not overlap we must *width of the first focal line* $r1^{6x} = 1.28$. Or, $50\theta^2 = 1.28$,

where θ is the semi-angle of the cone of the ions as they enter the magnetic field. So that,

$$\theta = \sqrt{\frac{1.28}{50}}/ = 0.61 \text{ radian or } \frac{0.61 \times 180}{\pi} = 9.17^\circ.$$

Example 16:

The Uranium isotopes of atomic, weights 235 and 238 may be separated by 180° magnetic focussing. Calculate the separation of their beams at the focus, brought in a magnetic field of 10^4 gauss if the radius of curvature of the heavier isotope be 100 cm when (i) the velocities (ii) the energies of the beams are equal.

Solution:

The separation of the two beams at the focus is clearly equal to the difference between the diameters of their semicircular paths in the magnetic field, *i.e.*, equal to $2(r_2 - r_1)$, where r_1 and r_2 are the radii of the two semicircular paths respectively. Now,

(i) in case the respective velocities of the two beams be v_1 and v_2 we have

$$r_1 = m_1v_1c/qB \text{ and } r_2 = m_2v_2c/qB,$$

whence, $r_1/r_2 = m_1v_1/m_2v_2$. Or, since the velocities of the two beams are equal, we have $v_1 = v_2$ and $\therefore\ r_1/r_2 = m_1/m_2 = 235/238$.

$$\text{Or,}\quad \frac{r_2 - r_1}{r_2} = \frac{238-235}{238} = \frac{3}{238}$$

$$\text{Or,}\quad r_2 - r_1 = \frac{3}{238}\,r_2 = \frac{3}{238}\times 100,$$

Hence *separation of the two beams*

$$= 2(r_2 - r_1) == 2\times\frac{3}{238}\times 100 = 2.522 \text{ cm.}$$

(ii) And, when the energies of the two beams are equal, we have

$$1/2m_1v_1^2 = 1/2m_2v_2^2,\ i.e.,\ v_1/v_2 = \sqrt{m_2/m_1}.$$

But, as seen under (i) above, $r_1/r_2 = m_1v_1/m_2/v_2$. So that,

$$\frac{r_1}{r_2} = \frac{m_1}{m_1}\sqrt{\frac{m_2}{m_1}} = \sqrt{\frac{m_1}{m_2}}$$

$$= \sqrt{\frac{235}{238}} = \left(1-\frac{3}{238}\right)^{\frac{1}{2}} = \left(1-\frac{3}{2\times 238}\right)$$

$$\therefore\quad r_2 - r_1 = \frac{3}{2\times 238}\,r_2 = \frac{3}{2\times 238}\times 100.$$

Hence *separation of the beams*

$$= \frac{2\times 3}{2\times 238}\times 100 = 1.261 \text{ cm.}$$

Example 17:

A stream of ions, of charge to mass ratio 1.44 × 10[14] esu gm, travelling horizontally in vacuum, encounters vertical electric and magnetic fields in the same region, on passing through which if falls on a fluorescent screen, showing a displacement of 2 cm both horizontally and vertically. What will be the horizontal displacement for a stream of ions of twice the above charge to mass ratio if the vertical displacement for it be 9 cm?

Solution:

Would it be possible to distinguish between streams of deutrons and α-particles in such an experiment?

Here, again, $$\frac{x^2}{y} = \frac{q}{m} \cdot \frac{B^2}{E} k.$$

For the stream of ions for which $q/m = 1.44 \times 10^{14}$ esu/gm and $x = y = 2$ cm,

we, therefore, have

$$\frac{(2)^2}{2} = 2 = (1.44 \times 10^{14}) \frac{B^2}{E} k,$$

Whence, $$\frac{B^2 k}{E} = \frac{2}{1.44 \times 10^{14}}.$$

∴ for ions for which $q/m = 2(1.44 \times 10^{14})$ esu/gm and for which $y = 9$ cm,

we have

$$\frac{x^2}{9} = 2\,(1.44 \times 10^{14}) \frac{B^2}{E} k = 2(1.44 \times 10^{14}). \frac{2}{1.44 \times 10^{14}} = 4.$$

Or, $$x^2 = 36,$$

whence, $x = 6$ cm.

Thus, the horizontal displacement of this stream of ions = 6 cm.

Since the charge to mass ratio (q/m) is the same for both a deutron and an α-particle, they will both suffer identical horizontal and vertical displacements in such an experiment and will thus be indistinguishable from the parabolas traced by them on the fluorescent screen.

Example 18:

An electric field, 10^3 volt/metre and a magnetic field, 3.2×10^{-3} weber/m^2 are parallel to each other along the z-direction. In the region of this combination of fields are introduced hydrogen ions of energy 200 electron volts along the axis of x. How long would it take the hydrogen ions to hit the fluorescent screen placed at a distance of 0.8 metres from the origin, perpendicular to their direction of motion and what will be *the coordinate of the point where the screen is hit ? (Mass of proton may be taken to be 1.6 × 10^{-27} kg and the charge on it equal to 1.6 × 10^{-19} coulombs).*

Solution:

Let v be the velocity of the hydrogen ions along the axis of x. Then, $1/2\ mv^2 = 200eV$,

$$\text{whence,} = \sqrt{\frac{2\times 200\times 1.6\times 10^{-19}}{1.6\times 10^{-27}}} = 2\times 10^{-5}\ \text{m/sec}.$$

(Because 1 eV = 1.6 × 10^{-19} joules).

Since no force is acting along this axis, the velocity remains unaffected and hence *time taken to cover the distance* x = 0.8 *metre to the screen* = 0.8/2 × 10^5 = 4 × 10^{-6} sec.

Now a moving charge constitutes an electric current and, therefore, the force acting on the hydrogen ions due to the magnetic field B is equal qcB* in a direction perpendicular to both B and v and, therefore, in the y-direction (in accordance with Fleming's left hand rule). So that, *acceleration of the ions in this direction* = qvB/m.

∴ *distance covered by the ions in this direction in time* t = 4 × 10^{-6} sec is given by

$$y = \frac{1}{2}\frac{qvB}{m}t^2$$

$$= \frac{1}{2}\frac{1.6\times 10^{-19}\times 2\times 10^5\times 3.2\times 10^{-3}}{1.6\times 10^{-27}}\times(4\times 10^{-6})^2 = 0.512 \text{ metre,}$$

because u = 0.

Similarly, distance covered by the ions in the z-direction in time t = 4 × 10^{-6} sec is given by

$$z = \frac{1}{2}\frac{qE}{m}t^2$$

$$= \frac{1}{2}\frac{1.6\times 10^{-19}\times 10^3}{1.6\times 10^{-27}}\times(4\times 10^{-6})^2 = 0.8 \text{ metre,}$$

Since the charge, the magnetic and the electric fields are all expressed here in RMKS (or SI) units, we need not divide the charge q by c as we do when using Gaussian units, *i.e.*, the charge in esu and the magnetic field in emu or gauss.

Thus, *coordinates of the point where the screen is hit are* 0.8, 0.512 *and* 0.8 *metre.*

Example 19:

A positive ion beam moving in the x-direction enters a region in which there is an electric field E_y = 3000 volts/cm and a magnetic field B_z = 300 gauss. Deduce the speed of those ions which may pass undeflected though the region. What will happen to ions which are (i) faster, (ii) slower than these?

Solution:

(ii) the velocity v_x of a charged particle, to remain undeflected in the crossed electric and magnetic fields E and B along the y-axis and z-axis respectively, must be given by the relation v_x = E/B. Or, if the Gaussian units be used (*i.e.*, E in *statvolts/cm* and B in gauss), we must have

$$v_x = \frac{cE}{B}.$$

Here, $E = E_x$ = 3000 volts/cm

= 3000/300 or 10 statvolts/cm,

$B = B_z$ = 300 gauss and $c = 3 \times 10^{10}$ cm/sec. So that,

$v_x = 3 \times 10^{10} \times 10/300 = 10^9$ cm/sec.

Thus, the *speed of the ions must be* 10^9 *cm/sec.*

At this speed, obviously, the deflection of the + ve ions in the downward direction in the plan of the paper, *i.e.*, in the - y direction, due to the magnetic field is just balanced by their deflection in the + y direction due to the electric field in that direction.

(i) If, therefore, the speed v_z of the ion-beam be higher that 10^9 cm/sec, it deflection in the - y direction due to the magnetic field will exceed that in the + y direction due to the electric field and *there will thus be a resultant deflection in the - y direction.*

(ii) If, on other hand, the speed v_x of the ion-beam be lower than 10^9 cm/sec, the deflection in the - y direction due to the magnetic field will be less than that in the + y direction due to the electric field and *there will, therefore, be a resultant deflection of the beam in the + y direction.*

Example 20:

A force P = 25i + 6j newtons acts on a particle of mass 2.5 kg for 5 sec. If the initial position of the particle is r_o = 6j + 8k metre and the initial velocity u = 2.5i + 3k metre/sec, calculate (i) the final velocity of the particle, (ii) the final position of the particle, (iii) the work done by the force on the particle.

Solution:

(i) Since the *initial velocity of the particle*, u= 2.5i + 3k m/sec and its *acceleration* a = P/m = (25i + 6j)/2.5 m/sec^2, we have *final velocity of the particle after time t* = 5 sec, *i.e.*, v = u + at,

Or, $v = 2.5i + 3k + \left(\dfrac{25i+6j}{2.5}\right)5 = 2.5i + 3k + 50i + 12j = 52.5i + 12j + 3k$ m/sec.

(ii) The *final position of the particle* is given by $r = r_o + ut + 1/2at^2$

$= (6j + 8k) + (2.5i + 3k)5 + 1/2\left(\dfrac{25i+6j}{2.5}\right)25 = 6j + 8k + 12.5i + 15k + 125i + 30j.$

Or, $r = 137.5i + 36j + 23k.$

(iii) Clearly, *work done by the force on the particle, i.e.*, W = (*force × distance through which the force is applied*) = $P(r - r_o)$

$= (25i + 6j)\,[(137.5i + 36j + 23k) - (6j + 8k)]$

Or, $W = (25i + 6j)\,(137.5i + 30j + 15k) = (3437.5 + 180) =$ 3617.5 joules.

Example 21:

The initial positions of two particles are – 2, 0 and 0, – 2 and they start simultaneously along the axes of x and y with uniform velocities 3i cm/sec and 4j cm/sec respectively. (i) Obtain the vector representing the position of the second particle with respect to the first as a function of time. (ii) When and where will the two particles be closest to one another?

Solution:

Let the velocities of the two particles be u_1 and u_2 respectively. Then, we have u_1 = 3j cm/sec and u_2 = 4j cm/sec. And, therefore,

(i) Position of the first particle after time t is given by

$$r' = r'_o + u_1 t = -2i + (3i)(t) = (3t - 2)i$$

and position of the *second* particle after the *same* time t is given by

$$r' = r''_o + u_2 t = -2j + (4i)(t) = (4t - 2)j.$$

$\therefore$ position vector of the second particle with respect to the first is given by

$$r = r'' - r' = (4t - 2)j - (3t - 2)i \text{ cm.}$$

(ii) Obviously, the two particles will be closest together when r has the minimum value, *i.e.*, when

$$\frac{d}{dt}(r^2) = 0. \quad \text{Or,} \frac{d}{dt}[ut - 2)^2 + (3t - 2)^2] = 0,$$

i.e., when $2(4t - 2)(4) + 2(3t - 2)(3) = 0$,

or when $t = 28/50 = 0.56$ sec.

At this time, the position of the second particle with respect to the first will be given by

$$r = (4 \times 0.56 - 2)j - (3 \times 0.56 - 2)i = (0.32i + 0.24j) \text{ cm.}$$

Thus, *the two particles will be the closest to one another after 0.56 sec when the position vector of the second particle with respect to the first is given by*

$$r = (0.32i + 0.24j) \text{ cm.}$$

Example 22:

At time t = 0, the velocity of an electron is $10^6 i$ cm/sec and the initial position vector $r_o = 100j$ cm. (i) Obtain the position vector after 0.1 sec. (ii) If the electric field be $10^{-2}i$ statvolt/cm, obtain the position and velocity vectors after 10^{-8} sec.

Solution:

(i) The position vector after t sec will be $r = r_o + ut$, where u is *velccity* of the electron (supposed uniform). So that,

$$r = 100j + 10^6 i\,(0.1) = 10^5 i + 10^2 j \text{ cm.}$$

(ii) In this case, the electron will have an *acceleration* a = (*force acting on it/its mass)* = eE/m, where e is the *charge on the electron* = -4.8×10^{-10} esu and E, the electric field = $10^{-2}i$ esu and m, the mass of the *electron* = 0.91×10^{-27} gm.

$\therefore$ *velocity of the electron after* $t = 10^{-8}$ sec, *i.e.*, v = u + at

$$= 10^6\,i - \frac{4.8\times10^{-10}(10^{-2}\,i)}{0.91\times10^{-27}}\times10^{-8}.$$

Or, $$v = \left(10^{-1} - \frac{4.8}{0.91}\right)10^7\,i = -5.2\times10^7\,i.$$

And *position vector* after t = 10^{-8} sec,

i.e., $$r = r_o + ut + 1/2at^2$$

$$= r_o + ut + \frac{1}{2}\frac{eE}{m}t^2$$

Or, $$r = 100j + 10^6i\,(10^{-8}) - 1/2\frac{4.8\times10^{-10}(10^{-2}\,i)}{0.91\times10^{-27}}\,(10^{-8})^2$$

$$= 10^2j + 10^{-2}i - 1/2\frac{4.8}{0.91}\times10^{-1}i$$

$$= 10^2j + 10^{-2}i\left(1 - \frac{24}{0.91}\right) = 10^2\,j + 10^{-2}\,i\left(-\frac{23.09}{0.91}\right)$$

$$= 10^2j - 25\times10^{-2}i.$$

Or, $$r = 10^2j - 0.25\times10^{-2}i \text{ cm}.$$

EXERCISES

1. An electron beam is accelerated through a P.d. of 1000 volts and then passed through a pair of deflecting plates 2 cm long and 0.5 cm apart. What should be the p.d between the plates in order that the electron beam may be deflected through 5° ?

   Ans. 4.3×10^2 volts.

2. The accelerating voltage in a cathode ray oscilloscope is 2400 volts, the effective length of the deflecting plates is 3 cm, their distance apart 0.5 cm and the distance of the screen from their mid-point 30 cm. If the p.d across the deflecting plates be 20 volts, obtain (i) the velocity of an electron as it enters the deflecting field, (ii) the time it takes to traverse the field, (iii) the angle of deflection due to the field and its displacement on the screen. (e/m for an electron = 5.2×10^{17} esu/gm).

   Ans. (i) 2.88×10^8 cm/sec, (ii) 1.04×10^{-9} sec, (iii) 1°26′, (iv) 0.75 cm.

3. An alternating potential difference $V = 2 \sin (2\pi \times 10^8)t$ is applied to two plane and parallel plates A and B arranged 0.5 cm apart along the of x. An electron is emitted from plane A with zero velocity just when plate B is acquiring positive potential. What will be the maximum speed acquired by the electron ?

Ans. 6.7×10^9 cm/sec.

[**Hint:** Here, obviously, $V_e = 2$ statvolts and $\omega = 2\pi \times 10^8$ so that, peak value of the electric field. *i.e.*, $E_0 = V_0/0.5 = 2/0.5 = 4$ statvolt/ cm. Therefore, $md^2x/dt^2 = - eE_0 \sin \omega t$. Or, $d^2 \times dt^2 = -(e/m) E_0 \sin \omega t$. Hence, $v = dx/dt = eE_0 (1 - \cos \omega t)/m\omega$. The value of v will be the maximum when $\cos \omega t = - 1$. Thus, $v_{max} = 2e E_0/m\omega = 2 \times 4 .8 \times 10^{-10} \times 4/9.1 \times 10^{-28} \times 2\pi \times 10^8 = 6.^7 \times 10^9$ cm/sec.]

4. An alternating electric field of peak value 3 volts/cm and frequency 1 mega hertz (*i.e.*, 10^8 c.p.s.) is established between two plane and parallel plates. Show that the electrons starting at different times with zero initial velocities in this field will acquire maximum velocities of $\pm 8.4 \times 10^8$ cm/sec.

[**Hint:** Proceed exactly as in problem 13 above and obtain t_{max}].

5. Show that the path of a charged particle in a uniform magnetic field is, in general, a helix. Under what condition is this path reduced to z circle?

6. A charged particle of mass M and charge q is moving in a constant magnetic field B. Show that (i) the component of the velocity of the particle along the axis of the magnetic field is constant, (ii) the magnetic field does not change the kinetic energy; of the particle, (iii) for a uniform circular motion in a plane perpendicular to the axis of the magnetic field, the angular velocity of the particle, α, given by $\omega = qB/Mc$, where c is the speed of light in free space.

7. Calculate the gyroradius and gyrofrequency of an electron moving in a uniform magnetic field of 10^8 gauss with a velocity 3×10^9 cm/sec in a direction perpendicular to that of the field.

Ans. 0.17 cm; $8.8 \times 10^9 \approx 9 \times 10^9$ *hertz* or c. p. s.

8. An electron describes a helix of radius 6.0 cm and pitch 2.0 cm in a magnetic field of intensity 30 gauss. Obtain the components of its velocity along and perpendicular to the magnetic field.

9. Particles of charge q and mass m are projected with velocity ai + bj in a magnetic field of B j units. Show that they follow a helical path. Deduce an expression for the pitch of the helix. What is the significance of the fact that this is independent of the component of velocity ?

10. An electron with a velocity 10^7 m/sec enters a region of uniform magnetic flex density of 0.10 weber meter^2, the angle between the direction of the field and the initial path of the electron being 25°. Find the pitch of its helical path. (e/m for an electron = 1.8 × 10^{11} coulomb/kg). **Ans.** 0.32 cm.

11. An electron, moving with a velocity equal to 4/5th of that of light in free space, enters a magnetic field of 300 gauss in a direction perpendicular to that of the field. Calculate the radius of the circular path described by the electron inside the field on non-relativistic as well as relativistic considerations.

    Ans. 4.55 cm; 7.58 cm.

    [**Hint:** Relativistically, the mass of the electron is given by m = $m_e/\sqrt{1 - v^2/c^2}$, where m_o = 9.1 × 10^{-28} gm.]

12. An alpha particle moves with velocity 10^8 cm/sec at an angle of 30° with a magnetic field of 10^8 gauss. Obtain the values of (i) the Lorentz force acting on it, (ii) the radius and the pitch of the helix described by it and (iii) its frequency of rotation. (take mass of α-particle = 4 × 1.6 × 10^{-24} gm and the charge on it = 2 × 4.8 × 10^{-10} esu).

    Ans. (i) 16 × 10^{-10} dynes, (ii) 10cm, 108.7 cm, (iii) 7.96 × 10^5 = 8 × 10^5 c.p.s.

    [**Hint:** Lorentz force F = q/c (v × B). So that, its magnitude = q/c v B cos θ.]

13. Explain clearly the principle underlying the working of a cyclotron. What are its drawbacks and how have they been remedied. Which is preferable, a lower or a higher radio-frequency voltage, to be applied to the dees?

14. What will be the energy of (i) a proton, (ii) an electron as it emerges from a cyclotron after making 100 revolutions inside it, if the p.d. across the dees be 10^8 volts ? Which of these will emerge with a higher velocity?

15. Calculate, ignoring relativistic effects, the amount of work done on an electron in a cyclotron if, starting from rest, it finally emerges with a velocity of 4×10^9 cm/sec. (Take mass of electron = 9×10^{-28} gm.) **Ans.** 4.5×10^2 eV.

16. (a) Show that the cyclotron frequency of a given kind of ions is independent of the energy. State the limiting condition, if any.

    (b) Explain the principle of magnetic focussing.

17. The radius of the dees of a cyclotron used to accelerate protons is 50 cm and they are connected to a radiofrequency supply of 10 mega cycles per second at maximum value 6000 volts. Calculate (i) the strength of the magnetic field applied, (ii) the velocity and the energy of the energy of the emerging protons, (iii) the minimum number of rotations made by them and (iv) the time spent by them inside the dees.

    Ans. (i) 6.557×10^3 gauss, (ii) 3.142×10^9cm/sec; 5.15×10^8 eV, (iii) 430, (iv) 4.3×10^{-5} sec.

18. A particle of charge q esu and mass m gm enters a magnetic field of 2000 i gauss with a velocity (2i + 3j) 10^8 cm/sec, Deduce (i) radius of the helical path, (ii) number of rotations made by the particle while travelling forward 10 cm along the axis of the helix.

    Ans. (i) $4.5 \times 10^{15}\left(\dfrac{m}{q}\right)$cm, (ii) $\left(\dfrac{q}{m}\right)(5.3 \times 10^{-16})$

19. Calculate the width of the focal line obtained by 180° — magnetic focussing if the radius of curvature of the path of the ion beam be 50 cm and the semi angle of the cone of the beam, as it enters the magnetic field, be 5°. **Ans.** 0.3807 cm.

20. The width of the focal line of an ion beam obtained by 180°— magnetic focussing is 0.3656 cm. If the angle of the cone of the ion beam entering the magnetic field be 4° , calculate the radius of curvature of the path taken by the beam in the magnetic field. **Ans.** 30 cm.

21. A beam containing ions of various charges q, masses m and speeds v (along the x-axis) enters a region of uniform electric field e and magnetic field B (both along the y-axis). The ions are received on a screen after travelling a distance L_x. Show that

(i) all ions of a given q/m value fall along a parabola irrespective of their speeds, (ii) all ions of a given speed fall along a straight line irrespective of their q/m value.

22. Ions with charge to mass ratio 2.9×10^{14} esu/gm travelling horizontally pass through a region of combined and parallel electric and magnetic fields along the vertical so as to hit a fluorescent screen on which they show horizontal and vertical displacements of 4 cm each. What will be the charge to mass ratio for charged particles which show horizontal and vertical displacements of 1 cm and 8 cm respectively?

Ans. 9.1×10^{12} esu/gm.

23. A beam at protons having a velocity of 2×10^7 cm/sec along the axis of x is introduced into a combination of parallel electric and magnetic fields of 100.2 volts/cm and 3.34×10^2 gauss respectively along the z-direction. Obtain the coordinates of the point where the beam will hit a photographic plate placed perpendicular to the direction of the beam at a distance of 6 cm from the origin. **Ans.** 6, 2.88, 4.32 cm.

24. A charged particle moves in the x-direction through a region in which there is an electric field E_y and a perpendicular magnetic field B_z. What is the condition necessary to ensure that the net force on the particle will be zero ? Show the v, E and B vectors on a diagram. What is the condition on v_x if E_y = 10 statvolts/cm and B_z = 300 gauss?

Ans. $F_{elec} + F_{mag} = 0$, *i.e.*, $qE + q/c\ V \times B = 0$, which leads to $v_x = cE_y/B_z$; $v_x = 1 \times 10^9$ cm/sec.

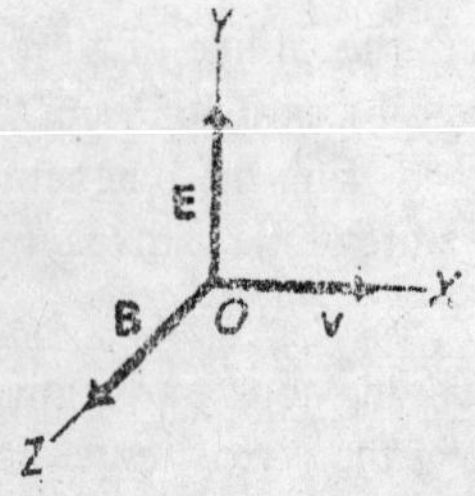

Fig. 3.19 : Vector diagram

25. A stream of charged particles, moving with a velocity of 6×10^9 cm/sec passed between two parallel plates 1 cm apart and having a p.d of 27 volts across them. What should be the magnitude and direction of the magnetic field that should be applied in the region of the electric field in order that the stream of particles may remain undeflected?

Ans. 0.45 gauss, perpendicular to both E and v.

26. What should be the velocity of a stream of (i) protons, (ii) deutrons in order that it may remain undeflected when passing through crossed electric and magnetic fields, E = 900 volts/cm and B = 600 gauss?

Ans. (i) 1.5×10^8 cm/sec; (ii) 1.5×10^8 cm/sec.

27. Explain the principle of a velocity selector for charged particles, using crossed electric and magnetic fields.

28. The initial position of a particle of mass 100 gm is r_0 = 5j + 7k cm and its initial velocity u = 30i + 10k cm/sec. If a force F = 100i + 75j dynes acts upon it for 4 sec, obtain its final velocity and final position. Also calculate the work done by the force: **Ans.** 34i + 3j + 10k; 128i + 11j + 47k; 13250 ergs.

29. A particle is projected upward in a direction inclined at 60° to the horizontal. Show that its velocity when at its greatest height is half its initial velocity. (Neglect the resistance of the air)

30. Find the greatest distance that a stone can be thrown inside a horizontal tunnel 10 ft high with a velocity of projection 80 ft/sec. Find also the corresponding time of flight.

Ans. 120 ft; 1.58 sec.

31. A particle P is projected with initial velocity 5i + 6j – 7k m/sec and it experiences a uniform acceleration 6j m/sec^2. Another particle Q is projected from the same position 20 see later with initial velocity 2i – 3j + 4k m/sec and it experiences a uniform acceleration –3j m/sec^2. Deduce the position of Q relative to P, 5.0 sec. after P is projected. **Ans.** 19i – 127.5j + 47k.

32. Show that when an electric field is applied at right angles to the direction of motion of a charged particle, it describes a path similar to that described by a particle thrown horizontally in the Earth's gravitational field.

33. A potential difference of 650 volts retards the motion of an electron proceeding normally from one electrode to another, parallel to it. If the initial energy possessed by the electron be 10^9 ergs, (i) will it be able to reach the second electrode ? (ii) What should be the value of the retarding potential so that the electron may just reach the second electrode?

Ans. (i) No; (ii) 625 volts.

34. Calculate (i) the force on a proton in an electric field of 100 statvolts/ cm, (ii) the intensity of the electric field that can just support the weight of an electron.

Ans. (i) 4.80×10^{-8} dynes;
(ii) 1.86×10^{-15} statvolts/cm.

35. Calculate the intensity of the electric field which will produce in an alpha particle an acceleration equal to that of g (*i.e.*, 980 cm/sec^2). (The mass of an alpha particle is four times that of a proton and the charge on it twice that on a proton).

Ans. 2.05×10^{-9} volts/cm

36. A proton is accelerated through a potential difference V and then allowed to cross a field-free evacuated space. How long would it take to cover the distance between two given points in this space, distance r apart. **Ans.** $r/\sqrt{2eV/m}$.

37. In an electric field e = 10^{-2}i statvolts/cm, the velocity of an electron is $v_e = 10^6$i cm/sec and its position vector $r_0 = 10^2$j cm at t = 0. Obtain its velocity and position vectors at t = 10^{-9} sec.

Ans. v = 5.4×10^2i cm/sec; r = 10^2j – 0.25i cm.

4

Conservation Laws–Laws of Conservation of Energy

CONSERVATION LAWS IN GENERAL

In Physics we come across a host of conservation laws and, even though all of them may not be equally exact or accurate, they nevertheless prove helpful is many ways. Thus, for instance,

(i) Without going into details of the trajectories or the forces involved in any particular case, they give us a broad and generalised picture of the significant facts that emerge in consequence of the equations of motion;

(ii) Even in cases where the nature of the force involved is not clearly known, the conservation laws have been successfully invoiced, particularly in the realm of what are called fundamental or elementary particles and nave, indeed, helped predict the existence of quite a few more of them;

(iii) They forewarn us of the impossibility of the occurrence of certain types of phenomena (like, for example, a perpetual action machine) and thus prevent wastage of time and effort that we might otherwise fed tempted to devote in tackling such problems;

(iv) They seem to have an intimate relationship with the concept of invariance and we may often use them, with success, in exploring and unravelling new and hitherto not well understood phenomena. Thus, as we have seen in chapter II, the principle of conservation of linear momentum has been obtained more or less as a direct consequence of Galilian invariance.

With these few examples regarding the importance of conservation laws, let us now proceed to deal with three important ones of them, viz., those of energy, linear and angular momentum, in proper detail.

CONCEPTS OF WORK, POWER OF ENERGY

Before dealing with the law of conservation of energy itself, we had better be quite clear about the concepts of *work*, power and *energy* involved in the law.

1. *Work :* Suppose a particle of mass m is acted upon by a force F, (Fig. 4.1), under the influence of which it is set into motion, not necessarily in the direction of the force, its actual path depending upon its initial velocity and the magnitude and direction of the force applied. Let it be along a curve C, joining the points A and B. It may be noted that the magnitude and direction of the force may vary from point to point along the curve.

To determine the work done in displacing the particle from A to B, let P and P be two neighbouring points such that their position vectors are r and r + δr and, therefore, PP' = δr (or δs if we prefer to work in terms of the length of curve C).

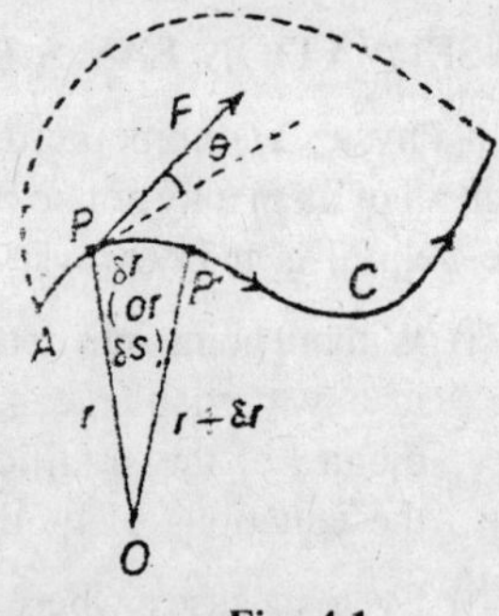

Fig. 4.1

Then, for a displacement of the particle (or the point of application of the force) from P to P', we have

$$\textit{work done by the force, } dW = F.ds$$

∴ *total work done by the force* in displacing the particle along the whole length of the curve from A to B, *i.e.*, W, is given by the relation

$$W = \int_A^B F.dr = \int_A^B F.ds = \int_A^B F\cos\theta\, ds, \qquad ...(i)$$

where θ is the angle that the direction of the force makes with the tangent to the curve at the point P and, therefore, F cos θ, the tangential component of the force along the curve at the point P.

Such an integral evaluated along the path of the particle is called the line *integral of the force F.*

We may, therefore, define work done by the force on the panicle as the *line integral of the ungential component of the force evaluated along the actual path of the particle.*

If the Cartesian components of F and dr (or ds) be respectively $F_x i$ $F_y j$, $F_z k$ and $d_x i$, $d_y j$ and $d_x k$, we have

$$W = \int_A^B (F_x i + F_y j + F_z k)\,(d_x i + d_y j + d_x k)$$

$$= \int_A^B F_x d_x + \int_A^B F_y d_y + \int_A^B F_z d_z \qquad \text{...(iii)}$$

Further, if there be a number of forces, F_1, F_2 etc., acting simultaneously on the particle, such that their resultant $F = F_1 + F_2 + F_3 + ...$, we have work done on the particle given by

$$W = \int_A^B F.dr = \int_A^B (F_1 + F_2 + F_3 + ...)dr$$

$$= \int_A^B F_1 dr + \int_A^B F_2\, dr + ..., \qquad \text{...(iii)}$$

i.e., equal to the sum of the work done by each individual force.

In case force *F be a constant or a nonvariable one and the displacement of the particle takes place along a straight line*, we have

$$W = \int_A^B F.dr \left(= \int_A^B F.ds\right) = F.r\ (= F.s) = F\cos\theta(r) = F\cos\theta(s) \quad \text{...(iv)}$$

where r (or s) is the displacement of the particle along the straight line.

This is in accordance with the elementary definition of work with which we are already familiar from our junior classes, viz., that work done is given by

W = component of the force along the displacement X the displacement.

Units of work : (i) In the *C.G.S. system*, the unit of work is the erg, which is the work done by a force of 1 *dyne* acting through a distance of 1 cm, *i.e.*, 1 erg = 1 dyne 1 cm.

(ii) In the F.P.S. system, the unit of work is the *foot-poundal*, which is the work done by a force of 1 poundal acting through a distance of 1 ft, *i.e.*, 1 *ft-poundal = 1 poundal × 1 foot.*

(iii) In the *RMKS (or SI) system*, the unit of work is the joule, which is the work done by a force of 1 *newton* acting through a distance of 1 *metre, i.e., 1 joule = 1 newton × 1 metre.*

(Since *1 newton = 10^5 dynes and 1 metre = 10^2 cm, 1 joule = $10^5 \times 10^2 = 10^7$ ergs*).

These are called the *absolute units*. The corresponding *gravitational* units are g *times the absolute units*. Thus, the gravitational units of work in the three systems are (i) the gm wt-cm = 981 ergs, (ii) the foot-pound = 32 foot-poundals and (iii) the kilogramme-metre = 9.81 joules.

2. *Power or Activity* : It is defined as the rate of doing work and is denoted by the letter P or A.

Thus, in Fig. 4.1, since F-dr is the work done by the force in moving the particle from P to P' in time dt, the *rate of working* of the force or its *power or activity* $P = F.<dr = F.v$, ...(v)

where v is the *instantaneous value* of the velocity acquired by the particle.

Relation V thus gives the *instantaneous value of the power of the force*. The *average value of power* is obtained by dividing the total work done (W) by the total time taken (t), *i.e.*, average power = W/t.

It will be easily seen from the above that the work done is given by

$$W = \int_{t_A}^{t_B} P.dt.$$

Units of power. (i) In the C.G.S. *system*, the unit of power is *1 erg/sec. (ii) In the F.P.S. system, it is 1 foot-poundal/sec*. The unit commonly used in this system is the *horse power* (H.P.), equal to 550 *ft-pounds/sec or 550 × 60 or 33,000 ft-pounds/min*. (iii) In the *RMKS* or (SI) system, the unit of power is 1 joule/sec, called the watt, usually used to measure *electrical power*.

3. *Energy* : We shall deal with it later

CONSERVATIVE FORCES

In the type of case discussed the force F generally varies from point to point and it may, therefore, appear that the work done by it (W) in moving the particle from point A to point B will depend upon the path taken between A and B. In actual fact it may or may not, depending upon the type of force we are dealing with. For, forces may, in general, be divided into two broad categories: *conservative and non-conservation.*

The work done by a conservative force in displacing a particle from one point to another depends only on the position of the two given points and is quite independent of the actual path taken between them. On the other hand, the work done by a non-conservative force does depend upon the actual path taken.

To the conservative class of forces belong most of what are called *central forces, i.e., forces which are directed towards, or away from, a fixed centre,* as in the case of gravitational and electrostatic forces respectively. Such forces are *position-dependent* forces, so to speak, since they depend only upon the instantaneous position of the particle with respect to the fixed centre and on nothing else.

Forces which are *velocity-dependent, i.e.,* whose value depends upon the magnitude and direction of velocity, like frictional and viscous forces, are, in general, *non-conservative,* though quite a few among them too may turn out to be conservative. Thus, for example, *Lorentz force* $\frac{q}{c}(v \times B)$, although velocity-dependent, is conservative because, acting as it does, in a direction perpendicular to both the magnetic field B and the velocity of the particle v, the work done by it, irrespective of the path taken, is always zero and thus independent of the path.

A region of space in which a particle experiences a conservative force at every point is referred to as a *conservative-force field* or a *conservative field of force.* In case the conservative force in the field happens to be a central force, it may be specified as a central-force field.

That a central force or a position-dependant force is a conservative force may be easily shown as follows:

Taking the origin as the fixed centre, a central force may be expressed as F = F(r), where F(r) is a function of r only and r, the unit vector along r.

Clearly, work done by such a force in displacing a particle from a point A to a point B is given by

$$W_{AB} = \int_A^B F.dr = \int_A^B F(r)_r\, dr.$$

As we know, $r^2 = r^2$, so that 2r.dr = 2rdr and dr = rdr.

$$\text{Hence} \quad W_{AB} = \int_A^B F(r)\, dr,$$

an expression dependent only on the functional form of f(r) and the two end-points A and B.

The force in question, viz, F is, therefore, a *conservative force.*

It is clear from the above that the work done by the force F in moving the particle from B to A along any path will be given by

$$W_{AB} = \int_B^A F.dr = \int_A^B F.dr = -W_{AB}.$$

It follows, therefore, that the total work done in displacing a particle from A to B and back to A along whatever path (as shown dotted in Fig. 4.1) is $W_{BA} - W_{AB} = 0$, indicating that the *work done by a conservative force along a closed path, i.e., along a closed curve, is zero.* This is denoted symbolically as f F.dr = 0, (the symbol $\oint$ standing for the integral over a closed path).

This is *characteristic property of a conservative force.* In fact, *we may perhaps best define a conservative force as owe by which the work done in a round trip is zero and, therefore, a non-conservative force as one by which the work done in a round trip is not zero.* In the first case, when the particle (or the body) comes back to its starting point, it has the same kinetic energy with which it started, so that its ability to do work is conserved. In the second case, the particle, as it comes back to its starting point, has less or more kinetic energy than the one with which it started, so that its capacity to do work is *not conserved* (*i.e.*, it is either less or more than before).

ENERGY

The energy of a particle is its *capacity to do work* and is measured by the amount of work it is capable of doing in virtue of (i) its motion or (ii) position or condition. The former is called the *kinetic energy* and the latter, the *potential energy* of the particle, represented by the symbols K (or now, more usually, T) and U (or F) respectively.

Since energy is really *work, its units are the same as those of work,* viz, the erg, the foot-poundal (or the foot-pound) and the joule in die three systems respectively.

We shall now proceed to consider the two types of energy separately and in necessary detail.

(i) *Kinetic energy :* Work-energy principle. Since kinetic energy (T) of a particle is due to its motion, let us obtain a proper expression for it for a particle moving with velocity y.

Let a force F be applied to the particle in a direction opposite to that of its motion. Then, clearly,

work done by the particle against the applied force. i.e., $W = -\int F dr$.
Since, in accordance with Newton's second law of motion,

$$F = mdv/dt,$$

where m is the mass of the particle, we have

$$W = -m\int \frac{dv}{dt}.dr = -m\int \frac{dr}{dt}dv = -m\int_v^0 v.dv = \frac{1}{2}mv^2$$

Since energy of the particle is its capacity to do work, we have

kinetic energy of the particle, $T = 1/2mv^2$.

It may as well be noted that $dT/dt = d\,(1/2\ mv^2)/dt = mv(dv/dt) = F.v = P$, *i.e.*, the *rate of change of energy is equal to the power or activity of the force.*

Again, invoking Newton's second law of motion, we have

$$mdv/dt = F.$$

Dot multiplying both sides by v or dr/dt and integrating with respect to t between the limits A and B (Fig. 4.1), we have

$$\int_A^B m\frac{dv}{dt}.vdt = \int_A^B F.\frac{dr}{dt}dt.$$

or $$\left[\frac{1}{2}mv.v\right]_A^B = \int_A^B F.dr$$

Since $v.v = v^2$ and v_A and v_B are the values of the initial and final velocities of the particle at points A and B respectively, we have

$$\frac{1}{2}mv_B^2 - \frac{1}{2}mv_A^2 = \int_A^B F.dr = W$$

where W is the work done by the force in displacing the particle from point A to point B.

Since $1/2\ mv_B{}^2$ is the kinetic energy of the particle at point B and $1/2mv_A{}^2$, its kinetic energy at point A, we have

work done in displacing the particle from A to B = its K.E. at B–*its K.E. at A = change in energy in proceeding from A to B, i.e., work done by a force on a particle is equal to the change in the kinetic energy of the particle.*

This is spoken of as the *work-energy principle* or the *work-energy theorem.*

As will be readily seen, the kinetic energy of the particle increases if work be done on it (by the force) and decreases if work be done by it (against the force).

(ii) *Potential energy :* We have already defined the potential energy of a particle or a body as its capacity to do work in virtue of its position (or condition). It is measured by the amount of work done by the force to restore the particle or the body from its present position (or condition) to a standard position (or condition) and is denoted by the symbol U or V. Thus, if the present or the existing position of the particle be specified by a vector r and its standard position by a vector r_0, we have

$$U = \int_{r}^{r_0} F.dr$$

It may be emphasised here that *this relationship exists only in the case of a conservative field of force*, for the work done by a conservative force alone depends on the position of the particle, irrespective of the path taken. In the case of a non-conservative force, the work done cannot be expressed in the form of potential energy since it depends on the direction of motion as also on the shape of the path taken and even, at times, on the magnitude of the velocity.

The standard position (or condition) chosen is the one in which the force acting on the particle or the body is zero, so that its P.E. in this position (or condition) may also be taken to be zero. The difference of P.E. in this position (or condition) and in the-existing position (or condition) of the particle then gives its P.E. in the latter position.

Thus, for example, the standard condition in the case of a spring is taken to be its normal, unstretched (or uncompressed) state. And, although for convenience we take the electrostatic and gravitational potential energies to be zero on the surface of the earth, the electrostatic and gravitational forces involved are actually governed by the inverse square law, so that the electrostatic force due to a charge or the gravitational force due to a material body is zero only at an infinite distance from it and the standard or the zero potential energy position is thus infinity. In such cases, therefore, $r_0 = \infty$ and we have

$$U = \int_{r}^{\infty} F.dr$$

or

$$U = -\int_{\infty}^{r} F.dr$$

i.e., the potential energy of a particle at a point r is given by the amount of work done in moving it from oo to that point.

In calculating the work we may, of course, use any convenient path since, as we know, the work done does not depend on the actual path taken but only on the positions of the end-points.

CONSERVATIVE FORCE AS NEGATIVE GRADIENT OF POTENTIAL ENERGY—CURL F = 0

We have just how

$$U = -\int_{\infty}^{r} F.dr, \text{ where F is a conservative force.}$$

Expressing F and r in terms of their components along the three coordinate axes, we have

$$F = F_x i + F_y j + F_z k,$$
$$r = xi + yj + 2k$$

and $$dr = d_x i + d_y j + d_y j + d_z k.$$

So that, $$U = -\int_{\infty}^{r} (F_x i + F_y j + F_z k)(d_x i + d_y j + d_z k)$$

$$= -\int_{\infty}^{r} (F_x dx + F_y dy + F_z dz),$$

which, on partial differentiation with respect to x, y, z, gives

$$F_x = -\frac{\partial U}{\partial x}, F_y = -\frac{\partial U}{\partial y}, \ F_z = -\frac{\partial U}{\partial z},$$

And ∴ $F = F_x i + F_y j + F_z k = -\left(\frac{\partial U}{\partial x} i + \frac{\partial U}{\partial y} j + \frac{\partial U}{\partial z} k\right) = -\text{grad } U$ or $-\nabla U$.

Thus, a conservative force F is equal to the negative gradient of potential energy U.

If instead of the three dimensions, we restrict ourselves to one dimension only, we have $F = -dU/dx$,

showing that a *conservative force is the negative of the space rate of change of potential energy,*

Now, we have Curl $F = \nabla \times F = -\nabla \times \nabla U$

$$= \begin{vmatrix} i & j & k \\ \partial/\partial x & \partial/\partial y & \partial/\partial z \\ F_x & F_y & F_z \end{vmatrix} = -\begin{vmatrix} i & j & k \\ \partial/\partial x & \partial/\partial y & \partial/\partial z \\ \partial U/\partial x & \partial U/\partial y & \partial U/\partial z \end{vmatrix}$$

$$= -\left[i\left(\frac{\partial^2 U}{\partial y \partial z} - \frac{\partial^2 U}{\partial z \partial y}\right) + j\left(\frac{\partial^2 U}{\partial z \partial x} - \frac{\partial^2 U}{\partial x \partial z}\right) + k\left(\frac{\partial^2 U}{\partial x \partial y} - \frac{\partial^2 U}{\partial y \partial x}\right)\right]$$

Since, with V taken as a perfect differential,

$$\frac{\partial^2 U}{\partial y \partial z} = \frac{\partial^2 U}{\partial z \partial y}, \frac{\partial^2 U}{\partial z \partial x} = \frac{\partial^2 U}{\partial x \partial z} \text{ and } \frac{\partial^2 U}{\partial x \partial y} = \frac{\partial^2 U}{\partial y \partial x}, \text{ we have}$$

$$\text{Curl F} = 0.$$

This condition (Curl F = 0) implies the existence of a potential function U such that F = –grad U. It thus follows that curl F = 0 is *a necessary and a sufficient condition/or a conservative field of force F.*

LAW OF CONSERVATION OF MECHANICAL ENERGY

It follows that

$$F.\delta r = -\left(\frac{\partial U}{\partial x}\right)\delta x - \left(\frac{\partial U}{\partial x}\right)\delta x - \left(\frac{\partial U}{\partial z}\right)\delta z = -\delta U.$$

∴ work done in displacing a particle from A to B is given by

$$W = \int_A^B F.dr = -\int_A^B DU = U_A - U_B$$

But, as (i)

$$W = \int_A^B F.dr = \frac{1}{2}mv_B^2 - \frac{1}{2}mv_A^2.$$

Obviously, therefore, $U_A - U_B = 1/2\ mvx^2 - 1/2mv_A{}^2$, indicating that *what the particle loses in P.E. it gains in K.E.*

So that, $1/2mv_A{}^2 + U_A = 1/2mv_B{}^2 + U_B$.

Or, putting T_A and T_B respectively for $1/2mv_A{}^2$ and $1/2mv_B{}^2$, we have

$$T_A + U_A = T_B + U_B.$$

Or, $\quad T + U = E$, a constant,

showing that *the sum total (E) of the kinetic and potential energies of a particle remains unaltered as the particle moves from one point A to another point B in a conservative force field. In other words, the mechanical energy of a particle remains constant in such a field.* This is referred to as the law of conservation of mechanical energy for conservative forces. We shall see later that it may be easily extended to include also other forms of energy when it is called the *general law af conservation of energy.*

This is perhaps by far the most accurate of all the conservation laws and there has been found not cue single exception to it so far.

It is easy enough to see that the tor holds good also for a *system of particles moving in a conservative field of force* for the simple reason that it holds good for each, individual particle.

The quantity E = (T + U) is often spoken of as the energy function of the particle and is obviously *invariant to both changes of time and position of the particle.* This may be seen from a simple illustration of the law in the case of a falling particle.

Let the particle of mass m be at rest at a height h above the ground, so that it has only potential energy U = mgh, its kinetic energy T being zero. Its total energy s thus (T + U) = (0 + mgh) = mgh.

Let it now be allowed to fall under the action of gravity, so that as it gathers speed, it acquires K.E. at the expense of its P.E.

When it has fallen to a height z above the ground, *i.e.*, through a distance (h – z), acquiring a velocity v, w have $v^2 - 0 = 2g(h - z)$ and the K.E. gained by it is, therefore,

$$T = 1/2mv^2 = 1/2m\ [2g(h - z) = mg(h - z)$$

And since its height from the ground is now z, the P.E. lost by it is $\int_h^z -mgdr = -mg(h-z)$ and ∴ its P.E. at this height (z) above the ground, *i.e.*, U = mgh – [–mg(z – h)] = mgz.

∴ *the sum total of ifs energy, E = T + U = mg(h – z) + mgz = mgh*, the same as at its initial height h.

And, when the particle has reached the ground, *i.e.*, when z = 0, its P.E., U = mgh = 0 and its K.E., T = mg(h – z) = mgh.

And ∴ its total energy E = T + U = mgh + 0 = mgh, the same as to start with.

Thus, the total energy of the particle remains the same mgh, in accordance with the law of conservation at any height from the ground between 0 and h, *indicating that the energy function E = T + U is independent of both time and the position of the particle.*

POTENTIAL ENERGY IN AN ELECTRIC FIELD—ELECTRIC POTENTIAL

We know that the intensity E of an electric field at a point is defined as the force experienced by a unit positive charge when placed at that point n the field. It follows, therefore, that the force experienced by 2 charge q placed at the point will be qE.

Again, the potential (V) at a point in an electric field is defined as *the amount of work done in moving a unit positive charge from ∞ (where the potential is assumed to be zero) to that point.* It is thus in ether words, the electrostatic potential energy of a unit positive charge placed that point. So that, if the potential energy of a charge of –q units b; U at a point r in an electric field, we have

$$V = U/q = -\int_{\infty}^{r} E.dr,$$

where E is the intensity of the field at the point.

Hence, the potential difference between two points r_1 and r_2 is given by

$$V_2 - V_1 = \int_{r_1}^{r_2} E.dr$$

where V_1 and V_2 are the potentials at the two point respectively.

This is obviously equal to the change in the electrostatic potential energy of a unit positive charge when moved from one point to the other. Clearly, therefore, if a charge q be moved between the two given points, the *change in electrostatic potential energy* will be

$$U_2 - U_1 = q\,(V_2 - V_1),$$

where U_1 and U_2 are the electrostatic potential energies of the charge q at the two points respectively.

Now, it may be noted that the expression for potential, $V = -\int_{\infty}^{r} E.dr$ is similar to that for the potential energy of a particle, viz., $U = -\int_{\infty}^{r} F.dr$ so that, taking

$$E = (E_x i + E_y j + E_z k)$$

and $$dr = (dxi + dyj + dzk)$$

and proceeding exactly as we obtain

$$E = -\left(\frac{\partial V}{\partial x} i + \frac{\partial V}{\partial y} j + \frac{\partial V}{\partial z} k\right) = -\text{grad } V \text{ or } -\nabla V,$$

and in one dimension, $E = -dV/dx$.

This enables us to define *electric potential* in two ways, viz.,

(i) *as the scalar quantity whose gradient with the negative sign gives the intensify of the electric field at the point in question.* Or,

(ii) *as the scalar quantity whose negative space rate of change in any given direction gives the component of the field intensity in that direction.*

Now, in the electrostatic system of units, we have electric intensity at a point distant r from a charge q in a medium of dielectric constant K given by $E = qr/Kr^2$, where r is the unit vector in the direction of r.

$$\therefore V = -\int E.dr = -\int \frac{dr}{Kr^2}.dr = -\int \frac{qdr}{Kr^2} = \frac{q}{Kr} + C$$

where C is a constant integration.

Since at $r = \infty$, $V = 0$, we have $C = 0$. And, therefore,

$V = q/Kr$ *esu or statvolt.*

In the RMKS (or SI) system, the unit of potential is the volt, where 1 *volt* = *l joule/1 coulomb* = $10^7/3 \times 10^9$ *esu or statcoulomb* = 1/300 esu.

Based on the volt, we have a very convenient unit for the measurement of energy, namely, the *electron volt which is equal to the work done by an electron moving through a potential difference of 1 volt.*

Thus, $1eV = 4.80 \times 10^{-10}$ esux $\times$ 1/300 esu = 1.6×10^{-12} ergs = 1.6×10^{-19} joules,

LINEAR RESTORING FORCE

As its very name indicates, a linear restoring force is the force which is proportional to the linear displacement of a particle from a fixed point—its mean position of rest or equilibrium—and is always directed towards that position (*i.e.*, opposite to that of its displacement), tending to restore the particle back to its original position. This is known as *Hooke's law.*

Thus, taking the mean position of the particle as the origin, if its displacement at a given instant be r, we have

$F \alpha (-r)$. or $F = -Cr$,

where C is a positive constant, called the *force constant.*

Or, since a small linear restoring force is often provided by a stretched or a compressed spring, C is also sometimes called the spring factor.

The negative sign of the expression for F indicates that the *force is directed oppositely to the displacement r from the mean position.* A graph between F and r will thus be a straight line passing through the origin and having a slope –C.

Obviously, work has to be done against this restoring force in displacing the particle from its mean position and this work is stored up in the particle in the form of potential energy.

To obtain the value (U) of this potential energy, we note that the linear restoring force is a conservative force as may be easily established by showing that curl F = 0, this being an essential condition for a conservative force. We thus have

curl F $= \nabla \times (-C\mathbf{r})$, where r = r = xi + yj + zk. So that,

$$\text{curl } F = -C \begin{vmatrix} i & j & k \\ \partial/\partial x & \partial/\partial y & \partial/\partial z \\ x & y & z \end{vmatrix} = 0,$$

because the partial differential coefficients ($\partial z/\partial y$ etc.) are all equal to zero.

And since it is a conservative force, it may be expressed as the negative of the gradient of potential energy U of the particle, acquired on displacement r, *i.e.*, F = –grad U, and, therefore,

$$U = -\text{j}\,(F_x dx + F_y dy + F_z dz)$$

$$= \frac{C}{2}\left(x^2 + y^2 + z^2\right) + A = \frac{Cr^2}{2} + A$$

where A is a constant of integration.

Since, with displacement zero, *i.e.*, r = 0, the P.E., U is also zero. we have A = 0. And, therefore,

$$U = Cr^2/2.$$

Thus, if the displacement be along the x direction, *i.e.*, if r = xi, where i is the unit vector along the axis of x, we have $U = Cx^2/2$. So that, plotting P.E., (U), against displacement x, we obtain a parabola, as shown in Fig. 4.2, clearly bringing put the fact that *the work done in displacing a particle against a linear restoring force, and hence the potential energy acquired by it, is proportional to the square of its displacement from its mean or equilibrium position.*

If the *maximum displacement* of the particle be a, the *maximum potential energy* acquired by it = $1/2ma^2$.

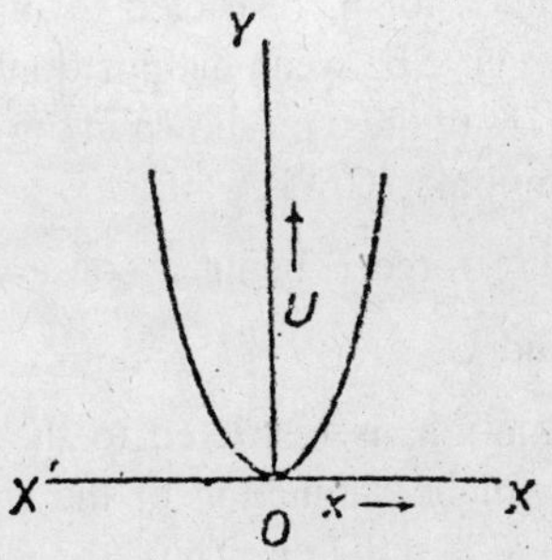

Fig. 4.2

Let us imagine the particle to be initially at its position of maximum displacement a, where the whole of its energy is present in the potential form and is equal to $1/2ma^2$. On releasing it, it will start moving towards its mean or equilibrium position under the action of the restoring force, thus acquiring kinetic energy at the expense of its potential energy. And since the restoring force is, as we have seen, a conservative force, the sum total of its energy E, at any point of its displacement, will remain constant. So that, at displacement x, if its velocity be v, we have *its kinetic energy*, $T = 1/2mv^2$ and its *potential energy*, $U = 1/2Cx^2$.

$\therefore$ its total energy, $E = T + U = 1/2mv^2 + 1/2Cx_2 = 1/2Ca^2$,

whence, $1/2mv^2 = 1/2C(a^2 - x^2)$ and $\therefore$ $v = \pm\sqrt{\frac{C}{m}(a^2 - x^2)}$

Again, as the particle reaches its mean or equilibrium position at $x = 0$, its P.E. = 0 and its kinetic energy, the maximum, equal to $1/2Ca_2$. Its velocity here will therefore be the maximum, say, V_{max}.

We, therefore, have $1/2mv^2_{max} = 1/2Ca^2$, whence, $y_{max} = \pm\sqrt{\frac{C}{m}}a$

POTENTIAL ENERGY CURVE

In a conservative force field, the potential energy of a particle is in general, a function of space and, therefore, changes from point to point. A curve, showing variation of the potential energy of a particle with its position in the field is spoken of as a *potential energy curve* and supplies a great deal of information about the motion of the particle without having to solve any equations of motion.

For the sake of simplicity, let us consider the case of a particle whose motion is only one-dimensional, restricted to only the axis of x, say. Then, a typical energy curve between the position (x) of the particle and its potential energy (U) is of the type shown in Fig. 4.3 with well-marked maxima (Q, N) and minima (C, G).

The linear restoring force F being a conservative one, we have

$$F = -\text{grad } U.$$

Since, here, the motion, is restricted to the .r-direction only, the potential energy (U) will be a function of the x-coordinate alone and we shall, therefore, have F = dU/dx. Thus, the slope of the carve (*i.e.*, dU/dx) at any given point gives the value of the force ($-\partial U/\partial x$) acting on the particle at that point.

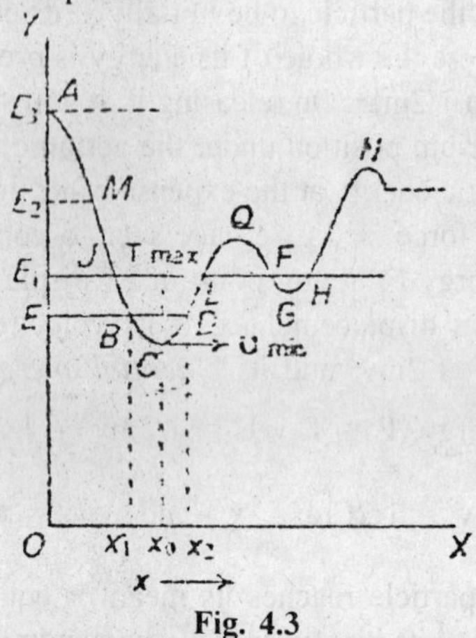

Fig. 4.3

It will be readily seen that for portions of the curve such as CQ and GN, where an increase in the value of x corresponds to an increase in the value of U, the slope of the curve at any point is positive and hence force F(= –dU/dx) acting on the particle there, negative. On the other hand, for portions of the curve, like AC and QG, where an increase m the value of x corresponds to a decrease in the value of 0, the slope of the curve at any point is negative and hence the force acting on the parade there, positive. *In either case, therefore, the force acting an the particle tends to pull it into a region of lower potential energy.*

And, clearly, at points on the curve representing minima and maxima of potential energy (U), such as C,G and Q,N respectively, the slope is zero and hence no force acts on the particle when lying at any one of these points which are, for this reason, referred to as positions of equilibrium. However, at paints such as Q (or N) corresponding to a maximum of P.E., the slightest displacement of the particle, either way, will result in a force acting on it, tending to move it along QDC or QFG to a position of lower potential energy C or G. These points, like Q (or N) of maximum potential energy are therefore, called positions of *unstable equilibrium*. On the other hand, at points such as C (or G) corresponding to minimum potential energy, any slight displacement of the particle, either way, results in a force acting on it, tending to bring it back- into its original position C (or G).

These points (like C or G) of minimum potential energy are, therefore, called positions of stable equilibrium.

In case the curve has also a horizontal portion, such as shown dotted in the figure, signifying a region of *constant potential energy*, the slope of the curve, and hence the force acting on the particle at any point on it, remains zero even if slightly displaced from its initial position, so that it stays in its new position, without either tending to go further away from it or to come back to its initial position. Such a region is, therefore, called a *region of neutral equilibrium*.

Bounded region or the potential well : Let us consider again the particle at the minimum potential energy or the stable equilibrium position, C, with its total energy E (= T + U) represented by the horizontal line shown, cutting the energy curve at B and D. Since the system is conservative and the energy is conserved, it follows that the P.E. of the particle at C being the minimum, its K.E. there must be the maximum, so that it moves towards either B or D. As it proceeds towards D, say, it comes under the action of the restoring force $F = -dU/dx$, so that its K.E. decreases and its P.E. increases (the sum total remaining the same, E), until when it reaches the point D, the whole of its energy is in the potential form, equal to E, its kinetic energy, and hence its velocity, now being reduced to zero. Here, therefore, the particle remains momentarily at rest and then starts moving back under the action of the restoring force and again reaches C where its potential energy is reduced to the minimum and its kinetic energy becomes the maximum. It, therefore, continues to move towards B, its potential energy now increasing and its kinetic energy correspondingly decreasing (the sum total remaining constant at E) until at B, the whole of its energy is again in the potential form and its kinetic energy, and hence its velocity, *zero*. Again, therefore, it is momentarily at rest at B and then starts moving back towards C under the action of the restoring force.

Thus, the particle remains confined to the region BCD, oscillating (simple harmonically) between the *turning points* B and D. Such a region (between the turning points B and D) is appropriately called a *bounded region or a potential well (or potential valley) and always exists about a point of minimum potential energy or-stable equilibrium.* The difference of energy between the top and the botton of the well is called the binding energy for the well and the particle confined to the well is said to be in the *bound state* or to perform a *bound motion* and, as we have seen, if its motion be one-dimensional, it is necessarily *periodic* and keeps on repeating itself after regular intervals, called its *time-period.*

It will be readily seen that

(i) If the total energy of the particle be E_1, the particle will have four turning points, J, L, F, and H and it can oscillate in either of the two potential wells JCL and F G H;

(ii) If its total energy be E_2, it will obviously have only one turning point at M and will always turn to the right from here, it? speed increasing and decreasing according as its potential energy (U) decreases or increases.

(iii) And, if its energy be greater than E_3, there will be no turning point at all, *i.e.*, the particle will not change direction and only its speed will vary as in case (ii).

Now, the potential energy function U = f(x) may, in general, be expanded about the position x_0 of stable equilibrium by means of *Taylor's theorem*. Thus,

$$U = U_{x0} + \left(\frac{dU}{dx}\right)_{x0}(x - x_0) + \left(\frac{d^2U}{dx^2}\right)^2_{x_0}\left(\frac{x-a}{2!}\right) + \ldots,$$

where $(dU/dx)x_0$, $(d^2U/dx^2)x_0$ etc. are the differentials at the stable equilibrium position (x_0) and, as such, are constants.

If we measure displacement from the position x_0, *i.e.*, if we take x_0 as the origin, so that $(x - x_0) = (x - 0) = x$, we have

$$U = U_0 + \left(\frac{dU}{dx}\right)_0 + \left(\frac{d^2U}{dx^2}\right)_0 \frac{x^2}{2!} + ...$$

Since x_0 is a point of stable equilibrium, and hence of minimum potential energy, we have $(dU/dx)_0 = 0$ and $(d^2U/dx^2)_0$, a positive quantity, say C. Then, denoting (d_2U/dx^3),.... etc. by C_1 C_2 ...., we have

$$U = U_0 + + \frac{Cx^2}{2!} + \frac{C_1x^3}{3!} + ...,$$

and, therefore, force acting on the particle, *i.e.*, $F = -\frac{dU}{dx}$.

or $$F = -Cx = \frac{C_1}{2}x^2 - \frac{C_2}{6}x^3 ...$$

In case the displacements be small, the higher powers of x become negligible and we, therefore, have

$$F = -Cx \text{ and } U = Uy_0 + 1/2Cx^2,$$

indicating that force F is a *linear restoring force*, proportional to displacement and directed towards the position of equilibrium. The positive constant C is thus the *force constant*.

The potential energy curve (or the graph between U and x) in this region of small displacements, x_1 and x_2, say, about the position of stable equilibrium (*i.e.*, within the potential well) thus voices out to be parabolic in form (as shown in Fig. 4.4), with its vertex at x_0, U_0. In case the minimum potential energy (U_0) be assumed to be zero, the apex of the parabola will touch the x-axis at the origin O,

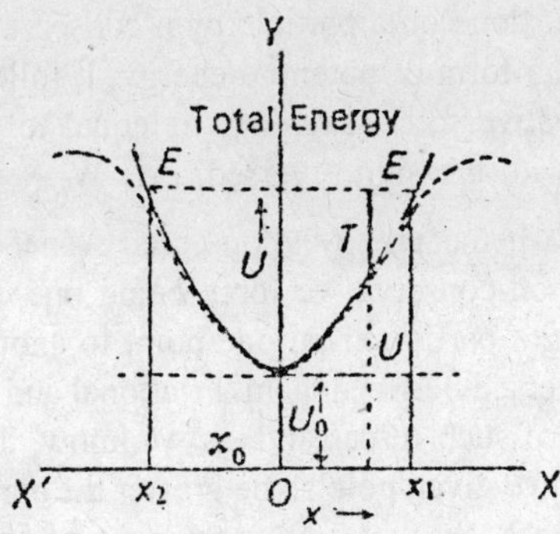

Fig. 4.4

The particle thus oscillates inside the potential well about the stable equilibrium (or minimum potential energy) position between the limits x_1 and x_2, so that its total energy E = T + U. It will be easily seen that the particle cannot cross the limits x_1 and x_2 for, within this region (x_1, x_2), the sum total of its energy is E = T + U. And at x_1 and x_2, T = 0 and, therefore, U = E, *i.e.*., the entire energy is present in the potential form. Beyond x_1 and x_2, the potential energy U would be greater than this total energy E, implying thereby that, in order that T + U may be equal to E, the kinetic energy T must be *negative*, which is simply absurd.

The motion of the particle thus remains restricted between x_1, and x_2 on either side of O, so long as the total energy is E and there are no non-conservative forces (like friction etc.) operating. In case the value of E be different, the values x_1 and x_2 too will naturally be correspondingly different. This is in fact the case of a simple harmonic oscillator which we shall study in some detail.

NON-CONSERVATIVE FORCESÙGENERAL LAW OF CONSERVATION OF ENERGY

We have seen how when conservative forces alone are concerned, the mechanical energy of a particle, or a system of particles, remains conserved, *i.e.*, the sum of its kinetic energy (T) and potential energy (U) gives its total energy E which remains constant or that the change in its total energy (ΔE) is zero *i.e.*,

$$T + U = E, \text{ a constant.}$$

$$\text{or } AT + \Delta U = \Delta E = 0.$$

Since the work done on a particle by a conservative force remains stored up in it in the form of potential energy, it follows that this work done by a conservative force, say, $W_{(C)}$ is equal to the increase in its potential energy, with the sign reversed, *i.e.*, $W_{(C)} = -\Delta U$.

Let us now see whether energy is conserved even if a non-conservative on a particle,—a non-conservative force being one, the work done by which in displacing a particle from one point to another depends upon the actual path traversed between them. Frictional and viscous forces are obvious examples of such forces and, as we know, the longer the path traversed between two given points, the greater the amount of work done by the frictional force. So that, unlike the case of conservative forces, the work done here in taking a particle from a point A to another point

B is *not equal and opposite* to the work done in taking it from B to A and the total work done from A to B and back is not, therefore, zero. In fact, there is loss of kinetic energy either way, *i.e.*, in proceeding from A to B as also in proceeding back from B to A, so that the resultant kinetic energy of the particle is less than its initial value by an amount equal 10 the work done in performing the round trip from A back to A.

Taus, if we have both, a conservative and anon-conservative, force acting on a particle and if the work done by the two be $W_{(C)}$ and $W_{(N)}$ respectively, then if ΔT be the loss suffered by the panicle in its kinetic energy, we have $W_{(C)} + W_{(N)} = \Delta T$. Or, putting—$\Delta U$ for $W_{(C)}$, we have $\Delta t + \Delta U = W_{(N)}$. Obviously, $\Delta t + \Delta E = \Delta E$ the change in the total energy of the particle. And, therefore, $\Delta E = W_{(N)}$.

Taus, the change in the total energy of the particle is no longer zero but $W_{(N)}$, the work done by the non-conservative force.

If the non-conservative force be a *frictional force*, the work done by it $[W_{(N)}]$ appears in the form of equivalent heat energy, say, equal to Q, where, clearly, $W_{(N)} = -Q$. We, therefore, have

$$\Delta E = -Q \text{ or } \Delta E + Q = 0,$$

i.e., the change in the total energy of the parade is zero or its total energy remains conserved.

Similarly, in the case of other non-conservative forces, the work done by them appears in some other forms of energy, like sound, light etc., but in all cases, without exception, the total energy is conserved. *The general law of conservation of energy thus holds good in the case of both conservative and KM-conservative forces.*

SOLVED EXAMPLES

Example 1:

Show that the following two forces are conservative:

(i) $F = (y^2 - x^2)i + 2xyj$ and (ii) $F = (2xy + z^2)i + x^2j + 2xzk$.

Also calculate the work down on a particle in case (ii) in moving it from the position 0, 1, 2 to the position 5,6,8.

Solution:

Clearly, if the curl of each of the two forces be zero, they will be conservative.

Now, in case (i), *curl* F = ∇ F = $\begin{vmatrix} i & j & k \\ \frac{\partial}{\partial x} & \frac{\partial}{\partial y} & \frac{\partial}{\partial z} \\ (y^2 - x^2) & 2xy & 0 \end{vmatrix}$

$$= \left[0 - \frac{\partial}{\partial z}(2xy)\right]i + \left[\frac{\partial}{\partial z}(y^2 - x^2) - 0\right]j + \left[\frac{\partial}{\partial x}2xy - \frac{\partial}{\partial y}(y^2 - x^2)\right]k.$$

$$= -\frac{\partial}{\partial z}(2xy)i + \frac{\partial}{\partial z}(y^2 - x^2)j + \left[\frac{\partial}{\partial x}(2xy) - \frac{\partial}{\partial y}(y^2 - x^2)\right]k$$

$$= 0 + 0 + (2y - 2y)k = 0.$$

Thus, curl F being zero, force F is a conservative force.

In case (ii), again, curl F = ∇F = $\begin{vmatrix} i & j & k \\ \frac{\partial}{\partial x} & \frac{\partial}{\partial y} & \frac{\partial}{\partial z} \\ 2xy + z^2 & x^2 & 2xz \end{vmatrix}$

$$= \left[\frac{\partial}{\partial y}(2xz) - \frac{\partial}{\partial z}(x)^2\right]i + \left[\frac{\partial}{\partial z}(2xy + z^2) - \frac{\partial}{\partial x}(2xz)\right]j + \left[\frac{\partial}{\partial x}(x)^2 - \frac{\partial}{\partial y}(2xy + z^2)\right]K.$$

Again, *curl F being zero, the force is a conservative one.*

Now, work done by force F in case (ii) in displacing a particle from A to B is given be $w = \int_B^A F.dr = \int_A^B (F_x dx + F_y dy + F_z dz)$

$$= \int_A^B [(2xy + z^2)dx + x^2 dy + 2xzdz]$$

$$= \int_A^B (2xydx + x^2 dy) + (z^2 dx + 2xzdx)$$

$$= \int_A^B d(x^2 y) + d(z^2 x)$$

$$= (x^2y + z^2x) = [x^2y + z^2x]_{0,1,2}^{5,6,8}$$

$$= 5\times5\times6 + 8\times8\times5 - 0 - 0.$$

Or, $W = 150 + 320 = 470.$

Thus, *work done by the force* = 470 units.

Example 2:

A particle lying in the x – y plane is acted upon by a force directed towards the origin whose magnitude is kr where r(equal to $\sqrt{(x^2+y^2)}$ is the distance from the particle to the origin, (a) Find the work that must be done to move the particle from the origin to the point x = 1, y = 1 along the radius vector to that point. (b) Find the work that must be done if the particle is first moved to x = 1, y = 0 and from there to x = 1, y = 1. (c) Is this particular force conservative or non-conservative?

Solution:

(a) Clearly, work done in moving the particle from 0 to r is given by

$$W = \int_A^B F.dr = \int_A^B F\cos\theta.dr = \int_0^r Fdr$$

[$\because$ tangentical component of F = F cos θ = F, cos θ being equal to 1.

Since F = kr, we have $W = \int_0^r kr.dr = \left[\frac{kr^2}{2}\right]_0^r = \frac{kr^2}{2}.$

Now, $r = \sqrt{x^2 + y^2} = \sqrt{1 + 1} = \sqrt{2}$. We, therefore, have

$$W = \frac{k\left(\sqrt{2}\right)^2}{2} k.$$

Thus, the work done = k units.

(b) In this case, the components of force F along the x and y-axes are F_x and F_y respectively [$\because r = \sqrt{x^2 + y^2}$]. And, clearly, no work is done against the component F_y in taking the particle from O to A (Fig, 4.5), along the axis of x, (where OA = x = 1), and similarly, no work is done against the component F_x in taking the particle from A to P along the axis of y, (where AP = y = 1).

$\therefore$ *work done in taking the particle from the origin* O to A (x = 1)

$$= \int_0^1 kx.dx = \frac{kx^2}{2} = \frac{k}{2}.$$

And, *work done in taking the particle from* A to P (y = 1)

$$= \int_0^1 ky.dy = \frac{y^2}{2} = \frac{k}{2}.$$

∴ *total work done in taking the particle from the origin 0 to* A (x = 1, y = 0) *and from* A to P (x = 1, y = 1) = k/2 + k/2 = k, the same as in case (i).

(c) It will be easily seen that if we take the particle to P by first moving it from the origin (0) to B(y = 1, x = 0) and from there to P(y = 1, x = 1), we shall again have to do the same work

$$\int_0^1 ky.dy + \int_0^1 kx / dx = \frac{k}{2} + \frac{k}{2} = k$$

as in cases (i) and (ii) .

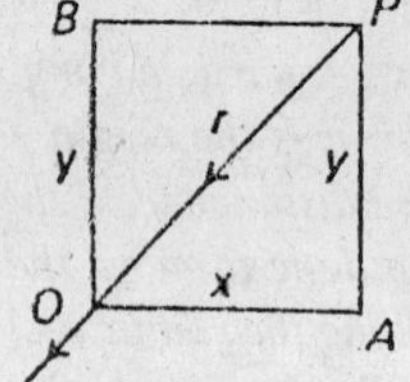

Fig. 4.5

Thus, the work done in taking the particle from the origin to P is the same (k) whatever the path taken, *i.e.*, it depends only upon the position of the point P and not upon the actual path taken. The force on the particle is, therefore, a *conservative force.*

Example 3:

A point-mass m starts from the point A of a curved track of the form shown in Fig. 5.6, with a horizontal velocity V_0. The track is frictionless except for the small horizontal portion DE of length L. (i) Calculate the velocity of the mass at points B and C. (ii) What constant deceleration is required to stop the mass at E if the retarding frictional force starts operating at point D?

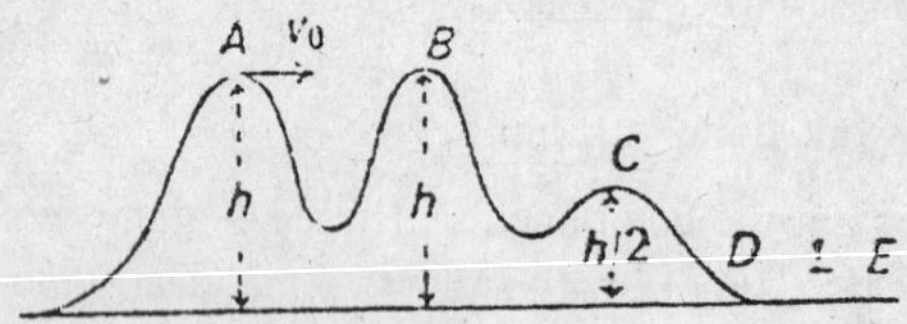

Fig. 4.6

Solution:

(i) Clearly, the total energy possessed by the mass at the point A = T + U = 1/2 mv_0^2 + mgh. Since point B is also at the same height h from the horizontal as the point A, the kinetic and potential energies of the mass there too are the same as at A, viz, $1/2mv_o^2$ and mgh respectively.

At the point C, however, which is at a height h/2 from the horizontal, the P.E. of the mass is reduced to 1/2 mgh. Since the total energy of the mass must remain the same as before, it follows that half the P.E. of the mass *i.e.*, 1/2 mgh is converted here into K.E. So that,

K.E. *of the mass* at C = $1/2\ mv_o^2 + 1/2$ mgh.

But if v_e be the velocity of the mass at C, its K.E. = $1/2\ mv_e^2$.

We ∴ have $1/2mv_c^2 = 1/2mv_o^2 + 1/2mgh$, whence, $v_c = \sqrt{v_o^2 + gh}$.

Thus, the *velocity of the mass at the point* C = $\sqrt{v_o^2 + gh}$.

(ii) By the time the mass reaches the point D, its P.E. is reduced to zero, so that the whole of its energy, $1/2\ mv_o^2$ + mgh is here converted into kinetic energy. If, therefore, v_D be the velocity of the mass at D, we have

$1/2mv_D^2 = 1/2mv_o^2 + mgh$, whence, $v_D = \sqrt{v_o^2 + 2gh}$.

Now, if – a be the constant acceleration of the mass between the points D and E which brings it to rest at E, we have

$0 - v_D^2 = 2(-a)(L)$. Or, $-a = -v_D^2/2L = -\sqrt{v_o^2 + 2gh}/2L$.

Thus, *the constant deceleration required at stop the mass at E* = $\sqrt{v_o^2 + 2gh}/2L$.

Example 4:

A particle of mass m slides along a track with elevated ends and a flat central part, as shown in Fig 4.7. The flat part has a length equal to 2 meters. The curved portions of the track are frictionless. For the flat part the coefficient of kinematic friction is $\mu = 0.20$. The particle is released at point A which is at a height h = 1.0 metre above the flat part of the track where does the particle finally comes to rest?

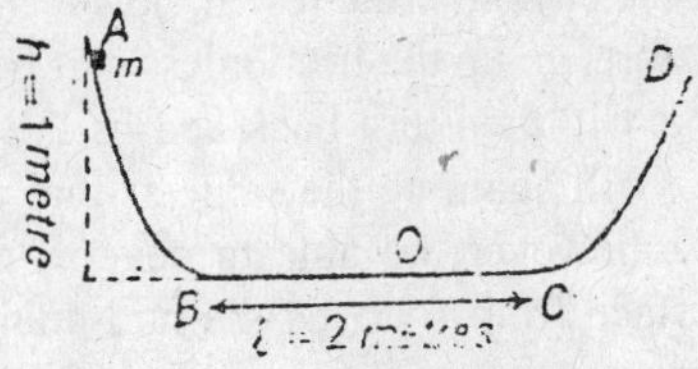

Fig. 4.7

Solution:

Here, clearly, the only energy that the particle possesses at the point A is potential energy = mgh, relative to the flat or the horizontal part BC. As it is released at A, this P.E. naturally starts converting itself into K.E. until at B, the vertical distance being reduced being reduced to zero, the whole of it is present in the kinetic from. So that, if v be velocity acquired by the particle on reaching B,

we have $1/2mv^2 = mgh$

As the particle now starts moving along the horizontal portion BC, it has to overcome the frictional force F and when the whole of its kinetic energy is used up in doing work against this frictional force over a distance s, say it will come to rest. We, therefore, have

$$1/2mv^2 = F.s.$$

Now, as we know, $F/R = \mu$ or $F = \mu R$, where R is the normal reaction of the horizontal surface over which the particle moves and is equal to mg, and μ, the coefficient of kinematic friction between the horizontal surface and the particle.

So that, $F = \mu R = \mu mg.$

$\therefore 1/2mv^2 = \mu mg.s.$ Or, $mgh = \mu mgs.$ whence, $s = h/\mu.$

Since h = 1 metre and $\mu = 0.2$ we have S = 1/0 2 = 5 metres, *i.e.*, the particle will come to rest after tranversing a 5 meter length of the flat surface.

The length of the flat surface being only 2 metres, the particle will still have some K.E. after travelling from B to C. It will, therefore, go up the frictionless, curved part CD until this K.E. is converted into P.E. It will then turn back its P.E. will get converted into K.E, until at C, it will again have the same K.E. as before when it went up along CD. It will, therefore, again cover a distance of 2 metres on the flat surface from C to B, with some more part of its K.E. used up in doing work against the frictional force encountered. It will, however, still have some K.E. left and will, therefore, go up the frictionless part BA until this K.E. is converted into P.E. It will then turn back and its P.E. converted into K.E. until at B its K.E. will again be the same as that when it went up along BA. It will now, however, be able to cover a distance of only 1 metre on the flat surface from B to O, the mid-point of the surface for by *then it will have covered a full 5 metre length of the flat surface that it could in virtue of its initial K.E.*

The *particle will thus finally come to rest at the mid-point O of the flat surface.*

Example 5:

The bob of a simple pendulum, initially held in the horizontal position at A, as shown in Fig. 5.8, is released such that it goes in a circle (shown dotted) round a point P at a vertical distance d below the point of suspension S of the pendulum. Show that the distance d should at least be 0.61, where l is the length of the pendulum.

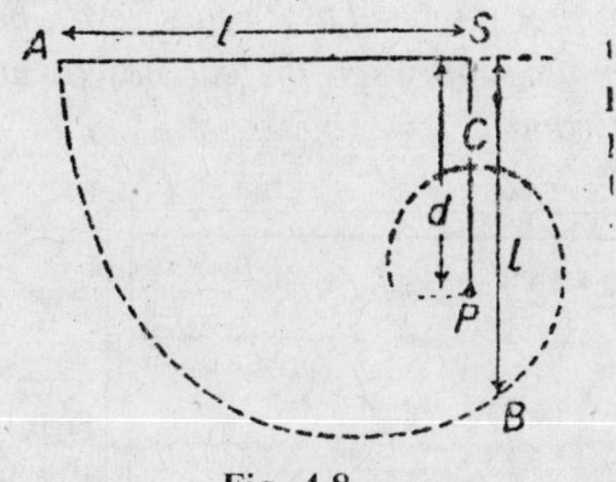

Fig. 4.8

Solution:

Clearly, K.E. *gained by the bob on reaching point* B = mg$l$, where m is the mass of the bob.

If therefore, v be the velocity of the bob at B, we have

$$1/2mv^2 = mgl, \text{ whence, } v^2 = 2gl.$$

In order that the bob should go in a circle round the point P, as shown dotted, its velocity v' at the uppermost point C to the circle should be such that the centrifugal force mv'^2/r just supports the weight mg of the bob, where r is the radius of the circle, equal to (l – d), *i.e.*.,

$$mv'^2/r = mg. \text{ Or, } v'^2 = rg.$$

Thus, K.E. *of the bob at* C = $1/2mv'^2$.

$\therefore$ loss of K.E. in proceeding from B to C = $1/2mv^2 - 1/2mv'^2$.

This must clearly be equal to the gain in P.E. of the bob in rising through a vertical distance BC(or 2r), *i.e.*, equal to mg (2r).

$\therefore$ $1/2mv^2 - 1/2mv'^2 = 2mgr$. Or, $v^2 - v'^2 = 4gr$.

Or, substituting the values of v^2 and v'^2, we have

$$2gl - rg = 4gr. \text{ Or, } 2l = 5r = 5(l - d) = 5l - 5d.$$

Or, $5d = 3l$, whence, $d = 3/5l = 0.6l$.

Thus, *the least distance of point P below the point of suspension S of the pendulum must be* 0.6 *l.*

Example 6:

A car carries a framework ABCD, shown below (Fig. 4.9) in which a 200 gm mass p is supported between two springs of force constant 5000 dyne/cm each. Side AB is kept horizontal and along the length of the car and the pointer attached to P reads zero when the car is at rest. What will the pointer read when the car has (i) uniform speed 20 m/sec on a straight road.

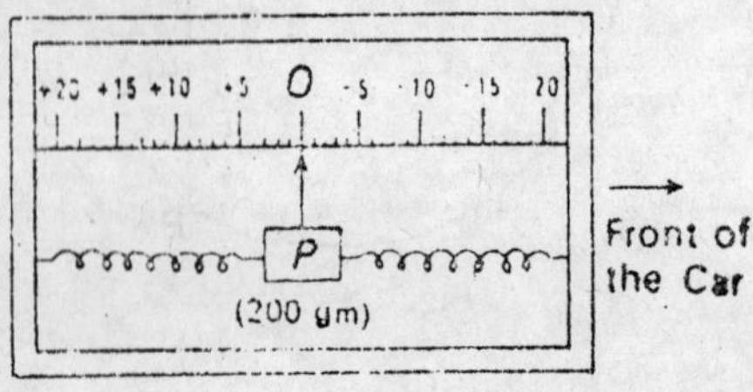

Fig. 4.9

(ii) uniform speed 10 m/sec on a circular road of radius 20 m, (iii) uniform acceleration of 0.5.m/sec² on a straight road (iv) uniform acceleration – 1.0m/sec² on a straight road?

Solution:

(i) In this case, the motion of the car being unaccelerated, there is no force acting on mass P and hence *the point r attached to it continues to rad zero on the scale.*

(ii) Here, since the car is moving with a uniform speed *along a circular path*, the centripetal force (mv^2/r) will act in a direction perpendicular to the length of the two springs, thus producing neither extension nor contraction in them. *The pointer will, therefore, again continue to read zero on the scale.*

(iii) The forward acceleration of the car being 0.5 m/sec² or 50 cm/sec², the force of reaction acting on the mass P = 200 × 50 dynes in the backward direction. The first spring (to the left of P) will thus get compressed and the second, stretched through a distance x say We shall, therefore, have

200 × 50 = – (C + C)x = – 2 Cx, where C is the *force constant* of each spring.

$$\text{or} \quad x = \frac{200 \times 50}{2C} = \frac{200 \times 50}{2 \times 5000} = 1\,\text{cm} = -10\ \text{mm}.$$

Thus, the mass will move 10 mm to the left and *the pointer will read* + 10 mm *on the scale, to the left of the zero mark.*

(iv) Here, the backward acceleration of the car being 1.0 m/sec^2 or 100 cm/sec^2, the force or reaction 200 × 100 dynes will act on the mass in the forward direction. As a result, the first spring (to the left of P) will get extended and the second compressed by the same amount x', say. We, therefore, have

200 × 100 = 2 × 5000 x', whence,

x' = 200 × 100/2 × 5000 = + 2 cm or + 20 mm.

The mass will thus move forward through 20 mm and *the pointer will thus read – 20 mm on the scale, to the right of the zero mark.*

Example 7:

An ideal massless spring S can be compressed 1.0 metre by a force of 100 newtons. This same spring is placed at the bottom of a frictionless inclined plane which makes an angle θ = 30° with the horizontal (Fig. 5.10). A 10 kg mass m is released from rest at the top of the incline and is brought to rest momentarily after compressing the spring 2.0 metres. (a) Through what distance does the mass slide before coming to rest? (b) What is the speed of the mass just before it reaches the spring?

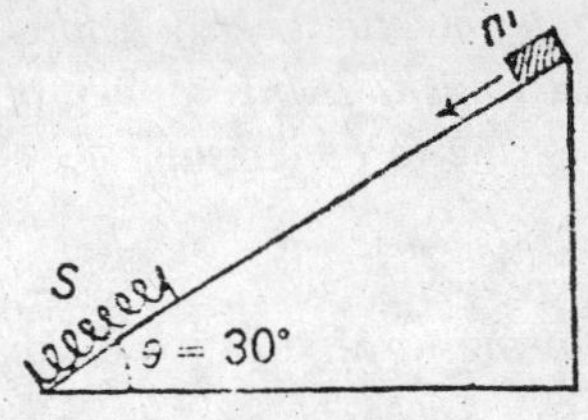

Fig. 4.10

Solution:

(a) Let the mass m slide a *total distance s along* the inclined plane before coming to rest on compressing the spring through 2 metres. Then, clearly,

gravitational potential energy lost by the mass = mg sin θ × S.

This must obviously be equal to the work done in compressing the spring through a distance of 2 metres, *i.e.*, equal to $1/2Cx^2 = 1/2 \times 100\,(2)^2 = 200$ joules, because C = 100 N/m (given).

We, therefore, have mg sin θ × s = 200 Or, 10 × 9.8 × 1/2 × s = 200,

i.e., 49s = 200. Or, s = 200/49 = 4.08 m, or 4.0 *metres*, say.

Thus, *the distance through which the mass slides before coming to rest* = 4 *metres.*

(b) Since this total distance of 4.08 meters also includes the 2 metres through which the spring is compressed, the distance through which the mass slides from the top of the plane until it reaches the spring is 4.08 – 2 = 2.08 *metres.*

And, since its initial velocity u = 0 (∵ it starts from rest), its velocity v on covering a distance of 2.08 metres along the incline is given by the relation $v^2 - u^2 = 2$ as. *i.e.*, $v^2 = 2 \times g \sin\theta \times 2.08 = 2 \times 9.8 \times 1/2 \times 2.08 = 20.38$, whence,

$v = \sqrt{20.38} = 4.514$ m/sec or 4.5 *metres per sec.*

Thus, *the speed of the mass just before it reaches the spring* = 4.5 metres/sec.

Example 8:

Obtain an expression for restoring force as a function of position for a particle moving in a potential energy field $U = A - Bx + Cx^2$. At that point does the force vanish? Is this a point of stable equilibrium? If so, obtain the value of the force constant.

Solution:

Clearly, *linear restoring force* = F = – dU/dx = B – 2 Cx.

This force will vanish when dU/dx = 0, or, B – 2 Cx = 0, *i.e.*, when x = B/2C.

Now, for stable equilibrium of the particle d^2U/dx^2, *i.e.*, 2C must be positive. Hence if C be a positive quantity, we shall have a *minimum potential energy position* of the particle at x = B/2C and this will, therefore, be a point of stable equilibrium.

From the relation F = B – 2 Cx, it is clear that the *force constant*=2C.

Example 9:

Two ideal and identical springs, S_1 and S_2, each of normal length l and force constant C are fixed at points (–l, o) and (+l, o) and connected together at the other end, as shown in Fig 5.11 (a).

Solution:

(a) Show that for the displacement, x, y, of their joined ends,

$$U = \frac{C}{2}\left[\left\{(x+l)^2 + y^2\right\}^{\frac{1}{2}} - l\right]^2 + \frac{C}{2}\left[\left\{(l-x)^2 + y^2\right\}^{\frac{1}{2}} - l\right]^2$$

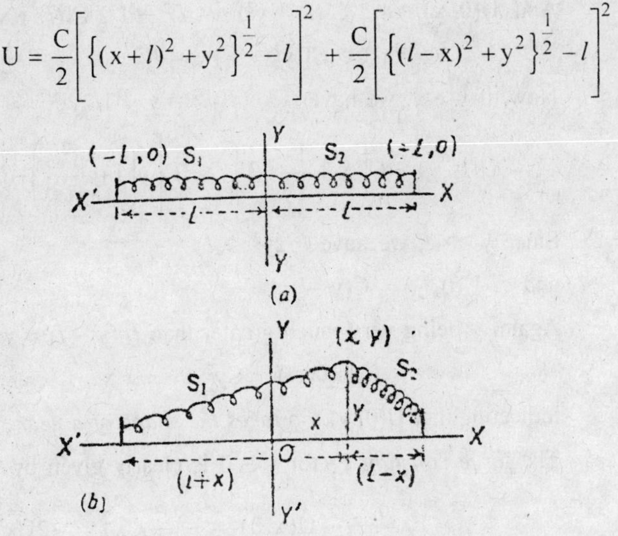

Fig. 4.11

(b) Obtain the values of the functions U (x, 0) and U(y, 0) show that U(x, 0) assumes the shape of a parabola for all values of x and that if y > > l, U(0, y) too assumes a shape which is a near parabola.

(c) Calculate the force components F_x for y = 0 and F_y for x=0.

(d) With the joined ends of the springs given the displacement (x, y), as shown in Fig. 4.11 (b) we find that spring S_1 gets stretched and spring S_2, compressed, their increased and decreased lengths now being.

$[(l + x)^2 + y^2]^{1/2}$ and $[(l - x)^2 + y^2]^{1/2}$ respectively. So that,

increase in length of spring $S_1 = [(l + x)^2 + y^2]^{1/2} - l$.

and *decrease in length of spring* $S_2 = [(l - x)^2 + y^2]^{1/2} - l$.

$\therefore$ $U(x, y) = 1/2\ C[\{(l + x)^2 + y^2\}^{1/2} - l]^2 + 1/2\ C[\{(l - x)^2 + y^2\}^{1/2} - l]^2$.

Obviously $U(x, 0) = 1/2\ C[\{(l+x)^2\} - l]^2 + 1/2C[\{(l-x)^2\}1/2 - l]^2$

$$= 1/2\ C(l + x - l)^2 + 1/2\ C(l - x - l)^2 = Cx^2,$$

indicating that U(x, 0) *assumes the shape of a parabola for all values of x.*

And, $U(0, y) = 1/2\ C[(l^2 + y^2)^{1/2} - l]^2 + 1/2\ C[(l^2 + y^2)^{1/2} - l]^2$

$$= C[(y^2 + l^2)^{1/2} - l]^2.$$

Now, if $y >> l$, we have $(y^2 + l^2)^{1/2} = y + 1/2y^{-1/2}\ldots \approx y + l^2/2y$

$$\therefore \quad U(0, y) \approx C\left(Y + \frac{1^2}{2y} - 1\right)^2 \approx C\left[y\left(1 + \frac{1^2}{2y^2}\right) - l\right]^2$$

Since $y >> l$, we have $l^2/2y^2 << l$

and $\therefore$ $U(0, y) \approx C(y - l)^2$

Again, y being very much greater than l, $(y - l) \approx y$. So that,

$$U(0, y) \approx Cy^2,$$

indicating that U(0, y) *too takes the shape of a near parabola.*

The *force constant* F_x for $y = 0$ is clearly given by

$$F_x = -\frac{\partial}{\partial x}U(x,0) = -\frac{\partial}{\partial x}Cx^2 = -2Cx.$$

And, *force constant* F_y for $x = 0$ is similarly given by

$$F_y = -\frac{\partial}{\partial y}U(0,y) = -\frac{\partial}{\partial y}C[(y^2 + 1^2)^{\frac{1}{2}} - 1]^2$$

$$= -2C\left[y - \frac{ly}{(y^2 + 1^2)^{\frac{1}{2}}}\right]$$

Example 10:

Fig. 4.12 shows the potential energy curve of a system in arbitrary units for U and r. State the possible r-values about which the system can oscillate, also the maximum energy of oscillations in each case. If the system just escapes from the potential well at smaller r, with what kinetic energy does it go to $r \to \infty$?

If 0.1 unit be the mass of the system, calculate its velocity when it escapes from the potential well at smaller r.

Solution:

As we know, the system can oscillate only about its position of stable equilibrium or minimum potential energy. As is clear from the graph, these positions occur at A and B, where the values of r are 2 units and 4.5 units respectively.

The *maximum energies of oscillation at* these position are obviously, 1 – (–2.5) or 3.5 units and 0.5 – (–1) or 1.5 units, respectively.

And, as the system just escapes at C from the potential well at smaller r = 2 units, its K.E. will obviously be 1 unit.

If y be the velocity of the system as it escapes from the potential well at smaller r, we have its K.E. = $1/2mv^2$ = 1 unit. So that,

$1/2\ (0.\ 1)v^2 = 1$. Or, $v = \sqrt{20} = 4.47$ units.

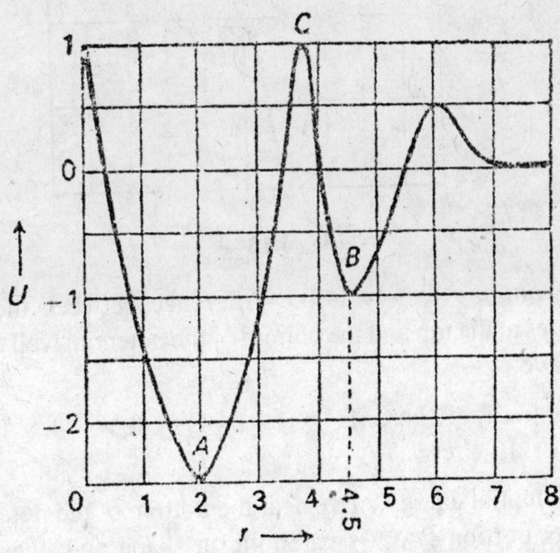

Fig. 4.12

Example 11:

The figure below (Fig. 4.13) shows the potential energy diagram of a system like a diatomic molecule, potential energy U being plotted against inter-nuclear distance r. Deduce values of the following:

(i) Equilibrium inter-nuclear distance r_o.

(ii) Binding energy E_b.

(iii) Force constant near the bottom of the well.

Solution:

(i) At the equilibrium inter-nuclear distance, the potential energy must be the minimum. As we can see from the graph supplied (Fig. 4.13), this happens at the point A which corresponds to r =2 Angstrom units. This, therefore, gives the values of r_o. Thus,

equilibrium inner-nuclear distance, $r_o = 2$ A. U. $= 2 \times 10^{-8}$ cm.

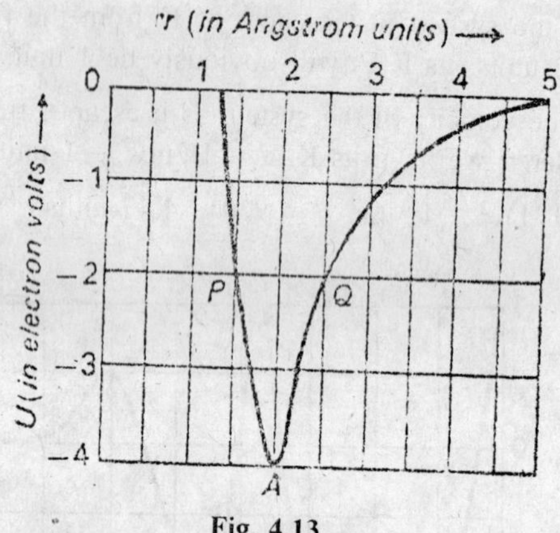

Fig. 4.13

(ii) Since *binding energy* is the difference between the potential energies at the top and the bottom of the potential well concerned, we have

binding energy, here, $E_b = 0 - (-4) = 4\text{eV} = 4 \times 1.6 \times 10^{-12}$ $= 6.4 \times 10^{-12}$ ergs.

(iii) The potential energy curve near the bottom of the potential well, *i.e.*, its portion PAQ is parabolic in shape and, therefore,

$U = U_A + 1/2\ C\ (r - r_o)^2$, whence, *the linear restoring force* F $= -\ dU/dr = -\ C(r - r_o)$

So that, $$C = \frac{dU/dr}{(r - r_o)} = \frac{dU}{dr(r - r_o)}$$

Considering the displacement from A to Q, therefore, we have

$dU = -2 - (-4) = 2eV = 2 \times 1.60 \times 10^{-12}$ ergs, dr = 0.5 A. U. $= 0.5 \times 10^{-8}$ cm and $(r - r_o) = (2.5 - 2.0) = 0.5$ A. U. $= 0.5 \times 10^{-8}$ cm.

Hence, *force constant* $= 12.8 \times 10^4$ dyne/cm.

$$C = \frac{2 \times 1.60 \times 10^{-12}}{(0.5 \times 10^{-8})(0.5 \times 10^{-8})} = \frac{3.20 \times 10^{-12}}{0.25 \times 10^{-16}}$$

Example 12:

The potential energy function for the force between two atoms in a diatomic molecule can approximately be expressed as $U_{(x)} = \dfrac{a}{x^{12}} - \dfrac{b}{x^6}$, *where a and b are positive constants and x is the distance between the atoms. (i) For what values of x, is* $U_{(x)}$ *equal to zero ? (ii) For what values of x, is* $U_{(x)}$ *minimum ? (iii) Calculate the force between the two atoms and plot* $F_{(x)}$ *against x. Show that the two atoms repel each other for x less than* x_o *and attract each other for x greater than* x_o*. What is the value of* x_o*? (iv) Assuming that one of the atoms remains stationary and the other moves along the x-axis, describe the possible motions. (v) Calculate the dissociation energy (the energy required to break the molecule into atoms) of the molecule.*

Solution:

(i) From the relation $U_{(x)} = \dfrac{a}{x^{12}} - \dfrac{b}{x^6}$, it is clear that for $u_{(x)}$ to be zero, we should have either (i) $a/x^{12} - b/x^6 = 0$ whence, $x^6 = a,b$ and, therefore,

$$x = (a/b)^{1/2}\ , \text{ or (ii) } x = \infty$$

Thus, *the two values of x for which* $U_{(x)}$ *will be zero are* $x = (a/b)^{1/2}$ and $x = \infty$.

(ii) For $U_{(x)}$ to be a minimum, we should have $dU_{(x)}/dx = 0$,

$$\text{i.e.,}\quad \frac{d}{dx}\left(\frac{a}{x^{12}}-\frac{b}{x^6}\right)=0.$$

Or, $-\frac{12a}{x^{13}}+\frac{6b}{x^7}=0$ whence, $x^6\frac{2a}{b}$ and, therefore, $x=(2a/b)^{1/6}$.

Substituting this value of x in the expression for $U_{(x)}$ above, we have $U_{(x)min} = -b^2/4a$.

Thus, $U_{(x)}$ will have the minimum value $-b^2/4a$ when $x = (2a/b)^{1/6}$.

If, therefore we plot a graph between x, the distance between the two atoms, and the potential energy $U_{(x)}$, we obtain a curve of the form show in Fig. 5.14 (a), which shows that $U_{(x)} = 0$ both at $x = (a/b)^{1/6}$ and at $x = \infty$ and that $U_{(x)}$ is a minimum at $x = (2a/b)^{1/6}$.

(iii) As we know, the force between the two atoms is given by $F = -d/dx\, U_{(x)}$, which gives $F_{(x)} = \frac{12a}{x^{13}}+\frac{6b}{x^7}$, whence, $F_{(x)} = 0$, when $x = (2a/b)^{1/6}$.

This is, therefore, the *point of stable equilibrium* and we denoted this value of x by x_o.

At values of x less than x_o, *i.e.*, between x = 0 and $x_o = (2a/b)^{1/6}$, F will be *positive*, indicating that the force is one of repulsion or that the atoms repel each other, and that at values of x higher than x_o *i.e.*, between $x_o = (2a/b)^{2/6}$ and ∞ F will be *negative*, indicating that the force is one of attraction or that the

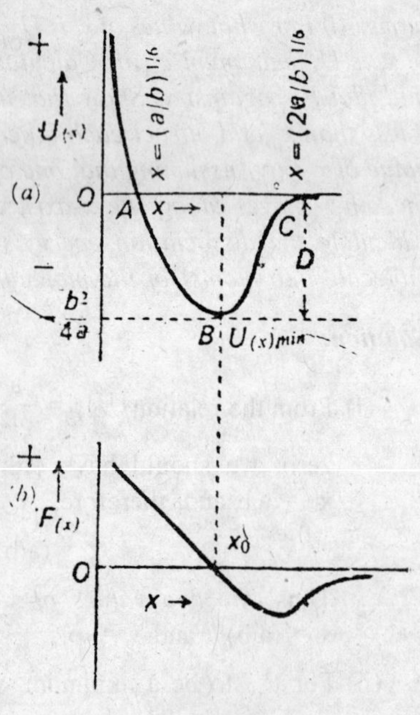

Fig. 4.14

atoms now attract each other.

A graph plotted between $F_{(x)}$ and x comes out to be of the form shown in Fig. 5.14 (b) showing $F_{(x)} = 0$ at $x_o = (2a/b)^{1/6}$, positive between x = 0 and $x = x_o$ and negative between $x = x_o$ and $x = \infty$.

(iv) If one of the atoms remains stationary and the other moves along the x-axis, it will have its minimum potential energy, or stable equilibrium, position at $x_o = (2a/b)^{1/6}$ from the stationary atom and will therefore oscillate simple harmonically about this position or within the potential well ABC.

(v) Once one of the atoms of the molecule acquires sufficient energy to get over the potential well, it will cease to be bound to the other atom and the molecule will thus get broken or dissociated. This energy to be able to get over the potential well gives the *dissociation energy* D of molecule.

The value of D is thus the difference between the potential energies at the top and the bottom of the potential well ABC, *i.e.*, between 0 and the minimum energy $- b^2/4a$ or equal to $0 - (- b^2/4a) = b^2/4a$.

Thus, *the dissociation energy of the molecule* $= b^2/4a$, meaning thereby that if the value of the kinetic energy at the equilibrium position be $b^2/4a$ (or more), the diatomic molecule will break up into two separate atoms.

Example 13:

The position of a moving panicle is at any instant given by r = A cos θi + A sin θj. Show that the force acting on it is a conservative one. Also calculate the total energy of the particle.

Solution:

The particle is obviously moving along a circular path of radius A. It must, therefore, be acted upon by a force directed towards the centre of its circular path, *i.e.*, the force acting upon it is a radial or a central one and hence necessarily *conservative.* However, this may also be show as follows:

The position vector of the particle at an instant t is

$$r = A\cos\theta\ i + A\sin\theta\ j = A\,(\cos\omega t i + \sin\omega t),$$

where ω is the angular velocity of the particle and $\omega t = \theta$.

The *linear velocity* of the particle at the given instant is

$$v = dr/dt = A\omega\,(-\sin\omega t\,i + \cos\omega t\,j)$$

and hence its *acceleration* $a = d^2r/dt^2 = -A\omega^2(\cos\omega t\,i + \sin\omega t\,j) = -\omega^2 r$,

∴ if m be the *mass* of the particle, the force acting on it is $F = -m\omega^2 r$.

Now, if this force (F) be a conservative one, we should have cur/ F = 0.

Let us see if this is so.

We have curl $F = \nabla \times F$

And, $\nabla \times F = -m\omega^2 \begin{vmatrix} i & j & k \\ \frac{\partial}{\partial x} & \frac{\partial}{\partial y} & \frac{\partial}{\partial z} \\ x & y & 0 \end{vmatrix}$, which is clearly equal to 0.

The *force acting on the particale is, therefore, a conservative one.*

Now, since F is a conservative force, we have $F = -grad\ U$, where U is the P.E. of the particle at the given instant.

We, therefore, have

$$F = -grad\ U = -\left(\frac{\partial U}{\partial x}i + \frac{\partial U}{\partial y}j + \frac{\partial U}{\partial z}k\right) = -\frac{dU}{dr}.$$

$$\therefore\ U = -\int F.dx = -\int -m\omega^2 r.dr = 1/2 m\omega^2 r.$$

And if v be the linear velocity of the particle at the given instant, we have

K.E. *of the particle,* $T = 1/2\ mv^2 = 1/2\ m\omega^2 r^2$ [∴ $v = r\omega$.

∴ *total energy of the particle,* $E = T + U = 1/2 m\omega^2 r^2 + 1/2 m\omega^2 r^2 = m\omega^2 r^2$.

Or, $E = m\omega^2 (A^2\cos^2\omega t + A^2\sin^2\omega t) = m\omega^2 A$.

Thus, the *total energy of the particle* $= m\omega^2 A$.

Example 14:

Calculate the distance of closest approach of (i) two protons, each of energy 500 MeV, approaching each other from opposite directions, (ii) a proton of energy V electron volts approaching a nucleus of atomic number Z.

Solution:

(i) As the two protons approach each other, they have to oppose the electrostatic force of repulsion between them and their kinetic energy is gradually converted into potential energy. Their distance of closest approach will, therefore, be that at which the whole of their kinetic energy is so converted into potential energy. Let it be r cm.

Then, since e is the charge on each proton, the potential energy of the two protons at distance r will be (e) (e)/r = e^2/r. This must be equal to their total kinetic energy, 2×500 MeV = $2 \times 500 \times 10^6$ eV = $2 \times 500 \times 10^6 \times 1.6 \times 10^{-12} = 1.6 \times 10^{-3}$ ergs.

Thus, $e^2/r = 1.6 \times 10^{-3}$,

whence, $r = e^2/1.6 \times 10^{-3} = (4.8 \times 10^{-10})^2/1.6 \times 10^{-3}$

$= 1.44 \times 10^{-16}$ cm.

Or, *distance of closest approach of the two protons*$=1.44\times10^{-16}$cm.

Or, distance of closest approach of the two protons = 1.44×10^{-16} cm.

(ii) Here, too, the proton will approach the comparatively heavier nucleus until the whole of its K.E. is converted into P.E.

Now, charge on the proton is e and that on the nucleus, Ze. So that, if r cm be the distance of closest approach, the P.E. of the proton and the nucleus = (Ze) (e)/r = Ze^2/r. This must be equal to the K.E. of the proton, given to be V electron volts = $V \times 1.6 \times 10^{-12}$ ergs, *i.e.*,

$$Ze^2/r = V \times 1.60 \times 10^{-12},$$

whence , $r = Ze^2/V \times 1.60 \times 10^{-12}$.

Or, *distance of closest approach,*

$$r = \frac{Z(4.8 \times 10^{-10})^2}{V \times 1.60 \times 10^{-12}} = \frac{14.4Z}{V} \times 10^{-8}.\text{cm}$$

$$= \frac{14.4Z}{V}\text{A.U.}$$

Example 15:

An electron is moving in a circular orbit of radius 2×10^{-8} cm about a proton supposed to be at rest. Calculate the velocity of the electron as also its kinetic and potential energies.

How much energy is required to ionise the system (i.e., to remove the electron to an infinite distance from the proton, so as to have zero kinetic energy)? What is the value of the ionization potential?

Solution:

For the electron to be moving in the circular orbit around the proton, the centrifugal force acting on it must just be balanced by the force of attraction between them. We, therefore, have $mv^2/r = (e)(e)/r^2$, whence, $v = \sqrt{e^2/mr}$, where m is the mass of the electron; v, its velocity; r *the radius of the orbit* and e, the charge on both the electron and the proton.

Substituting the values of e, m and r, therefore, we have *velocity of the electron,*

$$v = \sqrt{\frac{(4.8\times10^{-10})^2}{9.1\times10^{-28}\times2\times10^{-8}}} = \frac{4.8\times10^{-10}}{(18.2\times10^{-36})^{\frac{1}{2}}}$$

$$= 11.26 \times 10^7 \text{ cm/sec.}$$

$\therefore$ *kinetic energy of the electron*

$= 1/2mv^2 = 1/2 \times 9.1 \times 10^{-28} \times (11.26 \times 10^7)^2$

$$= 5.76 \times 10^{-12} \text{ ergs} = \frac{5.76\times10^{-12}}{1.6\times10^{-12}} = 3.6\text{eV}.$$

And, *potential energy of the electron*

$$\frac{(-e)(e)}{r} = \frac{-e^2}{r} = -\frac{(4.8\times10^{-10})^2}{2\times10^{-8}}$$

$$= -11.5 \times 10^{-12} \text{ ergs} = -\frac{11.5\times10^{-12}}{1.6\times10^{-12}} = -7.2 \text{ eV}.$$

Clearly, work done in removing the electron to an distance from the proton, *i.e.*, its *ionisation energy* = *total energy of the electron at infinity–its total energy at the given distance* r *from the proton, i.e.*, ionisation energy = 0 – (K.E. *of electron at distance* r *from the proton* + P.E. *of electron at distance* r *from proton)* = 0 – (3.6 – 7.2) = 3.6 eV = $3.6 \times 1.6 \times 10^{-12} = 5.76 \times 10^{-12}$ ergs.

Now, *ionisation potential* is the potential through which the electron must be accelerated to acquire its ionisation energy.

Since the ionisation energy of the electron here is 3.6 *electron volts,* it must have been accelerated through a potential of 3.6/1 = 3.6 *volts.*

$\therefore$ ionisation potential *of the electron* = 3.6 volts.

EXERCISES

1. (a) A particle at a point (2, 1, 3) is acted upon by a force F = 3i + 2xyj + xz^2k. Calculate the work done in moving it to a point (4, 1, 3).

   (b) A particle moves along one quarter of the circumference of a circle of radius 0.5 m. If the force applied be 0.2 N, inclined at an angle of 60° with the tangent to the circle at the point, calculate the work done.

   Ans. (a) 12 units; (b) 7.852×10^{-2} joule.

2. Investigate in two different ways whether the force F = (y^2 – x^2)i + 3xyj is a conservative one. **Ans.** non-conservative.

   [**Hint:** (i) Show that curl F ≠ 0. (ii) Evaluate the line integral from the point (0,0) to the point (x_o, y_o) along one path, made up of two straight sections (0,0) to (x_o, 0) and (x_o, 0) to (x_o, y_o), as shown in Fig. 5.15, and then along the other path made up of two similar straight sections (0, 0) to (0, y_o) and (0, y_o) to (x_o, y_o), as shown dotted in the Figure. Show that work done along the two paths is not the same.]

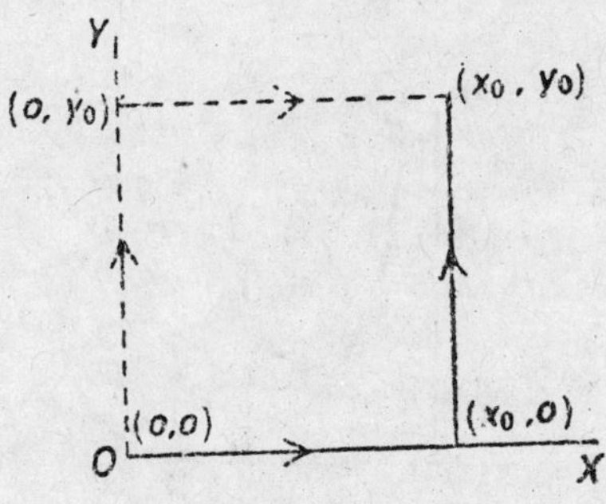

Fig. 4.15

3. (a) If the magnitude of the force of attraction between two particles of masses m_1 and m_2 respectively, and distance x apart, be given by F = k m_1m_2/x^2, where k is a *constant* , find (i) the potential energy function and (ii) work required to be done to increase the separation of the masses from x = x_1 to x = x_1 + d.

   Ans. (i) $U_{(x)}$ = –k m_1m_2/x, if V (∞) 0 = 0
   (ii) k m_1m_2 d/x_1 (d + x_1).

(b) Enunciate the principle of conservation of mechanical energy and illustrate by means of a homely example.

4. A particle of mass m is attached to a fixed point by means of a string of length l and hangs freely. Show that if it is projected horizontally with a velocity greater than $\sqrt{5gl}$ it will completely describe a vertical circle.

5. Show that for a given initial speed u, a projectile will have the same speed at the same elevation irrespective of its angle of projection.

6. A small block of mass m slides along the frictionless loop-the-loop track shown in Fig. 4.16. Calculate the height h at which the mass must be released on the track to be able to go round the loop of radius R. **Ans.** h = 5R/2.

[**Hint:** If v be the velocity acquired by the block in falling through height h, we have $v^2 - 0 = 2gh$. Or, $v^2 = 2gh$. As the mass reaches the top of the loops, let its velocity be v', so that centrifugal force exerted by the mass on the track upwards = mv'^2/R. In order that the block may be able to go round the loop, this must balance the weight mg of the block. So that, $mv'^2/R = mg$, whence, $v'^2 = gr$.

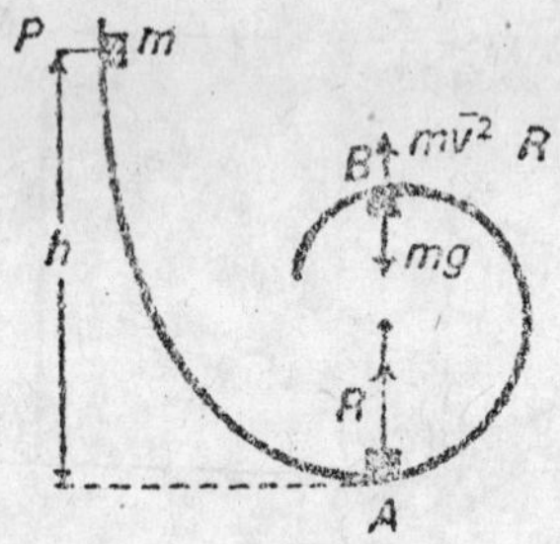

Fig. 4.16

Then, $1/2mv^2 - 1/2mv'^2$ = work done by the block in covering a vertical distance 2R against gravity from A to B, *i.e.*, 1.2m $(v^2 - v'^2) = 2mg$ R. Or, $v^2 - v'^2 = 4gR$.

Substituting the values of v and v'. we have $2gh - gR = 4gR$.

Or, $2h = 5R$, whence, $h = 5R/2$.]

7. A light rigid rod of length/ has a mass attached to its end, forming a simple pendulum. It is inverted and then released. What is its speed v at the lowest point and what is the tension T in the suspension at that instant? **Ans.** $2\sqrt{gl}$; 5mg.

8. What is the energy (in electron volts and in ergs) acquired by an alpha particle accelerated through a potential difference of 10^3 volts and what is then its velocity ? (Charge on an alpha particle = + 2e and its mass = 4 times the mass of a proton).

   Ans. 2000 eV or 3.2×10^{-9} ergs; 3.09×10^7 cm/sec.

9. (a) What is the electrostatic potential in volts due to an atomic nucleus of atomic number 10 at a distance of 2 Angstrom units from it?

   (b) What will be the energy of (i) an electron (ii) a proton, placed at that point ? (Charge on the nucleus = + 10e; 1 Angstrom unit = 10^{-8} cm.)

   Ans. (a) 72 volts; (b) (i) – 72 eV (ii) + 72 eV.

10. (a) Calculate (i) the intensity of the electric field (in volts/cm), (ii) the potential (in volts) at a distance of 1 A.U. from a proton. (Charge on a proton = + e = 4.80×10^{-10} esu; 1 esu of potential = 300 volts; 1 A.U. = 10^{-8} cm.)

    (b) A proton is released from rest at a distance of 1 A. U from another proton. What is the kinetic energy when the protons have moved infinitely apart from each other ? If one proton be kept at rest, what is the terminal velocity acquired by the other? (Mass of proton = 1.67×10^{-24} gm.)

    Ans. (a) (i) 1.44×10^9 volts/cm,
    (ii) 4.8×10^{-2} esu or 14.4 volts.
    (b) 23×10^{-12} ergs; 5.25×10^6cm/sec.

11. (a) Obtain the distance of closest approach of a particle of mass m and charge q, approaching with velocity v a massive particle carrying a charge Q and supposed to remain stationary.

    Ans. $2Qq/mv^2$.

    (b) Alpha particles of speed 2.0×10^9 cm/sec are shot into a thin film of gold (Z = 79). Deduce the distance of nearest approach between the a-particles and nuclei of gold atoms, assuming Coulomb's law to hold.

12. A mass of 1 kg is supported by a spring of force constant 10^5 dynes.cm. It is pulled down through a distance of 20 cm and then released. Calculate its highest velocity. **Ans.** 200 cm/sec.

13. A spring which does not conform to Hook's law obeys the force law $F = -Kx^2$. Calculate the potential energy at x, referred to U = 0 at x = 0. Also calculate the work done in stretching the spring from $x = x_1$ to $x = x_2$.

    Ans. $1/4\ Kx^4$; $1/2\ K(x_2^4 - x_1^4)$.

14. An object is attached to a vertical spring and slowly lowered to its equilibrium position. This stretches the spring by an amount d. If the same object is attached to the same vertical spring but permitted to fall instead, through what distance does it stretch the spring? **Ans.** 2d.

15. Show that a mass M falling freely from a vertical height h on a vertical spring, of force constant C, would produce a maximum compression y_o in the spring, where

$$y_o = \frac{mg}{C} + \frac{1}{C}\sqrt{m^2g^2 + 2mghC}.$$

16. What is the potential energy curve of a particle ? What significant information does it give about the behaviour of the particle ?

17. What is meant by a potential well (or a potential valley) ? Show that for small oscillations of a particle, the well of (or the valley) is parabolic in shape. What special significance attaches to this particular shape of the well (or the valley) ?

18. The energy of a particle is given by $U = A - \frac{B}{x} + \frac{C}{x^2}$; where A, B and C are all positive constants. What is the position of stable equilibrium of the particle? What is the force constant for small oscillations of the particle about this position ? For what values of the total energy can the motion of the particle be bounded (*i.e.*, for what values of the total energy will it be inside a potential well)?

    Ans. 2C/B; $B^4/8C^8$; for total energy $E < A$.

19. The accompanying figure (Fig. 4.17) shows the potential energy curve of a mass of 0.25 kg, with r in metres and U in joules. Supply the following information from the curve:

(i) The positions of stable equilibrium of the particle about which it can oscillate.

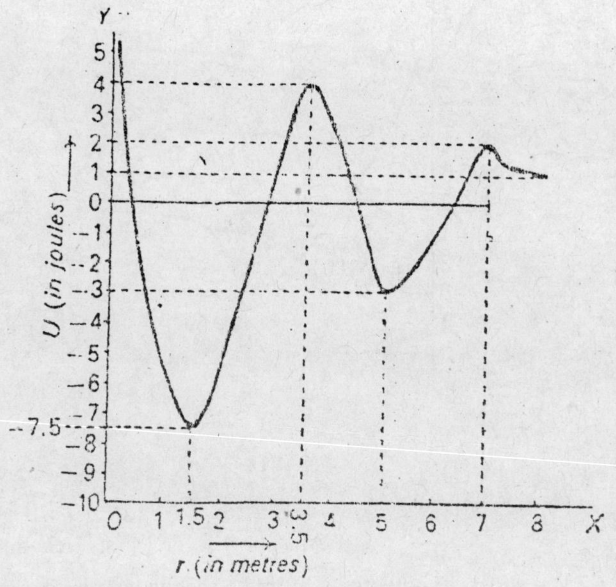

Fig. 4.17

(ii) The maximum energy of the oscillation in each case.

(iii) The kinetic energy with which the particle goes to $r \to \infty$ on just escaping from the potential well at the larger value of r and its velocity then.

(iv) The maximum total energy that would make the motion of the particle unbounded.

Ans. (i) r = 1.5 m and r = 50 m; (ii) 11.5 joules and 7 joules; (iii) 2 joules: 4 m/sec; (iv) 4 joules.

20. In Fig. 4.18, we have two potential energy curves of two diatomic molecules, with the interatomic distance (r) in *Angstrom units* and potential energy U in *electron volts*. What are the intra-atomic distances for equilibrium and the binding energies in the two cases?

What does the narrower potential well in the second case signify?

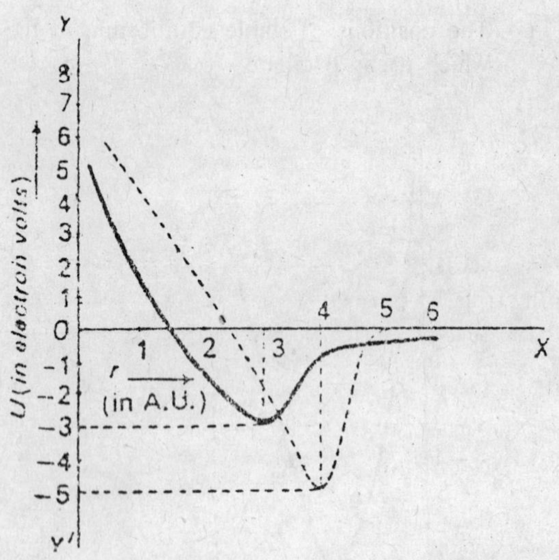

Fig. 4.18

Ans. 3 A.U or 3×10^{-8} cm and 4 A.U. or 4×10^{-8} cm; 3 eV and 5 eV.

The narrower potential will in the second case indicates a more rapid change of gradient dU/dr and hence a larger value of the force constant.

21. (a) What is meant by a conservative force ? Show that if force between two bodies is of a central kind, it is also conservative.

    (b) Show that in the case of a conservative force, the work done around a closed path is zero.

    (c) Show that Lorentz force is a conservative force.

22. Show that for a conservative force field F (r) we can define a scalar function U(r) such that $F(r) = -\nabla U(r)$.

    Also show that curl F(r) = 0.

23. State and explain the work-energy principle and write a short note on the utility of conservation laws in general.

24. Give a homely example or two of a conservative force and show that the force acting on a particle whose position at any instant t is given by $r = a \cos \omega t \, i + b \sin \omega t \, j$ is a conservative force.

25. (a) A centripetal force of 18 N is used to keep a 2 kg ball in a uniform circular motion at the end of a string 1 m long. How much work does the force do in each revolution of the ball ?

    (b) What is the work done by the earth's gravitational force of attraction on a satellite moving in a circular orbit around it?

Ans. (a) 0; (b) 0.

26. (a) A proton is released from rest at a distance of 1 A. U from another proton. What is the kinetic energy when the protons have moved infinitely apart from each other ? If one proton be kept at rest, what is the terminal velocity acquired by the other? (Mass of proton = 1.67×10^{-24} gm.)

    (b) Calculate (i) the intensity of the electric field (in volts/cm), (ii) the potential (in volts) at a distance of 1 A.U. from a proton. (Charge on a proton = + e = 4.80×10^{-10} esu; 1 esu of potential = 300 volts; 1 A.U. = 10^{-8} cm.)

5

Conservation of Laws *(Continued)*

LAWS OF CONSERVATION OF MOMENTUM

Despite its almost all-embracing character, the law of conservation of energy fails to provide complete or satisfactory solutions to a host of problems involving interaction of bodies. To take the familiar example of the firing of a rifle, all that the law tells us is that the kinetic energies of the forward moving bullet and the backward moving or recoiling rifle, along with the heat and sound energies liberated, must together be equal to the chemical energy of the explosive detonated in the rifle. It does not tell us how this total energy is distributed among the bullet, the rifle and the atmosphere. And further, because energy is a scalar quantity, the fact of its remaining conserved does not even suggest that the bullet and the rifle must move in opposite directions.

For the proper solution of many a problem in Dynamics, such as the one mentioned above, in regard to which we have no knowledge of the active forces actually involved, some additional principle of a vector *nature* is, therefore, obviously needed and the law of conservation of momentum is the answer.

As we have seen earlier, the momentum of a particle is a vector p, defined as the product of its mass and velocity. Being the product of a vector and a scalar, it is itself a *vector*. More correctly, it must be called *linear momentum* to distinguish it from *angular momentum*, a related but different quantity with which we shall deal a little later.

Thus, linear momentum p = mass × velocity = mv.

Since in accordance with Newton's second law,

$$F = dp/dt = d\,(my)/dt,$$

we have $F\,dt = d\,(mv)$.

integrating which with respect to t, we have $\int_{t_1}^{t_2} F\,dt = = \int_{mv_1}^{mv_2} d(mv) =$

$mv_2 - mv_1$, where $(mv_2 - mv_1)$ is the change in the momentum of the particle in the interval of time from t_1 to t_2 and $\int_{t_1}^{t_2} F\,dt$, dt, the impulse of the force during the *same* interval (the integral being taken because the force may vary during the time interval). So that, we have

Impulse = change in momentum.

Now, in case the external force applied to a particle (or a body) be *zero*, we have

$$F = dp/dt = d\,(mv)/dt = 0,$$

whence, $p = mv =$ *a constant*,

showing that *in the absence of an external force, the momentum of the particle {or the body) remains constant.*

We have already discussed the *law of conservation of linear momentum* in the case of a system of two mutually interacting particles, so that, in the absence of an external force on the system,

$$p_1 + p_2 \text{ or } m_1v_1 + m_2v_2 = \text{constant},$$

i.e., the total momentum of the particles remains conserved.

It would be interesting to give here an example to show that the law of conservation of momentum enables us to solve a problem which cannot be solved by a straight application of the relation $F = ma$.

Suppose a particle of mass m, initially at rest, suddenly explodes into two fragments of masses m_1 and m_2 which fly apart with velocities v_1 and v_2 respectively. Obviously, the forces resulting in the explosion of the particle must be internal forces since no external force has been applied. In the absence of the external force, therefore, the momentum must remain conserved and we should have $mv = m_1v_1 + m_2v_2$.

Since the particle was initially at rest, $v = 0$ and, therefore, also $m_1v_1 + m_2v_2 = 0$,

whence $v_1/v_2 = -\,m_2/m_1$,

showing at once that the velocities of the two fragments must be inversely proportional to their masses and *in apposite directions along the same line*. This result could not possibly be arrived at from the relation F=ma, since we know nothing about the forces that were acting during the explosion. Nor could we derive it from the law of conservation of energy, as already pointed out.

The law may be easily extended to any number of particles, either in fixed relative positions so as to form a rigid body or in the Form of a loose conglomeration and thus being free to move into all sorts of different positions.

Let m_1, m_2 ..... m_n be the masses of the different particles and, in addition to their own interaction with each other, let there also be external forces acting on them, so that they acquire velocities v_1, v_2 ....v_n respectively. Then, clearly, their total momentum P, say, is the vector sum of their individual momenta p_1, p_2, p_3 ..... etc., *i.e.*,

$$P = P_1 + P_2 + p_3 + ... + p_n = m_1v_1 + m_2v_2 + m_3v_3 + ... m_nv_n.$$

Differentiating with respect to t, we have

$$\frac{dp}{dt} = \frac{dp_1}{dt} + \frac{dp_2}{dt} + ... \frac{dp_n}{dt} = \frac{d}{dt}(p_1 + p_2 + p_3 + ... p_n),$$

i.e., $F = F_1 + F_2 + ...F_n$,

where F_1, F_2, F_3......F_n are the forces acting on the panicles of masses m_1,m_2......m_3, respectively

Now, the internal forces alone cannot bring about any changes in the momentum of the body since, forming pairs of equal and opposite forces, they give rise to equal and opposite changes of momentum which cancel out. The above forces F_1, F_2..... F_n thus actually represent only the external forces acting on the panicles and F, their resultant: So that if F = 0, we have dp/dt = 0 and, therefore, $P = p_1 + p_2 + p_3 + ... p_n$, a constant, implying that even though the individual values of p_1, p_2 etc. may change, *their total sum* P remains unaltered. This is the *law of conservation of linear momentum* and may formally be stated thus:

When the vector sum of the external forces acting upon a system of particles equals zero, the total linear momentum of the system remains constant.

The conservation of linear momentum of a system is *a direct* consequence of the translational invariance of the potential energy of *the system* which, as we know is one of the results of Galilian invariance.

Thus, considering, for the sake of simplicity, the case of two particles in one dimension, say, along the axis of x, such that their coordinates are x_1 and x_2 we have their potential energy U (x_1, x_2), depending upon their positions.

Now, Galilian invariance demands that this P.E. should remain constant even if each particle is displaced through the *same* distance d, *i.e.*, the potential energy $U[(x_2 + d), (x_2 + d)]$ after the displacement must be the same as before the displacement, viz, $U(x_1, x_2)$. This means that if $U(x_1, x_2) = (x_2 - x_2)^2$, we should also have $U[(x_1 + d), (x_2 + d)\}$ equal to $(x_1 - x_2)^2$. Let us see if this is so.

We have $U[(x_1 - d), (x_2 - d)] = [(x_1 - d) - (x_2 + d)] = (x_1 - x_2)^2$

$= U(x_1, x_2)$, clearly showing that *the potential energy is independent of the displacement d suffered by each particle.*

In general, therefore, if the potential energy $U(r_1, r_2)$ is a function of only $(r_1 - r_2)$ it is translationally invariant.

Since the force on a particle is the negative of $\partial u/\partial x$ have

$$F_1 = \partial u/\partial x_1$$

and $$F_2 = \partial u/\partial x_2$$

But being a function of $(x_1 - x_2)$ only, we have

$$\frac{\partial U}{\partial x_1} = \frac{\partial U}{\partial x}\frac{\partial x}{\partial x_1} = \frac{\partial U}{\partial x}$$

and $$\frac{\partial U}{\partial x_2} = \frac{\partial U}{\partial x}\frac{\partial x}{\partial x_2} = -\frac{\partial U}{\partial x}$$

And, therefore, $$\frac{\partial U}{\partial x_1} = -\frac{\partial U}{\partial x_2}$$

or $$F_1 = -F_2$$

So that, the total force acting on the particles which interact with each other is $F = F_1 + F_2 = 0$

The total force acting on the system of the two particles thus being two, we have, by Newton's second law,

$$\frac{d}{dt}(m_1v_1 - m_2v_2) = 0.$$

or $$m_1v_1 + m_2v_2) = \text{constant},$$

i.e., the total momentum of the two particles remains conserved.

The same argument can be extended to any number of particles, the one and only condition being that the potential energy of the system must be translationally invariant.

The law of conservation of linear momentum is one of the fundamental conservation laws and is also applicable to atomic and nuclear physics even though 'Newtonian mechanics is not. It is applicable even at relativistic velocities, provided we take the relativistic value of the mass of the particles, given by $m = m_0 \sqrt{1 - v^2/c^2}$.

CENTRE OF MASS

When we consider the motion -of a system consisting of a large number of particles, say, N, there is one point in it which behaves as though the entire mass of the system (*i.e.*, the sum of the masses of all the N individual particles) were concentrated there and its motion is the same as would ensue if the external forces acting on the system were applied directly to it. This point is called the *centre* of mass of the system. As we shall see as we proceed along, this concept of centre of mass has proved most useful in tackling many is problem and, m particular, those concerned with collision of particles

Taking first the simple case of a system of just two particles of masses m_1 and m_2 and of a total mass M, with r_1 and r_2 as their position vectors with respect to some origin, if R be the position vector of the centre of mass, we have

$$R = \frac{m_1 r_1 + m_2 r_2}{m_1 + m_2} = \frac{m_1 r_1 + m_2 r_2}{M}$$

or $$MR = m_1 r_1 + m_2 r_2,$$

i.e., the product of the total mass of the system and the position vector of the centre of mass (c.m.) is equal to the sum of the products of the individual masses and their respective position vectors.

The position of the c.m. of the system may thus be easily obtained.

To take a general case, if we have a system consisting of N particles, of masses $m_1, m_2, \ldots\ldots m_n$, with $r_1, r_2 \ldots\ldots r_n$ as their position vectors at a given instant, the position vector R of the centre of mass of the system at that instant is given by the relation $MR = m_1 r_1 + m_2 r_2 + \ldots\ldots m_n r_n$, where M is the *total mass of the system.*

Clearly, $M = m_1 + m_2 + \ldots\ldots m_n = \sum_{k=1}^{N} m_k$, where mn is the mass of the k_{th} particle. If, therefore, r_k be the position vector of the k_{th} particle, we have

$$R = \frac{m_1 r_1 + m_2 r_2 + \ldots . m_N r_N}{m_1 + m_2 + \ldots . m_N} = \frac{\sum_{k=1}^{N} m_k r_k}{\sum_{k=1}^{N} m_k} = \frac{\sum_{k=1}^{N} m_k r_k}{M},$$

where $\Sigma m_k r_k$ is called the *first moment of mass for the system.*

Now, $r_k = r_k i + y_k j + z_k k$ and $R = xi + yj + zk$. So that, if X, Y and Z be the Cartesian coordinates of the centre of mass, we have

$$X = \frac{m_1 x_1 + m_2 x_2 + \ldots . m_N x_N}{m_1 + m_2 + \ldots . m_N} = \frac{\sum_{k=1}^{N} m_k x_k}{\sum_{k=1}^{N} m_k} = \frac{\sum_{k=1}^{N} m_k x_k}{M},$$

or $$X = \frac{1}{M} \sum_{k=1}^{N} m_k x_k,$$

$$Y = \frac{m_1 y_1 + m_2 y_2 + \ldots . m_N y_N}{m_1 + m_2 + \ldots . m_N} = \frac{\sum_{k=1}^{N} m_k y_k}{\sum_{k=1}^{N} m_k} = \frac{\sum_{k=1}^{N} m_k y_k}{M},$$

or $$Y = \frac{1}{M} \sum_{k=1}^{N} m_k y_k,$$

and $$Z = \frac{m_1 z_1 + m_2 z_2 + \ldots . m_N z_N}{m_1 + m_2 + \ldots . m_N} = \frac{\sum_{k=1}^{N} m_k z_k}{\sum_{k=1}^{N} m_k} = \frac{\sum_{k=1}^{N} m_k z_k}{M},$$

$$Z = \frac{1}{M} \sum_{k=1}^{N} m_k z_k$$

Since a rigid body consists of a large number of particles, compactly packed together, it must also have a *centre* of mass. In view, however, of the large number of particles constituting it and the extremely small spacing between them, the distribution of mass may be deemed to be continuous throughout the body. The sign of summation used above is thus replaced by that of integration taken over the whole volume of the body.

If ρ be the density of the body at a point whose position vector is r, the mass of an element of volume dV there is clearly dm = ρdV. So' that, proceeding as above, if R be the position vector of the centre of mass of the body, we have

$$R = \frac{1}{M}\int rdm = \frac{\int v \; rdm}{\int v dm}$$

and its cartesian coordinates,

$$X = \frac{\int v \; xdm}{\int v dm} = \frac{1}{M}\int_v x \; dm = \frac{\int v \; x\rho dv}{\int v \; \rho dv},$$

$$Y = \frac{\int v \; ydm}{\int v dm} = \frac{1}{M}\int_v y \; dm = \frac{\int v \; y\rho dv}{\int v \; \rho dv},$$

and

$$Z = \frac{\int v \; zdm}{\int v dm} = \frac{1}{M}\int_v z \; dm = \frac{\int v \; z\rho dv}{\int v \; \rho dv},$$

For homogeneous bodies of regular geometrical shapes, the centre of mass obviously lies on the point or the line of symmetry as the case may be, as, for example, at the centre in the case of a sphere and on the axis in the case of a cone. In fact, it is located at a point for which $\int dmr = 0$ and consequently R = 0. This integral also, like the summation $\Sigma m_k r_k$ (mentioned earlier) is called the first moment of mass for the system.

MOTION OF THE CENTRE OF MASS

Let us consider the motion of a system of N particles of individual masses $m_1, m_2....m_n$ and total mass M, *it being assumed that no mass enters or leaves the system during its motion, so that M remains constant, i.e., leaving out, for the moment, cases like that of a rocket* where the mass goes on decreasing (due to expulsion of the hot gases) throughout its motion. Then, as we have seen, we have the relation

$$MR = m_1 r_1 + m_2 m_2 + ... m_N r_N,$$

where R is the position vector of the centre of mass and $r_1, r_1, ...r_N$, the position vectors of the individual particles.

Differentiating this expression with respect to t, we have

$$MdR/dt = m_1 dr_1/dt = m_2 dr_2/dt + ...m_N d_{rN}/dt \qquad ...(i)$$

Clearly dR/dt = V, the *velocity of the centre of mass and* dr_1/dt, dr_2/dt ....d_{rN}/dt, the velocities v_1, v_2...v_N of the individual particles. So that,

$$MV = m_1v_1 + m_2v_2 + ...m_Nv_N = \Sigma m_kv_k = P, \qquad ...(ii)$$

indicating that *the total linear momentum of the system is equal to the product of the total mass of the particles and the velocity of the centre of mass of the system.*

It follows, therefore, that the velocity of the centre of mass, *i.e.*

$$V = \Sigma m_kv_k/M.$$

Since, in the absence of any external force, the momentum remains conserved, we have MV = a constant

or V = a constant, [M being constant.

i.e., the velocity of the centre of mass of a system remains constant if no external force be applied to it.

This is a remarkable property indeed, seeing that during the motion of a system, the relative positions of the particles with respect to each other may be changing in a complex or complicated manner. Thus, for example, in the case of a decaying radioactive nucleus in flight, the particles of decay may move with different-velocities in different directions but the velocity of then-centre of mass remains unaltered.

Now, differentiating expression (ii) with respect to r, we have

$$MdV/dt = m_1dv_1/dt + m_2dv_2/dt + ...m_Ndv_N/dt,$$

$$= m_1a_1 + m_2a_2 + ...m_Na_N,$$

where a_1, a_2......a_N are the accelerations of the individual panicles of the system.

In accordance with Newton's second law of motion, $m_1a_1 = F_1$, $m_2a_2 = F_2$.....m_Na_N.....F_N, the threes acting on the different particles.

Since $dV/dt = a_{c.m.}$ the Acceleration of the centre of mass, we have

$$Ma_{c.m.} = F_1 + F_2 + ...F_N,$$

i.e., the product of the total mass of all the particles and the acceleration of the centre of mass of the system is equal to the vector sum of all the forces acting on the individual particles of the system.

All these forces must obviously be *external forces* since, as we know any internal forces only exist in pairs of equal and opposite forces cancelling each other out and producing no effect on the system. We can, therefore, write the above as

$$Ma_{c.m.} = F_{ext.}$$

$$\text{And } \therefore \quad F_{est} = MdV/dt = dP/dt = Ma_{c.m.},$$

where F_{ext} is the vector sum of all the external forces applied to the system.

Thus, as pointed out earlier, *the centre of mass of a system of particles moves as though it were a particle of mass equal to that of the whole system, with all the external forces acting directly on it.*

CENTRE OF MASS FRAME OF REFERENCE

A frame of reference carried by the centre of mass of an *isolated system* of particles (*i.e.*, a *system not subjected to any external forces*) is called the *centre-of-mass* or the *C-frame of reference*. In this frame, obviously, R (the position vector of the centre of mass) is equal to zero and hence the velocity of the centre of mass, *i.e.*, V = dR. dt also equal to zero. In consequence, the *linear momentum* P of the system too, given by $P = MV = \Sigma m_k v_k$, is zero. The C-frame of reference is, therefore, also called the *zero-momentum frame*. And since in the absence of external forces, the centre of mass, and hence the C-frame, moves with constant velocity, it is also an *inertial frame*. This makes it particularly useful in solving many a problem which are difficult to solve in the laboratory frame of reference, as we shall presently see. Indeed, it has become almost customary to deal with all collisions in Nuclear Physics in this frame of reference alone.

COLLISION

Contrary to the meaning of the term 'collision' in our everyday partance, in Physics it does not necessarily mean one particle (or body) '*striking*' against another. Indeed, two particles may not even 'touch' each other and may still be said to collide. All that is implied is that as the particles approach each other, (i) *a large force far a relatively short time (i.e., an impulse) acts on each colliding particle, (ii) the motion of the particles (or, at feast, of one of the particles) is chatted rather abruptly and (iii) the total momentum (as also the total energy) of the panicles remains conserved.*

The essence of a collision is, in fact, a *redistribution of the total momentum of the particles*. This is the reason why the how of conservation of momentum is simply indispensable in dealing with the phenomenon of collision between particles.

Now, there are two limiting cases of collision:

(i) *A perfectly elastic collision,* in which the kinetic energy of the particles is fully conserved. Some of the collisions between atomic, nuclear and fundamental particles belong to this class.

In such a collision between two particles, therefore, we have $m_1u_1 + m_2u_2 = m_1v_1 + m_2v_2$ as also $1/2m_1u_1^2 + 1/2m_2u_2^2 = 1/2m_1v_1^2 + 1/2m_2v_2^2$, where m_1 and m_2 are the respective masses of the two particles and u_1, u_2 and v_1, v_2 their velocities *before* and *after* the collision, *i.e.*, initially, before the interacting forces between the particles have yet started acting and finally, after these forces have again been reduced effectively to zero.

(ii) *A perfectly inelastic collision, in which the particles stick permanently together on impact and the loss of kinetic energy is the maximum consistent with the requirements of conservation of momentum.* A bullet remaining embedded in a target is one such example.

In either case, however, since the interacting forces after the collision become effectively zero, *the potential energy of the system remains the same both before and after the collision.*

In the case of inelastic collision' between larger particles, the loss of kinetic energy occurs mostly in the form of heat energy due to the increased vibrations of the constituent atoms of the particles. On the other hand, in the case of inelastic atomic collisions, an atom may absorb some energy and get into what is called an *'excited state'*, with one or more of its electrons moving into a higher orbit or a higher energy level, the energy so absorbed (and remaining in the potential form) being referred to as *'excitation energy'* (ξ). The final kinetic energy of the system is thus less by this much amount than the initial one, *i.e.*, *final kinetic energy of the system* + ξ = *initial kinetic energy of the system* or, $1/2m_1v_1^2 + 1/2m_2v_2^2 + \xi \; 1/2m_1u_1^2 + 1/2m_2u_2^2$.

In case the atom be already in an excited state before collision, it may come into its normal state on collision, *i.e.*, the electron (or electrons) may fall back to the original orbit, thereby-releasing the potential energy ξ absorbed during excitation, resulting in an increase in the kinetic energy of the system after collision. Thus, in this case, we have *final K.E. of the system* $-\ \xi$ = *initial kinetic energy of the system.*

or $$1/2m_1v_1^2 + 1/2m_2v_2^2 - \xi = 1/2m_1u_1^2 + 1/2m_2u_2^2.$$

CALCULATION OF FINAL VELOCITIES OF COLLIDING PARTICLES

The velocities of the particles after a collision can be easily calculated by solving their equations of motion if we know their velocities before collision and the forces acting during the collision. Unfortunately, these interacting forces are not always known to us. Even so, we can obtain the velocities of the particles after collision from those before collision in case *(i) the collision w perfectly inelastic and (ii) the collision is elastic but one-dimensional, i.e.,* in which the relative motion of the particles before and after the collision is along the *same* line.

In case the elastic collision be two or three-dimensional, the velocities of the particles after collision can be determined only if we also know the direction of motion of the particles after the collision. (See case (iii) below).

(i) *Inelastic collision* : In the case of a perfectly inelastic collision, if m_1 and m_2 be the masses of the two colliding particles and u_1 and u_2 their velocities before collision, we have

momentum of the particles before collision $= m_1u_1 + m_2u_2$.

After the collision, the two particles stick together and move with a velocity v, say. So that,

momentum of the particles after collision $= (m_1 + m_2)v$.

Since momentum is always conserved in a collision, we have

$$(m_1 + m_2)v = m_1u_2,$$

whence $$v = \frac{m_1u_1 + m_2u_2}{m_1 + m_2}$$

Thus, knowing the masses of the two particles and their initial velocities, we can easily obtain their common velocity (v) after the collision.

In case the second particle be at rest, $u_2 = 0$. So that, we shall have $(m_1 + m_2)v = m_1u_1$, whence, $v = m_1u_1/(m_1 + m_2)$.

That there will be a loss of K E. may be easily seen from the fact that K.E. *of the particles before collision* = $1/2m_1u_1^2$

[the second particle being at rest, $u_2 = 0$],

and *K.E. of the particles after collision* = $1/2(m_1 + m_2)v^2$,

$$\therefore \frac{\text{K.E. after collision}}{\text{K.E. before collision}} = \frac{1/2\,(m_1 + m_2)v^2}{1/2\,m_1u_1^2}$$

$$= \frac{1}{2}(m_1 + m_2)\left(\frac{m_1u_1}{m_1 + m_2}\right)^2 \frac{1}{1/2\,m_1u_1^2}.$$

$$= \frac{m_1}{m_1 + m_2},\ \text{i.e.,} < 1,$$

showing that K.E. after collision < K.E. before collision.

As already pointed out, this loss of K.E. on collision manifests itself in the form of heat energy in the system due to increased vibrations of the atoms of the two particles.

Let us now consider the problem if the two particles be in the *centre-of-mass frame of reference* in which the centre of mass remains at rest and hence the momentum both before and after the collision is zero.

Since the velocity of the centre of mass in the laboratory frame of reference is $Y = m_1u_1/(m_1 + m_2)$ in accordance with relation it follows that the centre of mass frame moves this velocity (V) relative to the laboratory frame. The initial velocities of the two particles in this frame arc, therefore,

$$u'_1 = u_1 - V = u_1 - m_1u_1/(m_1 + m_2) = m_2u_2/(m_1 + m_2)$$

and $\quad u'_2 = u_2 - V = 0 - V = -m_1u_1/(m_1 + m_2)$

After the collision, the two particles stick together, their combined mass being $(m_1 + m_2)$ but they must beat rest since their momentum in this frame must remain zeros Relative to the laboratory frame, however the, velocity of the combined mass in the centre-of-mass frame must naturally be the same as that of the frame itself, viz., $V= m_1u_1/(m_1 + m_2)$ which, it will be readily seen, is the *same* as the velocity T of the combined mass in the laboratory frame.

(ii) *Elastic one-dimensional collision* : Again, if mi and mi be the masses of two particles, u_1, u_2 and v_1, v_2, their respective velocities before and after a *head-on collision along the line joining their centres,* we have *momentum equation,* $1/2m_1u_1^2 + 1/2m_2u_2 = m_1v_1 + m_2v_2$ ...(i)

and *K.E equation,* $1/2m_1u_1^2 + 1/2m_2u_2 = 1/2m_1v_1^2 + 1/2m_2v_2^2$ ...(ii)

Or, writing these two equations as

$$m_1(u_1 - v_1) = m_2(v_2 + u_2) \quad \text{...(iii)}$$

and $$m_1(u_1^2 - v_1^2) = m_2(v_2^2 + u_2^2) \quad \text{...(iv)}$$

and dividing the latter by the former, we have

$$u_1 + v_1 = v_2 + u_2.$$

or $$u_1 - u_1 = v_2 - v_2. \quad \text{...(v)}$$

showing that *in an elastic one-dimensional collision, the relative velocity with which we two panicles approach each other before collision is equal to the relative velocity with which they recede from each other after collision.*

As for the values of the velocities v_1 and v_2 of the particles after collision, we have from relation (v) above

$$v_2 = v_1 + u_1 - u_2$$

and $$v_1 = v_2 - u_1 + u_2$$

Substituting this value of v_2 in relation (iii) above and solving for v_1 and similarly, substituting the value of v_1 in (iii) above and solving for v_2, we have

$$v_1 = \left(\frac{m_1 - m_2}{m_1 + m_2}\right)u_1 + \left(\frac{2m_2}{m_1 + m_2}\right)u_2 \quad \text{...(vi)}$$

and $$v_2 = \left(\frac{2m_1}{m_1 + m_2}\right)u_1 + \left(\frac{m_2 - m_1}{m_1 + m_2}\right)u_2$$

Now, the following special cases at interest arise:

(a) *When the colliding particles hare the same mass :* In this case, $m_1 = m_2$. So that, from relations (vi) and (vii) we have $v_1 = v_2$ and $v_2 = u_1$, *i.e., in one-dimensional elastic collision of two particles of equal moss, the particles simply interchange their velocities on collision.*

(b) *If one of the particles be also initially at rest :* Let the second panicle be at rest. Then, $u_2 = 0$ and, therefore, from relations (vi) and (vii), we clearly have

$$v_1 = \left(\frac{m_1 - m_2}{m_1 + m_2}\right) u_1 \quad ...(viii)$$

and $$v_2 = \left(\frac{2m_1}{m_1 + m_2}\right) u_1 \quad ...(ix)$$

And, since $m_2 = m_2$,

we have $v_1 = 0$

and $v_2 = u_1$,

i.e., the first particle comes to rest on collision and the second particle acquires the initial velocity of the first.

(c) If the particle at rest be very much more massive than the other. Let the second panicle, at rest, be much more massive than the first, so that m_1 is negligible compared with m_1. Then, we have, from relations (viii) and (ix),

$$v_1 = -u_1 \text{ and } v_2 = 0$$

indicating that when a tighter particle collides against a much more sessile particle at rest the latter continues to remain at rest and the velocity of die former gets reversed.

A common example of this type is the dropping of a steel on an equally hard horizontal surface on the ground. The collision thus occurs between the ball and the earth. Since the latter is very much more massive and is at rest, the velocity of the ball gets reversed on impact and it rises to the sane height from which it was dropped.

(d) If the particle at rest be very much lighter than the other. Again, let the second panicle be at rest and very much lighter than the first, so that now m_2 is negligible compared with m_1. Then, we have, from relations (viii) and (ix) above,

$$v_1 = u_1 \text{ and } v_2 = 2u_1$$

i.e., the velocity of the massive particle remains practically unaltered on collision with the lighter particle at rest and the lighter particle acquires, nearly twice the initial velocity of the massive particle.

We can see from the above why in a reactor, hydrogen in used as a moderator for slowing down neutrons, produced by the fission of Uranium atoms, so as to enable them to produce more fissions. It is obviously because the hydrogen nucleus (*i.e.*, the proton) has nearly the same mass as a neutron and hence in a head-on cofission with a hydrogen nucleus at rest, the neutrons are thus almost brought to rest or greatly slowed down. If a massive nucleus like that of lead were to be used as the target, the neutrons will simply bounce back from it with the same velocity with which they impinge on it; and if the target were a lighter one, like an electron, for example, the neutrons will simply continue to move with their original velocity after collision with it. There are, of course, other factors too in the choice of a suitable moderator but from considerations of momentum and energy, a hydrogen nucleus as a stationary target is one of the best bets.

(iii) *Elastic collision in two or three dimensions* : In the case of a two or three-dimensional elastic collision between two particles, the initial velocities of the particles and the conservation laws of momentum and energy alone are not enough for the determination of the velocities of the particles after collision, because the velocity of each particle has three components but we have only four known relations between them, viz., three for the conservation of momentum for each of the three dimensions and one for the conservation of energy. More information is, therefore, needed to be able to calculate the final velocities of the particles and the simplest one is the angle of deflection of one of the particles.

Let us discuss the problem in both the laboratory frame of reference and the centre-of-mass frame of reference.

(a) *In the laboratory frame of reference* : Let a particle of mass m_1, moving with velocity u_1 collide with a particle of mass m_2, initially at rest (*i.e.*, $u_2 = 0$) in the laboratory frame of reference (Fig. 5.1). Let it be deflected or *scattered* at an angle θ_1 with its initial direction, after collision, with velocity v_1, where u_1 and v_1 lie in the X-Y plane. Obviously, then, the velocity v_2 of m_2, after collision, making an angle θ_2 with the original direction, will have no component in the z-direction for the simple reason that m_1u_1 and m_1v_1 have no z-components.

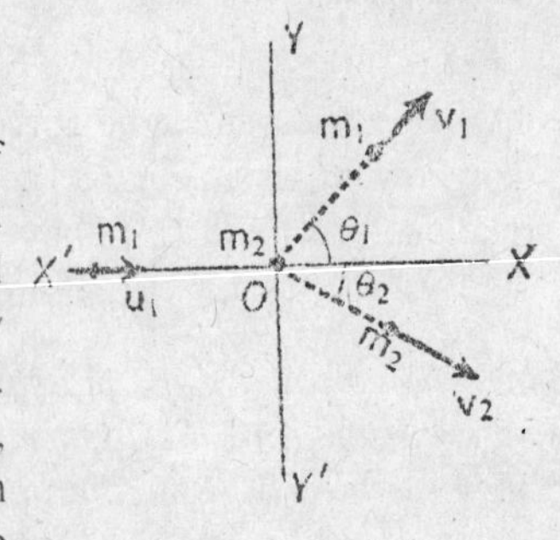

Fig. 5.1

Now, from the law of conservation of momentum, we have *for the x-component of the motion,*

$$m_1u_1 = m_1v_1 \cos \theta_1 + m_2v_2 \cos \theta_2 \qquad \text{... (i)}$$

and for the y component of the motion $0 = m_1v_1 \sin \theta_1 - m_2v_2 \sin \theta_2$...(ii)

And, from the law of conservation of energy, we have

$$1/2m_1u_1^2 = 1/2m_1v_1^2 + 1/2m_2v_2^2. \qquad \text{...(iii)}$$

If we know the initial conditions, viz., m_1, m_2 and u_1 we have to determine four unknown quantities, *i.e.*, v_1, v_2, θ_1 and θ_2 and have only three equations connecting them. We must, therefore, be given the value of one more quantity, such as θ_1, to be able to determine the motion after collision.

Solving these equations, then, Tor v_1, v_2 and θ_2, in the particular case when $m_1 = m_2$, *i.e.*, when the two masses are equal, we obtain $v_1 = u_1 \cos \theta_1$ and from relation (iii), therefore, $v_2^2 = u_1^2 - v_1^2$, whence v_2 can be obtained.

Finally, $\sin \theta_2 = v_1/v_2 \sin \theta_1$. So that, knowing v_1, v_2 and θ_1 (given), we can also determine θ_2, where the sum of θ_1 and θ_2 is always 90°, showing that the two equal masses move, after collision, in directions perpendicular to each other.

This method is somewhat lengthy and laborious. If, however, we view the problem in the centre-of-mass frame, its solution becomes very much simpler and perhaps also a little more instructive, as will be seen from the following.

(b) *In the centre-of-mass frame of reference :* As pointed out earlier under case (i) above, the centre-of-mass frame of reference moves with velocity $V = m_1u_1/(m_1 + m_2)$ relative to the laboratory frame of reference, so that the centre of mass remains at rest in this frame. The initial and final velocities of the two particles (m_1 and m_2) in this frame of reference are thus given by

$$u'_1 = u_1 - V,\ u'_2 = -V,\ v'_1 = v_1 - V \text{ and } v'_2 = v_2 - V.$$

Since the centre of mass throughout remains at rest in this reference frame, the total momentum is always zero and hence the momentum of the two particles must always be equal and opposite, clearly implying that the velocities v'_1 and v'_2 of the two panicles, after collision, must be oppositely directed and inclined at the same angle to the initial direction of motion of the particles, as shown in Fig. 5.2. We thus have

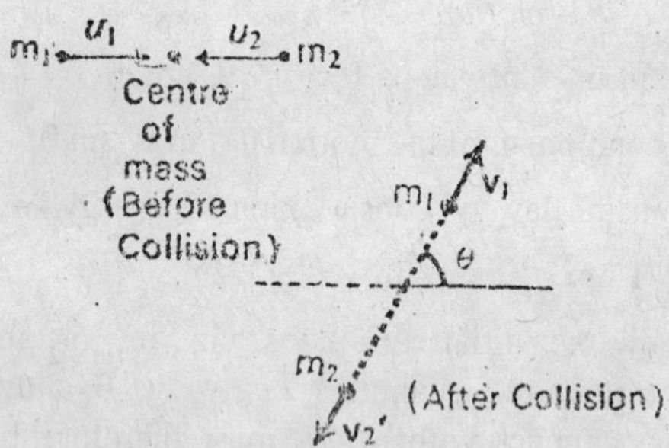

Fig. 5.2

$$m_1u'_1 = m_2u'_2 \quad \text{...(i)}$$

and $$m_1v'_1 = -m_2v'_2 \quad \text{...(ii)}$$

whence, $$v'_2 = \frac{m_1}{m_2}v'_1 \text{ and } u'_2 = \frac{m_1}{m_2}u'_1.$$

Since total energy remains conserved *on collision, we have

$$1/2m_1u_1'^2 + 1/2m_2u_2'^2 = 1/2m_1v_1'^2 + 1/2m_2v_2'^2 \quad \text{...(iii)}$$

Substituting the values of u'_2 and v'_2, obtained above, we have

$$1/2m_1u_1'^2 + 1/2m_2\left(\frac{m_1^2u_1'^2}{m_2^2}\right) + 1/2m_1v_1'^2 + 1/2m_2\left(\frac{m_1^2v_1'^2}{m_2^2}\right)$$

or $$1/2m_1u_1'^2\left(1+\frac{m_1}{m_2}\right) = 1/2m_1v_1'^2\left(1+\frac{m_1}{m_2}\right)$$

which gives $$1/2m_1u_1'^2 = 1/2m_1v_1'^2$$

or $$u'_1 = v'_1$$

And, hence from energy relation (iii) above, $u'_2 = v'_2$.

Thus, *in the centre-of-mass frame of reference, the magnitudes of the velocities of the particles remain unaltered m an elastic collision.*

VALUE OF THE SCATTERING ANGLE

(i) *In the centre-of-mass frame of reference* : In this frame of reference, there are absolutely no *limitations* on the value of the scattering angle θ, so that it can haw all possible values.

(ii) *In the laboratory frame of reference* : In this frame of reference, there are some restrictions 03 the value of the scattering angle, θ_1, (Fig. 5.1), as will be dear from the following:

We have $\tan \theta_1 = v_1 \sin \theta_1 / v_1 \cos \theta_1$.

Now, the y-component of the final velocity of the first particle being the same in the laboratory reference frame as well as the center-of-mass frame, we have $v_1 \sin \theta_1 = v'_1 \sin \theta$. And, since the x-components of this velocity differ by V in the two frames of reference, we have $v_1 \cos \theta_1 = v'_1 \cos \theta + V$. Substituting these values in the relation above, we, therefore, have

$$\tan \theta_1 = \frac{v_1' \sin \theta}{v_1' \cos\theta + V} = \frac{\sin\theta}{\cos\theta + (V/v_1')} \qquad \text{...(iv)}$$

As we know, $$V = \frac{m_1 u_1}{m_1 + m_2} = \frac{m_1 (u_1' + V)}{m_1 + m_2}$$

$$= \frac{m_1 u_1'}{m_1 + m_2} = \frac{m_1 V}{m_1 + m_2} \qquad [\because u_1 = u'_1 + V.$$

or $$V\left(1 - \frac{m_1}{m_1 + m_2}\right) = \frac{m_1 u_1'}{m_1 + m_2}$$

or $$V\left(\frac{m_2}{m_1 + m_2}\right) = \frac{m_1 u_1'}{m_1 + m_2}$$

whence, $$V \frac{m_1}{m_2} u_1' = \frac{m_2}{m_2} v_1' \qquad [\because u'_1 = v'_1.$$

Substituting this value of V in relation (iv) above, we have

$$\tan \theta_2 = \frac{\sin \theta}{\cos \theta + (m_1/m_2)} \qquad \text{...(v)}$$

This shows that

(a) If $m_1 > m_2$, so that m_1/m_2 a little > 1, the denominator can never be zero and hence $\tan \theta_1$ can never be ∞. And, therefore, θ_1 must be less than 90°;

(b) If $m_1 = m_2$, so that $m_1/m_2 = 1$, the denominator can be zero for $\cos \theta = -1$ and, therefore, $\tan \theta_1$ can be ∞. Thus, in this case, θ_1 can have any value up to the limiting value of 90°;

(c) If $m_1 < m_1$, so that $m_1/m_2 < 1$, the value of $\tan \theta_1$ can also be negative. In this case alone, therefore, all values of θ_1 can be possible.

If follows from case (a) that in an elastic collision, *if a massive particle collides against a lighter one at rest, it can never bounce back along its ordinal path.* On the other hand, it follows from case (c) that *if a lighter particle collides against a massive one at rest*, it may well *bounce back along its original path.* And, it follows from case (b) that in an elastic collision between two particles of equal mass, one of which is initially at rest, the two particles move at right angles to each other after the collision.

SYSTEMS OF VARIABLE MASS—THE ROCKET

In our discussion of the conservation of linear momentum, we have, so far dealt with systems whose mass remains constant. We now propose to consider those whose mass is *variable*, *i.e.*, those in which mass enters or leaves the system. A typical case is that of a *rocket* from which hot gases keep on escaping, thereby continuously decreasing its mass,

A rocket may use either a liquid or a solid fuel. In the former case the fuel (like liquid hydrogen or liquid paraffin) and a suitable oxidiser (like oxygen, hydrogen peroxide or nitric acid), stored up in separate chambers, are injected into s *combustion chamber* where the fuel is burnt (Fig. 5.3). In the latter case, the fuel itself carries its own oxidiser (as, for example, gun powder) and hence a separate chamber for it is not necessary. In either case, the large quantity of the heat of combustion produced greatly raises the pressure inside the chamber resulting in the burnt up gases (like CO, steam etc.) issuing out of given instant t in the inertial Further, let a be the rate of change of mass of the rocket due to the escape of the hot gases through the orifice at its tail end with an *exhaust velocity* –v relative to it. Them, clearly, *rate of change of mass of the rocket* = *dM/dt* = $-\alpha$, the negative sign indicating that the change of mass is a *decrease* in mass.

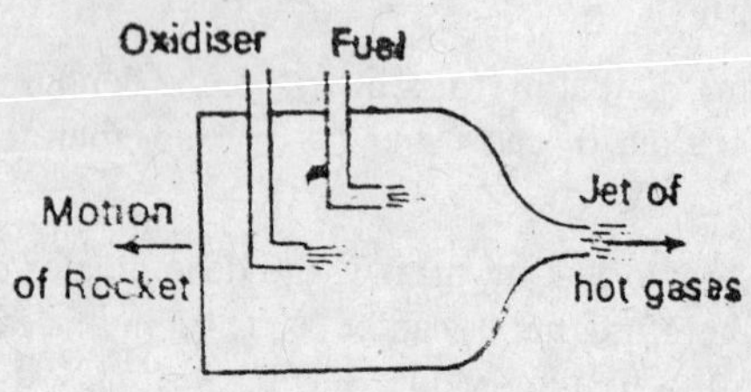

Fig. 5.3

Assuming the forward velocity of the rocket to be along the same line with the oppositely directed velocity v is the hot gases or the jet, we have

velocity of the jet in the laboratory frame = $-v + V = (V - v)$

$\therefore$ *rate of change of momentum of the jet issuing out of the rocket,*

or *force acting on it* = $-\alpha\,(V - v) = dM/dt\,(V - v)$

Hence, in accordance with Newton's third law of motion, this must also be equal to the force acting on the rocket, tending to move it forwards.

So that, *force acting on the rocket, propelling it forward* = $dM/dt\,(V - v)$, neglecting the weight (Mg) of the rocket, or the force of attraction on it due to the earth, in comparison with it.

Since M is the *mass* and V, the *velocity* of the rocket at the instant considered, we have also, (in accordance with Newton's second law), force acting on the rocket given by d(MV)/dt. We therefore, have

$$\frac{d}{dt}(MV) = \frac{dM}{dt}(V - v) \qquad \text{...(i)}$$

or $$\frac{dM}{dt}V + M = \frac{dV}{dt} = \frac{dM}{dt}V - \frac{dM}{dt}v$$

or $$M\frac{dV}{dt} = \frac{dM}{dt}v \qquad \text{...(ii)}$$

which gives the force (*i.e., mass* × *acceleration*) or the thrust on the rocket at the instant t, *i.e.*, force or thrust F = dM/dt v.

We may re-write equation II above as $dV = -\dfrac{dM}{M}v$

when, on integrating it with respect to t, we have

$$\int \frac{dV}{dt}\,dt = -\frac{1}{M}\int \frac{dM}{dt}\,v\,dt.$$

or $$V = -v \log_e M + C,$$

where C is a constant of integration.

If M_0 be the mass of the rocket and V_0, its velocity at t = 0, (where V_0 would obviously be zero if the rocket just takes off from its launching pad), we have

$$V = -v\log_e M_0 + C,$$

whence $C = V_0 + v\log_e M_0$

$\therefore \quad V = -v \log_e M + V_0 \log_e M_0.$

or $\quad V = V_0 + v \log_e M_0/M$ ...(iii)

This gives the value of the velocity V of the rocket at instant t in terms of the mass of the rocket initially and after time t.

Now, since mass of the rocket (along with the fuel and the oxidiser left in it) is M after time l and since the mass decreases at the rate α, clearly, $M = (M_0 - \alpha t)$, *i.e.*,

$$M = M_0 \left(1 - \frac{1}{M_0} \alpha t\right)$$

Denoting α/M_0, the *rate of change of mass in terms of the initial mass* M_0, by β, we have

$$M = M_0 (l - \beta t).$$

Substituting this value of M in relation III above, we have

$$V = V_0 + v \log_e \left(\frac{M_0}{M_0 (1 - \beta t)}\right) = V_0 + v \log_e \left(\frac{1}{1 - \beta t}\right)$$

or $\quad V = V_0 - v \log_e(1 - \beta t)$ ...(iv)

which gives the velocity of the rocket at the instant t in terms of β (the rate of change of mass of the rocket in terms of its initial mass).

The distance covered by the rocket in time t can be easily obtained by integrating expression (iii) or (iv) with respect to t.

In case the weight of the rocket is taken into account, the equation of motion I for a rocket moving vertically upwards becomes

$$\frac{d}{dt}(MV) = \frac{dM}{dt}(V - v) - Mg$$

or $\quad \dfrac{dM}{dt} V + M \dfrac{dV}{dt} = \dfrac{dM}{dt} V - \dfrac{dM}{dt} v - Mg$

or $\quad M \dfrac{dV}{dt} = - \dfrac{dM}{dt} v - Mg$

whence $\quad \dfrac{dV}{dt} = - \dfrac{dM}{Mdt} v - g,$

integrating which with respect to t, we have

$$\int \frac{dV}{dt} dt = -v \int \frac{dM}{Mdt} dt - \int gdt.$$

or $$V = -v \log_e M - gt + C.$$

Since at t = 0, M = M_0 and V = V_0, we have $V_0 = -v \log_e M_0 + C$.

or $$C = V_0 + v \log_e M_0. \text{ So that}$$

$$V = -v \log_e M - gt + V_0 + v \log_e M_0.$$

or $$V = V_0 + v \log_e M_0/M = gt. \quad ...(v)$$

Again, since M = M_0 (l – βt), we have

$$V = V_0 - v \log_e (1 - \beta t) - gt. \quad ...(vi)$$

It will be noted that both relation (v) and (vi) for the velocity of the rocket are essentially the same as (iii) and (iv) respectively, with the additional term – gt (in view of the downward pull due to gravity), where t is the time of combustion of the fuel.

In case t be small, the addition of the term – gt hardly makes any difference, *i.e.*, there is hardly any appreciable reduction in the value of V

It t be large, however, *i.e.*, if the fuel is burnt at a slower rate, the term – gt may become appreciably large and may slow down or even stop the ascent of the rocket. The rate of fuel combustion has, in fact, to be judiciously regulated, since both a slower and a more rapid rate of combustion are, for various reasons, equally undesirable.

A glance at the expressions for velocity (v) of the rocket will show that to obtain a higher value of the rocket velocity we must have (i) *a higher exhaust velocity v and* (ii) a *higher value* of M_0/M (or of ρ) *i.e., a higher loss of mass, so that the final mass of the rocket may be very much smaller than its initial mass.*

Now, the exhaust velocity depends upon the design of the nozzle as also upon the temperature and pressure developed inside the combustion chamber, the highest theoretical value possible being the *root mean square velocity* of the gas molecules at the temperature prevailing in the combustion chamber. Since, however, the temperature can hardly ever exceed 3000 C, the practical limit to the value of the exhaust velocity works out to 2 km/sec.

Then, again, the ratio M_0/M can at best be 10% of the initial value M_0, (even if the whole of the fad is burnt, so that At is only the mass of the empty containers), but, its advantage is more than offset by a smaller exhaust velocity in this case.

As can be easily calculated from expression (iii) above, with an exhaust velocity of 2 km/sec, initial velocity (V_0) of the rocket *zero* and the ratio $M_0/M = 10$, the velocity (V) of the rocket works out to $0 + 2 \log_e 10 = 2 \times 2.3026 = 4.6052$ km/sec.

This is less by far than the escape velocity 11.2 km/sec necessary to be able to escape from the gravitational field of the earth, or even the orbiting velocity 8 km/sec necessary to be able to orbit around the earth close to its surface.

It can be easily seen that, with the exhaust velocity $v = 2$ km/sec, in order that a rocket may attain a velocity V equal to the escape velocity 11.2 kms.sec, the mass ratio M_0/M must be 270 even a bit more, actually, .in view of the fact that the rocket will encounter, and will have to overcome, the resistance of the air. This means that a little more than 270 kg of the fuel will be consumed for every 1 kg mass of the rocket that leaves the earth. In the present stage of its development no single stage rocket is capable of such a performance and we have, therefore, to use a *multistage, i.e., a two or three stage rocket* to achieve the purpose.

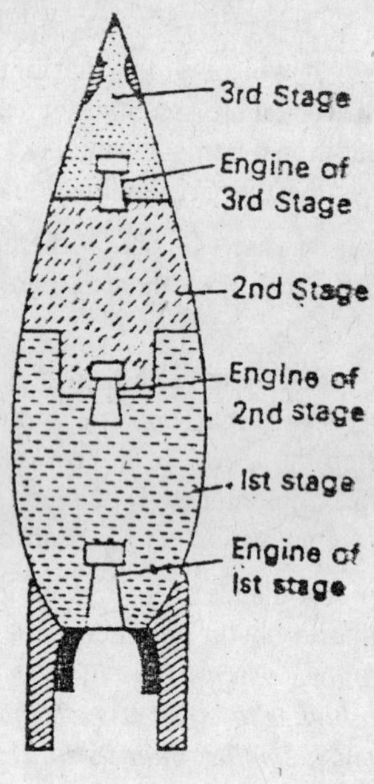

Fig. 5.4

A multistage rocket is just a combination of rockets, either (i) joined consecutively, *i.e.*, in series, (ii) *one inside the other*, or (iii) with the rear part of one inside the nozzle of the other, as shown diagrammatically in Fig. 5.4. In all the three types, the first stage is the largest in dimensions as well as in weight, and the last stage the smallest and the lightest.

Naturally, the first stage rocket is used first and when its fuel is ail burnt up and it has done its job it gets detached and is discarded, with the second stage taking over the task of producing further acceleration. This too, in its turn, is discarded when its fuel is burnt up and the third stage rocket takes over. The velocity thus goes on increasing at each stage by the same amount as it does in a single stage rocket. The fuel consumption and the thrust for the first stage are about 100 times more than for the third stage and the fuel stock carried by it about 60 times that carrried by the third stage.

Finally, a word may as well be slid here as to the shape of the rocket. As can be readily seen, a rocket is subjected to intense air pressure during its upward flight through the atmosphere and a lot of heat is generated due to the viscous friction of the air. Both these factors are taken into account in designing it. Its frame is made of a beat-resisting material, with its velocity during its flight through the denser regions of the atmosphere kept sufficiently low and its body is given a cylindrical and *tapering, stream line* form to reduce the air pressure on each individual part of it to the very minimum, its overall shape resembling that of a cigar.

ANGULAR MOMENTUM—TORQUE

The angular momentum of a particle about a point fixed in an inertial frame is the *moment of its linear momentum about that point.* That is why it is also called *moment of momentum*. It is usually denoted by the symbol J and is measured by the *product of the linear momentum p (=mv) of the particle and its vector distance r from the fixed or the reference point in the inertial frame*. Thus,

$$J = r \times p = m(r \times v) \qquad \text{...(i) [where } p = mv.$$

It is obviously a *vector quantity* and its direction, perpendicular to both r and p (or v), is given by the *right hand screw rule*. So that, in the case of a particle describing a circular motion its value is equal to $m\omega^2 r$ = mvr (where v is the linear velocity of the particle) and is perpendicular to the plane of the circle described, in the same direction with ω.

A component of the angular momentum about any axis (*i.e.*, a line) passing through the fixed or the reference point is referred to as the angular momentum of the particle about that axis (or line)

The units of angular momentum in the C.G.S. and the RMKS (or SI) systems are respectively

gm-cm^2/sec

or erg-sec

and kg-m²/sec

or joule-sec.

Relation I holds good even for a system of particles, each free to move independently of the other or compactly arranged to form a rigid body. Thus, if J_1, J_2, J_3 etc. be the angular momenta of the different particles of the system about the fixed or the reference point, the angular momentum of the whole system about that point is given by

$$J = J_1 + J_2 + J_3 + ... = (r_1 \times mv_1) + (r_2 \times mv_2) + (r_3 \times mv_3) + ...$$

or
$$J = \Sigma(r \times mv) = \Sigma(r \times p),$$

i.e., the angular momentum of the whole system about the point is the vector sum of is angular momenta of the individual particles about that point.

Torque. Differentiating relation I above with respect to t, we have

$$\frac{dJ}{dt} - \frac{d}{dt}(r \times p) = \frac{dr}{dt} \times p + r \times \frac{dp}{dt}.$$

Now, $dr/dt \times p = v \times mv = 0$ and dp/dt = rate of change of momentum or the force F acting on the particle. So that,

The vector product $r \times F$ is called *torque* (*or moment of the force*) about the fixed or the reference point and is represented by the symbol T.

Thus, the time rate of change of angular momentum of a particle about a point gives the torque T about that point, where

$$\tau = r \times F = dJ/dt \qquad ...(ii)$$

In the case of a system of particles, independent of each other or forming a rigid body, we have seen above how $J = \Sigma(r \times mv) = \Sigma(r \times p)$. It follows, therefore, that

$$\tau = dJdt = \Sigma r \times d/dt\,(mv) = \Sigma(r \times F),$$

i.e., the torque or the time rate of change of angular momentum of the system about the fixed or the reference point is the sum of the torques of all the external forces acting on the system. The reason why we explicitly say *external* forces is that the internal forces all form collinear action-and-reaction pairs of equal and opposite forces, having equal and opposite moments about the given point and their sum is, therefore, zero, *i.e.*, they produce no effect

As will be readily seen, *angular momentum is the rotational analogue of linear momentum.* Just as the rate of change of linear momentum of a particle gives the force acting upon it, so also the rate of change of angular momentum of the particle gives the torque acting upon it. And, as we shall shortly see, just as *linear momentum =(mass) (linear velocity)* or mv, we have, for a rigid body, *angular momentum = (moment of inertia) (angular velocity) or $I\omega$.*

Another expression for angular momentum of a system. Sometimes it is found to be more convenient to express the angular momentum of a system of particles in terms of the velocity of the centre of mass and the velocities of the particles of the system relative to the centre of mass.

Thus, if R and V be the *position vector* and *velocity* respectively of the centre of mass of a system of particles relative to a fixed or a reference point, and r_c and v_c, the position vector and velocity of a particle of mass m of the system, relative to the centre of mass, the position vector and velocity of the particle *relative to the fixed or* reference point will clearly be $r = R + r_c$ and $v = V + v_c$ respectively. Hence, the *angular momentum of the system about the fixed or the reference point* will, as just explained above, be given by

$$J = \Sigma m(R + r_c) \times (V + v_c) = \Sigma m(R \times V) + \Sigma m(R \times r_c)$$
$$+ \Sigma m\,(r_c \times V) + \Sigma m(r_c \times v_c)$$

Now $\quad r_c = (r - R)$ and $\therefore\ mr_c = mr - mR$

or $\quad \Sigma mr_c = \Sigma mR - \Sigma mR = \Sigma mr - MR$

[$\because$ $\Sigma m = M$, mass of the system..

But, as we know, the inherent property of the centre of mass demands that

$$MR = m_1r_1 + m_2r_2 + ... = \Sigma mr$$

We, therefore, have $\quad \Sigma mr - MR = 0$

or $\quad \Sigma mr_c = 0$

Similarly, $\quad \Sigma mv_c = 0$

So that, the above relation for J simplifies to

$$J = R \times MV + \Sigma(r_c \times mv_c).$$

Here, clearly, $\Sigma(r_c \times mv_c)$ is the angular momentum of the system about the centre of mass, say, $J_{c.m}$ and $MV = P$, the momentum of the centre of mass or the total linear momentum of the system. We, therefore, have

$$J = R \times P + J \qquad ...(iii)$$

i.e., the total angular momentum of the system about the fixed or the reference point is the vector sum of the angular momentum of the centre of mass about that point and the angular momentum of the system about the centre of mass. The former depends upon the fixed or the reference point chosen and is referred to as the *orbital angular momentum* and the latter, which is independent of the fixed or the reference point, is called the *spin angular momentum*. So that, the *total angular momentum of the system is the vector sum of its orbital and spin angular momenta.*

Differentiating expression III above with respect to t, we have

torque acting on the system, $\frac{dJ}{dt} = \frac{dJ_{c.m.}}{dt} + \frac{d}{dt}(R \times P)$

If the centre of mass be chosen as the origin, so that it remains stationary, we have P = MV = 0 (since V = 0). The second term in the expression for τ above, therefore, vanishes and we nave torque $\tau = dJ_{c.m.}$...../dt and the body is said to be in a *state of spinning.*

Thus, each planet in the Solar system possesses both an orbital and a spin angular momentum and its total angular momentum about the sun is the vector sum of the two. In the case of our own planet. Earth, the two momenta sire inclined to each other at an angle of 23.5°. The sun alone in the solar system (considered in isolation from the rest of the universe) possess *zero orbital angular momentum* (because F = 0) and it, therefore, has only spin angular momentum.

The total angular momentum of the solar system A thus the vector sum of all these orbital and spin angular momenta of the various planets, the contribution o f the spin angular momentum of the sun itself being just 2%.

If, however, we consider the universe, as a whole, with the centre of the galaxy as the origin, or the reference point, the sun too has an orbital angular momentum about the origin and thus possesses both types of angular momenta.

CONSERVATION OF ANGULAR MOMENTUM

We know that the external torque applied to a system of particles is given by the relation τ = dJ/dt = r × F. If, therefore, the external torque be zero, we have dJ/dt = 0, *i.e.*, J = *constant*. This is the prinoiple or the law of conservation of angular momentum and may be formally stated thus:

When the external torque (or the sum of external torques) applied to a system of particles is zero, the total angular momentum of the system remains conserved.

The law is not restricted to only closed orbits but applies equally well to open orbits as well as collisions.

Thus, if J_1, J_2, J_3 etc. be the angular momenta of the different panicles of a system, their sum total $J = J_1 + J_2 = J_3 + ...$ remains constant in the absence of any external torque applied to the system, though the particles may exchange momenta among themselves.

Similarly, in the case where the centre of mass is taken to be the origin, we have $\tau = dJ_{c.m.}/dt$. If, therefore, the external torque τ be zero, we have $J_{c.m.}$ = *constant*. This may be called the *law of conservation of spin angular momentum.*

It will be seen at once that just as the external force applied determines the motion of the centre of mass of a system; the external torque applied determines the rotation of the system about the centre of mass.

Now, from the relation for τ, mentioned above, viz, $\tau = dJ/dt = r \times F$, it is abundantly clear that *the torque acting as a particle will be zero, and hence its angular momentum consented, when (i) its position vector r is zero or (ii) the force F applied to it is zero or (iii) the direction of both r and F is the same, i.e.*, the line of action of force F passes through the fixed or the reference point.

The first two conditions are obvious by themselves and the third condition is satisfied in the case of *central forces, i.e.*, forces which are always directed towards (or away from) a fixed reference point. This will be seen from the following:

The magnitude of a central force on a particle depending only on the distance from the fixed or the reference point, we may represent it by $F = rf(r)$, where r is the unit vector along the direction of r and is equal to r/r and f(r), a *scalar function* of distance r.

$\therefore$ torque acting on a particle under the action of a central force is given by

$$\tau = \frac{dJ}{dt} = r \times F = r \times rf(r) = f(r)\ (r \times r/r) = 0, \ [\because r \times r = 0.$$

showing that $J = r \times mv$ = *constant,*

i.e., the angular momentum of a particle moving under the influence of a central force always remains conserved.

It follows, therefore, that J, being constant, must be perpendicular to the plane containing r and v, *i.e., the path of a particle under the influence of a. central force must lie in a plane.*

Thus, since electrostatic forces are central forces, the angular momentum of the electron moving in its orbit around the proton in a hydrogen atom remains conserved.

Again, since gravitational forces too are central forces, the angular momenta of planets orbiting around the sun remain conserved (although they move in elliptical orbits) and so also of the satellites moving around a planet.

Let us examine the case of a planet moving around the sun *in an elliptical orbit* in a little more detail.

MOTION OF A PLANET IN AN ELLIPTICAL ORBIT AROUND THE SUN

The planets, as we know, move in elliptical orbits around the sun, with the sun at one of the foci. In fact, it is the centre of mass of the sun and the planet which should lie at this locus but, in view of the enormously greater mass of the sun compared with that of the planet, the centre of mass of the system (*i.e.*, of the sun and the planet) lies close to the centre of the sun itself which may, therefore, be assumed to lie there.

Thus, let O be the centre of the sun lying at one focus of the ellipse and P and P', the two positions of the planet at instants t and r + Δt, given by the position vectors r and r + Δr respectively, with the gravitational force on it due to the sun always directed towards the latter, (Fig. 5.5).

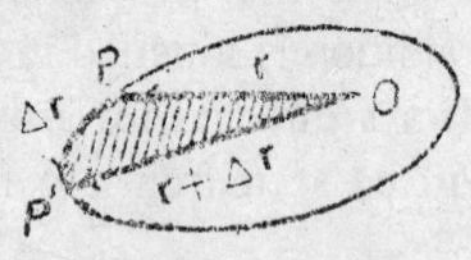

Fig. 5.5

In this interval of time Δt the planet sweeps out, on account of its displacement Δr, a vector area ΔS (shown shaded) = 1/2r × Δr.

∴ *rate at which the radius vector sweeps out the area*

$$= \Delta S/\Delta t = 1/2 r \times \Delta r/\Delta t.$$

or, in the limit $\Delta t \to 0$,

we have $dS/dt = 1/2r \times dr/dt = 1/2r \times v$,

where dS/dt may be called the *areal velocity of the radius vector.*

Now, *angular momentum* $J = r \times mv$;

so that, $r \times v = J/m$,

and, therefore, *areal velocity of the radius vector, dS/dt = J/2m.*

Since J/2m is a constant, it follows that *the areal velocity of the radius vector remains constant under the influence of the central gravitational force.*

This is really Kepler's second law which states that the *radius vector from the sun to any planet sweeps out equal areas in equal time.*

A consequence of this is that *a planet moves fatter at the point of its closest, than that of its farthest, approach to the sun.* This is obviously so because at these points the vector r is perpendicular to v (Fig. 5.6) and the angular momentum is thus equal to mvr. For the conservation of momentum, therefore, the angular momentum mv_2r_2 at the point of nearest approach must be equal to the angular momentum mv_1r_1 at the point of farthest approach. We, therefore, have $mv_1r_1 = mv_2r_2$. Or $v_1r_1 = v_2r_2$, showing that *the larger velocity of the planet is associated with ifs shorter distance from the sun and vice versa.*

And since the angular momentum (J) remains constant, J/2m or the areal velocity of the radius vector too remains constant.

In the case of the earth, its orbital velocity is 18.8 miles/sec when it is nearest the sun and 18.2 miles/sec when it is farthest from it.

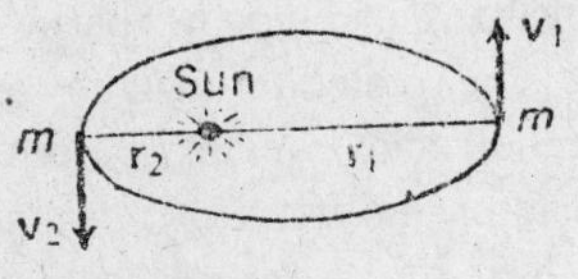

Fig. 5.6

This result, arrived at by *Kepler* from his long and laborious calculations and most careful observations, is thus just a natural consequence of the law of conservation of angular momentum.

Interestingly enough, *Newton* used the reverse argument, half a century later, to show that the areal velocity dS/dt of the radius vector joining a planet and the sun being constant, the gravitational force is a central force.

CONSERVATION OF ANGULAR MOMENTUM OF A SYSTEM, A CONSEQUENCE OF ROTATIONAL INVARIANCE OF POTENTIAL ENERGY OF THE SYSTEM

Just as the conservation of linear momentum of a system of particles is a consequence of translational invariance of the potential energy of the system also the conservation of angular momentum of a system of particles is a consequence of rotational invariance of the potential energy of the system, *i.e., the angular momentum of the system remains conserved if its rotation does not bring about any change in its potential energy.*

This follows from the fact that the rotation of a system would normally require work to be done against any external torque that may be acting on the system; thus necessarily resulting in some change in its potential energy. In case, therefore, the rotation of the system brings about no change in its potential energy, the obvious inference is that no work is being done against any external torque or that, in other words, no external torque is acting on the system. And, as we know so well, in the absence of an external torque, the angular momentum of the system mass remain conserved.

Analytically, we may proceed in a manner similar to that adopted in the case of linear momentum.

Thus, let us consider first, for the sake of simplicity, a system consisting of only two particles Having position vectors r_1 and r_2. Let these be rotated through any arbitrary angle about any arbitrary axis and let the rotated vectors be represented by Ωr_1 and Ωr_2 (where Ω is an operator and not 4 mere number), such that their lengths are the same as those of r_1 and r_2 respectively, indicating that, on rotation, the vectors r_1 and r_2 undergo for change in magnitude but only a change in direction.

Now, the rotational invariance of potential energy demands that the potential energy $U(r_1, r_2)$ *before rotation* = the potential energy $U(\Omega r_1, \Omega r_2)$ *after rotation.* This can obviously be possible only if $U(r_1, r_2) = U(r_1 - r_2)$ and $U(\Omega r_1, \Omega r_2) = U(\Omega r_1, \Omega r_2)$, *i.e.*, if the potential energy depends only on the magnitude of the distance between the two particles, this magnitude being the *same* before and after the motion. In order that this be so, the force acting on the panicles must be directed along the Sine $(r_2 - r_1)$, *i.e.*, it must be a *central force*. And since a central force gives rise to no torque on the system, its angular momentum remains conserved.

What is true for a system of two particles is equally true for a system consisting of any number of them, provided the potential energy depends on the magnitude of the distance between the various particles.

The conservation of angular momentum of a system is thus a natural corollary of the rotational invariance of its potential energy.

SOME ILLUSTRATIVE EXAMPLES OF CONSERVATION OF ANGULAR MOMENTUM

(i) *Motion of a planet or a satellite in its orbit* : This has already been dealt with under where it has been shown:

(a) how to be able to conserve its angular momentum, a planet must move faster at its point of nearest approach to the sun than at the point of its farthest approach and also:

(b) how the principle of conservation of angular momentum has led to the deduction of Kepler's second law.

(ii) *Scattering of a positive particle by a massive nucleus* : Suppose a positive particle, like a proton or an alpha particle, approaches a massive nucleus N of ao atom, (Fig. 5.7), and that its velocity is not appreciably affected due to its passage into the atom. Then, clearly, as it approaches the (positive) nucleus of the atom, it experiences an electrostatic force of repulsion, varying inversely as the square of its distance from the nucleus. So that as it gets closer to the nucleus, this repulsive force on it goes on increasing. As a result, it gets defected or scattered from its straight path (towards the nucleus) into a hyperbolic path, PAP', as shown, with the nucleus N at its external focus.

Let the particle in question be a *proton*, of mass m and carrying a charge + e, and let the atomic number of the nucleus be Z, so that the charge on it is + Ze.

Suppose at an infinite distance from the nucleus where the repulsive force on it is zero, the particle travels with velocity it and enters the atom (with no appreciable change in its velocity) along the direction PO and that, on reaching the point A, the apes of the hyperbola, it proceeds along AP' such that PA and AF make equal angles with the straight line from N passing through A.

The perpendicular distance NB = b between the nucleus and the initial direction PO of the particle is called the *impact parameter* or the collision parameter and the distance NA = r_s is called the distance of closest approach of the particle to the nucleus.

Since electrostatic forces (like gravitational ones) are central forces, the angular momentum of the particle must remain conserved and we, therefore, have

$$mv_0b = mv_AT_A,$$

whence, $v_A = v_0b/r_A$, where v_A is the velocity of the particle at the point of closest approach, *i.e.*, at A. Also, since the energy of the particle must also remain conserved, we have

$$1/2mv_A^2 + Ze^2/r_A = 1/2mv_0^2,$$

where $1/2mv_A^2$ is the kinetic, and Ze^2/r_A, the potential energy of the particle at A and $1/2mv_0^2$, its *initial kinetic energy*, when its potential energy is zero (its distance from the nucleus being infinite).

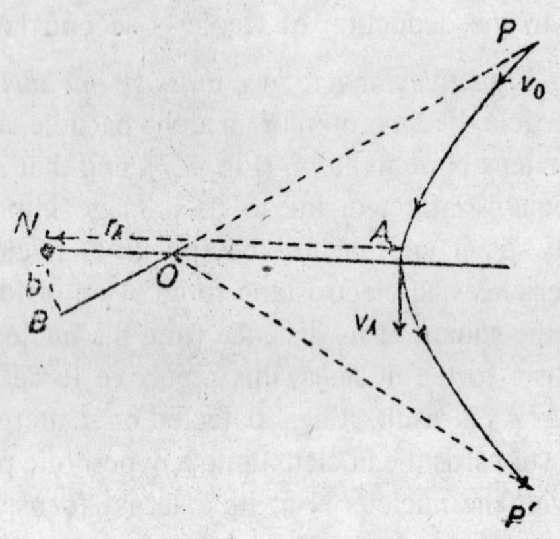

Fig. 5.7

Or, substituting the value of v_A, obtained above, we have

$$Ze^2/rA = 1/2mv_0^2 - 1/2m(v_0b/r_A)^2.$$

or $$\frac{Ze^2}{r_A} = \frac{1}{2}mv_0^1\left[1-\left(\frac{b}{r_A}\right)^2\right]$$

So that, knowing the *impact parameter b, the distance of closest approach* of the particle (r_A) can be easily obtained.

(iii) *Effect on linear and angular speeds of a particle on contraction of its orbit:* Let a particle of mass m be carried at one end of a light string such that it can be rotated in a horizontal circle, *i.e.*, made to move in circular orbit. Let the free end of the string pass through a small hollow tube, as shown in Fig. 5.8, so that the radius of the orbit of the particle can be decreased by simply applying a downward pull F on the free end of the string.

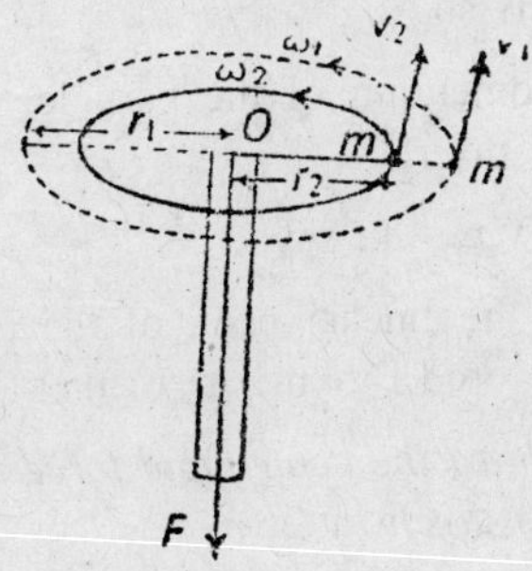

Fig. 5.8

Let the *initial radius* of the circular orbit described by the particle be r_1, its *linear speed* v_1 and *angular speed* $\omega_1 = v_1/v_1$, so that its *angular momentum* about the centre O of the orbit = mv_1r_1 and its *kinetic energy* $1/2 = m_1v_1^2$.

And, when the free end of the string is pulled down a little by a force F, so that the radius of the circular orbit of the particle is reduced to r_2, let the linear and angular speeds of the particle be v, and ω_2 respectively and hence its *angular momentum* about 0 = mv_2r_2 and its kinetic energy = $1/2mv_2^2$.

Clearly, the downward pull on the free end of the string is transmitted to the particle as a *radial force* (viz., the centripetal force), directed towards the centre O of its orbit. As a radial force, therefore, it exerts no torque on the particle (its line of action passing through O) and its angular momentum thus remains conserved, so that we

have $\qquad mv_1r_1 = mv_2r_2,$

whence, $\qquad v_2 = v_1r_1/r_2$

Since, $r_2 < r_1$, it follows that $v_2 > v_1$ and hence the *K.E. of the particle in the second case*, *i.e.*, $1/2mv_2^2 = 1/2m(v_1r_1/r_2)^2 = 1/2mv_1^2\ (r_1/v_2)^2$ is *greater than its kinetic energy* $1/2mv_1^2$ in the first case.

Thus, by shortening the radius of the orbit of the particle from r_1 to r_2, its K.E. has increased by $1/2mv_2^2 - 1/2mv_1^2$.

$$1/2mv_1^2\,(r_1/r_2)^2 - 1/2mv_1^2 = 1/2mv_1^2\left(\frac{r_1^2}{r_2^2} - 1\right) \quad ...(i)$$

This additional energy must obviously be supplied by the work done on the particle against the centrifugal force in pulling it inwards towards O. Let us see if this is so.

Since angular momentum of the particle remains conserved, we have $mv_1r_1 = mv_2r_2$

or $v_1r_1 = v_2r_2 = k$, say.

This means that the linear speed of the particle, $v = k/r$. And, therefore, centrifugal force $F = mv^2/r = (m/r)\,(k^2/r^2) = mk^2r^3$

∴ *work done against the centrifugal force in decreasing the orbit radius from r_1 to r_2 is given by*

$$W = \int_{r_1}^{r_2} -F dr = \int_{r_1}^{r_2} -\frac{mk^2}{2r^2} dr = \left[\frac{mk^2}{2r^2}\right]_{r_1}^{r_2} = \frac{mk^2}{2}\left[\frac{1}{r^2}\right] = mk^2\left[\frac{1}{r_2^2} - \frac{1}{r_1^2}\right]$$

Or, substituting v_1r_1 for k, we have

$$W = \frac{mv_1^2 r_1^2}{2}\left(\frac{1}{r_2^2} - \frac{1}{r_1^2}\right) = \frac{1}{2}mv_1^2\left(\frac{r_1^2}{r_2^2} - 1\right)$$

which is the same as expression I for gain in K.E., above, *showing that the gain in K.E., of the particle on shortening its orbit, is indeed supplied by the work done against the centrifugal force acting on the particle.*

Obviously enough, in the absence of any such supply of energy to the particle, it would not be possible for it to conserve its angular momentum.

This may be easily seen if we simply attach the end of the string to a fixed vertical stick and set the particle in rotation, as before, at the end of the string. It will be found that as the particle moves in its circular orbit, the radius of the orbit goes on progressively decreasing and the string goes on winding itself around the stick in the form of a spiral.

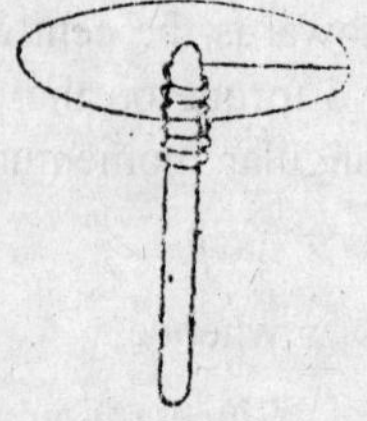

Fig. 5.9

Clearly, no work is done here against the centripetal force, as in the first case, and hence no energy is supplied to the particle. As a result, despite the continuous decrease in the radius of the orbit, there cannot be any increase in the kinetic energy, and hence, in the linear speed of the particle which therefore continues to remain the same throughout, although the angular speed $\omega = v/r$ goes on increasing with decrease in the value of r. The constancy of the linear speed indicates that the tension in the string is always perpendicular to the direction of v at any point is the *spiral path* of the particle and is no longer directed towards the centre of its initial circular path. This is so because, in the absence of energy supplied to the particle, its angular momentum mvr cannot possibly remain conserved, but must go on decreasing along with the radius r of the orbit.

(iv) *The shape of the galaxy :* Galaxies are physical systems, containing anything from 10^9 to 10^{12} stars with a lot of free gas, bound by gravitational attraction, sad free to rotate about their own axes of symmetry. One galaxy is separated from another by a huge distance of about 3×10^{24}cm. The galaxy which contains our own sun is referred to as *the galaxy* and has a radius of about 10^{23}cm.

A conjecture as to the formation of galaxies, which some regard to be oversimplified, is due to *Hubble*, according to which there was, to start with, just a gaseous mass spread throughout space, some portion of which acquired angular momentum due to gravitational interaction within itself and the rest, free from angular momentum, condensed in the form of spheres to ultimately form stars.

A galaxy is supposed to have come to acquire its more or less lens-like or pancake shape due to the condensation of a great mass of the gas inside it under gravitational interaction. This may be explained somewhat in the following manner.

Suppose there is a large mass of gas in a more or less spherical form, having an initial *angular momentum* J about any axis, as shown in Fig. 5.10(a). As the gas contracts under gravitational interaction, the forces of interaction being pairs of equal and opposite forces, there is no external torque acting on the system and its angular momentum, therefore, remains conserved. Since however, its volume is greatly reduced, energy has to be supplied to it to enable it to conserve its angular momentum, as we have seen under (iii) above. This energy can obviously come only from the gravitational potential energy of the gas itself.

Thus, if m be the mass of a particle of the gas in the outer regions of the galaxy, initially at a distance r, from its centre and rotating with velocity re about it, its K-E. $1/2mv_0^2$.

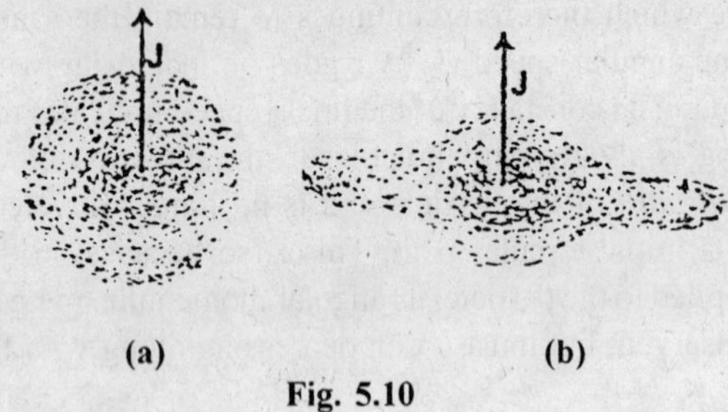

Fig. 5.10

After gravitational interaction, if the particle lies at a distance r from the centre of the galaxy and if its velocity be r, its K.E. = $1/2mv^2$.

Since the angular momentum remains conserved, we have

$$rnv_0r_0 = mvr,$$

whence, $v = v_0r_0/r$

and $\therefore$ K.E. of the particle $1/2mv_0^2\left(\frac{r_0}{r}\right)^2$

And its gravitational P.E. at this distance r from the centre of the galaxy = –GMm/r,, where M is the *mass of the galaxy*.

$\therefore$ *total energy of the particle* $= \frac{1}{2}mv_0^2\left(\frac{r_0}{r}\right)^2 - \frac{GMm}{r}$

The contraction of the gas will stop, or equilibrium attained, when the total energy of the gas or the particle is an *extremum* (*i.e.*, either a maximum or a minimum), *i.e.*, when

$$\frac{d}{dr}\left[\frac{1}{2}mv_0^2\left(\frac{r_0}{r}\right)^2 - \frac{GMm}{r}\right] = 0$$

or $$-mv_0^2\frac{r_0^2}{r^3} + \frac{GMm}{r} = 0$$

or $$\frac{GMm}{r^2} = \frac{mv_0^2r_0^2}{r^3} = \frac{mv^2}{r}$$ [Substituting vr/r_0 for v_0].

Thus, the contraction of the gas continues, with its angular momentum conserved, until *equilibrium is attained when the inward pull due to gravitational force on it is just balanced by the outward pull due to centrifugal force,* the value of r, then being $v_0^2r^2/GM$.

Any further reduction in the value of r beyond this is not possible, because the rapidly increasing centrifugal force opposes any further contraction of the gas.

There is, however, no such limitation to the contraction of the gas with its angular momentum conserved *along the direction* of the axis of rotation, any K.E. acquired during contraction being dissipated presumably in the form of radiation. As a result, the shape ultimately acquired by the galaxy is that of a *pancake*, with a small spherical, or nearly spherical, portion at the centre, as shown in Fig. 5.10(a).

(v) *Diving, skating, ballet-dancing etc. :* In all these processes, the performer utilises the principle of conservation of angular momentum. For, as mentioned earlier, the angular momentum of a body free to rotate about an axis is equal to $I\omega$ where I is the *moment of inertia* of the body about the axis and ω, its angular velocity. Now, I depends upon the square of the *radius of gyration* of the body about the axis of rotation, *i.e.*, upon the *effective distance* of the particles of the body from the axis of rotation and hence upon the *distribution of mass* about the axis. A large variation in the value of the moment of inertia (I) of the performer and hence a corresponding variation in the value of his angular velocity ω is, therefore, possible on his stretching his limbs out or pulling them in.

Thus, a diver, as he springs from the diving board, has a certain angular velocity about a horizontal axis passing through his centre of mass to enable him to spin about the axis or to make a somersault or two before striking water.

Let his initial M.I. about the axis be I_0 and his angular velocity about it, ω_0, so that his initial angular momentum about the axis is $I_0\omega_0$. The only external force acting on the diver is the gravitational force due to the attraction of the earth, but since it is a *central force* it produces no torque about his centre of mass and his angular momentum thus remains conserved. He can, therefore, double or triple his angular velocity and thus increase the number of his somersaults before touching water by correspondingly reducing his M.I. about the axis of rotation to say, I, by simply pulling in his hands and feet towards the centre of his body. If ω be his angular velocity corresponding to the value I of his moment of inertia, we have

$$I_0\omega_0 = I\omega,$$

Since $I < I_0$, we have $\omega > \omega_0$ and $\therefore$ $1/2I\omega^2 > 1/2I_0\omega_0^2$, *i.e.*, the rotational K.E. of the diver increases in the process.

The source of this additional K.E. acquired by the diver is obviously the work done by him in pulling in his limbs towards the centre of his body.

In a similar manner, a skater or a bullet dancer, (Fig. 5.11), can increase or decrease his or her angular velocity and thus skin faster or slower about a vertical axis through the centre of mass by pulling in the hands and the feet or by extending them as far out as possible, thereby decreasing or increasing the M.I. about the axis of rotation, as desired.

Fig. 5.11

ANGULAR MOMENTUM OF ELEMENTARY PARTICLES— ITS QUANTISATION

We have seen how a planet moving in its elliptical orbit around the sun possesses (i) *an orbital angular momentum* due to its motion around the sun and (ii) *a spin angular momentum* due to its rotating or spinning about an axis passing through its own centre of mass. The same is true for almost all the fundamental or elementary particles, including the *photon*, which may possess one or both these types of angular momenta. Thus, for example, the electron in a hydrogen atom possesses orbital angular momentum on account of its motion in its orbit around the proton and also spin angular momentum on account of spinning about its own axis (passing through its centre of mass).

Now, the student is, in all probability, already aware that in *Bohr's* atomic model, an electron can revolve round the nucleus, without emitting any radiant energy, in only certain specified or permissible orbits for which its orbital angular momentum is an integral multiple of $h/2\pi$ *i.e.*, $nh/2\pi$, where n is an integer, 1, 2, 3, etc. and h, the *Planck's constant* (6.25×10^{-27} erg-sec). These orbits are, for this reason, referred to as stationary orbits or Bohr's orbits.

The orbital angular momentum of the revolving electron is thus restricted to certain fixed or discrete values (viz,, integral multiples of $h/2\pi$) and is, therefore, said to be quantised.

Bohr, of course, had no theoretical basis for making this assumption except that it could satisfactorily explain the various spectral series of hydrogen.

A mathematical basis for this hypothesis was, however, provided later by *De Broglie* who suggested that every particle has a wave associated with it and that the smaller the particle, the longer this wave. Thus, the ware associated with an electron is longer than that associated with a proton. The wavelength of this associated wave, called *de Broglie wavelength*, may be easily obtained. For, in accordance with *Einstein's mass-energy equation* $E = mc^2$ and from *Planck's equation* $E = h\nu$, coupled with the fact that $c = \nu\lambda$ and, therefore, $\nu = c/\lambda$, (where ν is the frequency of the wave), we have $hc/\lambda = mc^2$, whence, $\lambda = h/mc$.

If, therefore, v be the velocity of an electron, the De Broglie wavelength associated with it may be obtained by substituting v for c in the above relation, when we have $\lambda = h/mv$, a fact tested experimentally by *Davisson* and *Germer*.

So that, instead of an electron moving m a stationary orbit, we must now consider a series of waves moving in the orbit and, in order that they may not cancel each other due to interference they must move in such a manner as to produce a stationary or standing wave in the orbit. The necessary condition for this is that the length of the path, *i.e.*, the circumference of the orbit, must be equal to an integral multiple of the wave length. Thus, if r be the radius of a stationary orbit, we must have $2\pi r = m\lambda = nh/mv$. whence, the *orbital angular momentum* of the electron, $mvr = v\lambda/2\pi$, as assumed by *Bohr*.

Now, the *centripetal force acting on the electron* moving with speed v in an orbit of radius r around the nucleus is clearly $-mv^2/r$, the –ve sign indicating that it is directed *inward* towards the centre of the orbit or the nucleus.

Since the charge on the nucleus is Ze, where Z is its atomic number and e, a charge equal in magnitude to that on an electron, the force acting on the electron inwards, towards the nucleus, is also $-Ze.e/r^2$, (the charge on the electron being negative).

In the case of hydrogen, Z = 1 and, therefore,

∴ *force acting on the electron, inwards* = e^2/r^2.

We, therefore, have $\frac{e^2}{r^2} = \frac{mv^2}{r}$,

whence, $mv^2 = \frac{e^2}{r}$ ...(i)

And, since angular momentum of the electron, *i.e.*, $mvr = nh/2\pi$, we have $m^2v^2r^2 = n^2h^2/4\pi^2$ ...(ii)

∴ dividing relation (ii) by (i), we have $\frac{m^2v^2r^2}{mv^2} = \frac{n^2h^2r}{4\pi^2e^2}$,

whence, $r = \frac{h^2}{4\pi^2me^2}n^2$

or r e n^2.

Thus, *the radii of the stationary or remissible orbits arc proportional to n^2.*

Further, *potential at a distance r from the charge Ze on the nucleus* = Ze/r. We, therefore, have

P.E. of the electron = (Ze) (–e)/r

$= Ze^2/r = e^2/r$ [∵ Z = 1.

And, since K.E. of *the electron* = $1/2mv^2$, we have *total energy of the electron*, E = K.E. + P.E. = $1/2mv^2 - e^2/r$

$$= \frac{1}{2}\frac{e^2}{r} - \frac{e^2}{r} = -\frac{1}{2}\frac{e^2}{r}$$

Or, substituting the value of r from above, we have

$$E = -\frac{1}{2}\frac{e^2 4\pi^2me^2}{n^2h^2} = -\frac{2\pi^2me^4}{n^2h^2} \quad \text{...(iii)}$$

It may be pointed out that *Bohr* had originally taken the electron orbit to be circular. Later *Sommerfield* extended the theory to include also elliptical orbits but showed that the value of E = K.E. + P.E. for the electron orbit is still given by relation (iii) above, deduced for circular orbits.

Here, obviously, the motion of the nucleus (or thy proton) has been ignored. More correctly, we can imagine the proton (of mass M, say) to be stationary and obtain *relative motion* of the electron about it if we replace the mass m of the electron by the *reduced mass* $\mu = mM/(m + M)$ of the electron and the proton. In that, case, clearly,

$$r = \frac{h^2}{4\pi^2 \mu e^2} n^2$$

and $$E = -\frac{4\pi^2 \mu e^2}{n^2 h^2}$$

Correct value of orbital angular momentum of an electron and other fundamental particles. A full-Hedged *quantum theory* of the atom based on wave mechanics, was developed in the year 1925, thanks in main to the efforts of *Heisenberg, Schroifinger, Born* sod others, according to which the correct value of the orbital angular momentum of an electron or, in fact, any elementary particle, is given not by nh/ $2\pi \frac{nh}{2\pi}$ but by $\sqrt{l\,(l+1)}\,\frac{h}{2\pi}$ where l is called the orbital or

Azimuthal quantum number and may have values 0, 1, 2, 3,... (n – l), where n is the principal quantum number.

A significant departure from Bohr's theory is at once noticeable in that here, with $l = 0$, we can have angular momentum (J) = 0 which is surely not consistent with Bohr's theory which demands that an electron moving in its orbit around the nucleus must necessarily possess angular momentum.

Now, angular momentum is a vector quantity and is represented by a vector along the axis of rotation (or perpendicular to the plane of rotation). If, therefore, we express it in units of $h/2\pi$, *i.e.*, if $h/2\pi$ be taken as unit *angular momentum*, the orbital angular momentum may be represented by a vector 1, whose magnitude is not l but $\sqrt{(l\,(l+1)}$ units, where 1 unit = $h/2\pi = 1.0542 \times 10^{-7}$ erg-sec or $\approx 10^{-34}$ kg-m²/sec, usually denoted by the symbol $\hbar$ (read as h bar).

In a complex atom, where there may be more orbiting electrons than one, the total orbital angular momentum (L) is the vector sum of the orbital angular momenta of the individual electrons and will always have an integral value.

It may also be mentioned here that any *observable component* of the orbital angular momentum, such, for example, as the one in the direction of a magnetic field, is also *quantised* and is given by h/m_l where m_l is called the *magnetic orbital quantum number* and may have any of its $(2l + 1)$ possible values, l, $(l - 1)$, $(l - 2)$, 0, –1, –2, –l.

Spin angular momentum : In order to explain the *fine structure* of the spectral lines of hydrogen and some other elements (*i.e.*, the splitting up of their spectral lines, when examined by an instrument of high resolving power), as also to explain *anomalous Zeeman effect*, *Uhlenbeck* and *Gondsmit* mid also *Dirac* suggested, in the year 1925, that an election, whether in an orbit or free, spins about an axis just like a top and has, therefore, also a *spin angular momentum*, whose magnitude is $sh/2\pi$, where s is called the spin quantum number. The same is true for all other elementary particles except π or K-mesons

The spin angular momentum too is thus *quantised* and expressed in $h/2\pi$ or ft units Its correct value, however, based on wave mechanics (when measured in terms $h/2\pi$ as the unit) is represented by the vector s, whose magnitude is not s but $\sqrt{s(s + 1)}$ units.

The value of the spin quantum number is not the same for all elementary particles. Thus, while it is 1/2 for the electron and for most of the other elementary particles, it is 1 for the *photon* and 0 for π and K-masons. So that for the electron, proton and neutron etc., the vector s (*i.e. , the spin angular momentum vector*) has the magnitude $\sqrt{(1/2\ (1/2 + 1)} = \sqrt{3}/2$ units, where, of course, 1 unit = $h/2\pi$; or $\hbar$.

Since a free and independent electron too spins about its axis, and so do most of the other elementary particles, they all possess spin angular momentum. For this reason, the spin angular momentum of an electron or an elementary particle is also referred to as its *intrinsic angular momentum*, which it preserves under all circumstances.

The vector sum S of the spin angular momenta of more than one electron must thus be an odd multiple of 1/2 if the number of electrons

be odd and must be an even integer, or an even multiple of 1/2 if the number of electrons be even, *i.e.*, being quantised, it must always be a multiple of $1/2\hbar$ or $1/2h/2\pi$. It also follows that the values of spin angular momenta for different directions of spin must differ from each other by a whole unit h or h/2π. This means, in other words, that the vectors representing the spin angular momentum vectors (or the spin, for short) must always be *either parallel* (*i.e.*, similarly directed), as shown in Fig. 5.12 (i) *or anti parallel* (*i.e.*, oppositely directed), as shown in Fig. 5.12 (").

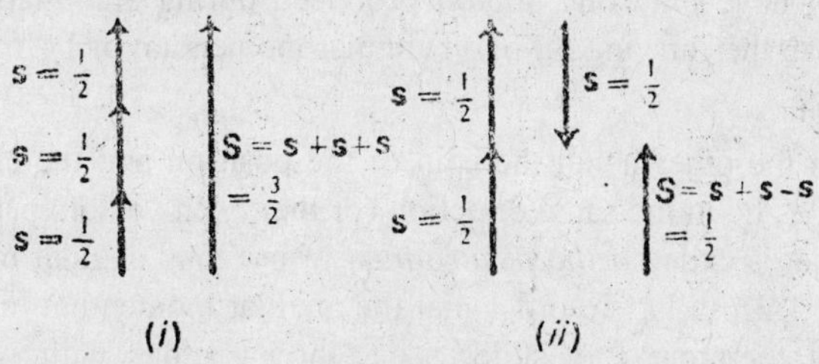

Fig. 5.12

Fundamental particles, like *electrons, nucleons* (*i.e.*, protons and neutrons), nutrinos μ-mesons etc. for which the spin angular momentum (or spin) is 1/2 are called *fermions*, after *Fermi*,–the heavier ones among them (like nucleons) being called *baryons* and the lighter ones (like *electrons*, *μ-mesons* or *mucus*), *leptons.*

A very important and far-reaching property of fermions is that no *two fermions inhabiting the same region, can possess identical properties*, thus *no two electrons in an atom can exist in the same state.*

This was first stated by *Pauli*, in the year 1925, and is known as *Pauli's exclusion principle*. As can be easily understood, it is this principle which accounts for the arrangement of electrons in an atom,

Particles for which the spin is zero or an integral multiple of $\hbar$ (or h/2π) are called Bosons (after *Base*), as for example, *α-particles*, *spin zero, photons, spin 1, deuteron* (nucleus of deuterium), *spin 2.*

Example of conservation of spin angular momentum : Such an example is furnished by a *positron slowing* down in a material and capturing an electron to form an '*atom*' of *positronium*, for a brief period, with the positron as nucleus. Since both positron and electron have the same mass, the two are really distributed about their centre of mass. So that, when they capture each other, the only angular momentum involved is the spin angular momentum of $1/2\hbar$ of each. If their spins be opposed to each other, *i.e.*, if their spin vectors be anti- parallel, we have what is called para positronium whose total spin is

$1/2\hbar - 1/2\hbar = 0$. On decay, therefore, the annihilated mass appears as electromagnetic energy in the form of two *photons*, with their spins of 1 each in opposite directions, so that the spin angular momentum carried off is zero. The spin angular moment us thus remains conserved, being zero both before and after decay of the 'parapositronium atom.' Obviously, we cannot have only one photon produced during annihilation, for it would mean the carrying off of an angular momentum of $1\hbar$ from where it is zero.

If, on the other hand, the spin of the positron and the electron be aligned, *i.e.*, be in the same direction (or their spin vectors parallel), we obtain what is called *ortho positronium* whose total angular momentum is $1/2\hbar + 1/2\hbar = 1\hbar$. In order that the angular momentum may remain conserved, therefore, it must decay into three photons, with two of them having opposite spins, so that the total angular momentum carried off is $1\hbar + 1\hbar - 1\hbar = 1\hbar$, the same as that of the 'ortho positronium atom'.

It need hardly be pointed out that we can observe the quantum nature of angular momentum only in the interaction of small entities like the fundamental particles because its unit $\hbar$ or $h/2\pi$ is much too small (equal to 1.0542×10^{-7} erg-sec). It, God forbid, we were living in a world in which $\hbar$ or $h/2\pi$ were equal to I, large objects of human size would start behaving the way these small panicles do and very strange things will happen to us. Thus, for example, a 1 kg mass whirled at the end of a metre long string would revolve only at frequencies 0, $1/2\pi$, $2/2\pi$, $3/2\pi$ etc. and the swing would occur in directions such that the change, in angular momentum from me direction to the next equal to $1\hbar$. And, our bicycles would then pick up speeds which arc integral multiples of some basic speed (depending upon their size and weight), their wheels and bandies, would turn only through certain discrete angles, and so on.

Total angular momentum : The total angular momentum of a revolving electron would obviously he equal to the sum of its orbital angular momentum and its spin angular momentum and would be given by j $(h/2\pi)$, where j is the *total angular momentum quantum number*. So that, j be the *total angular momentum vector*, we have j = l + s, such that the vector sum has always a half odd integral value, *i.e.*, it is always an odd multiple of 1/2. Since s in always equal to 1/2, it means that j can have only two values for a given value of *l*, viz., *l* + 1/2 and *l* = 1/2, except, of course, when 1 = 0, in which case it will have only the value 1/2.

Again, on the basis of wave mechanics, the magnitude of j must be $\sqrt{j(j+1)}$, that of l must be $\sqrt{l(l+1)}$ and that of s must be $\sqrt{s(s+1)}$, all in units h or $h/\partial\pi$.

In the event of there being two or more electrons in an atom, it is found that in an overwhelming number of cases we have what is called an L – S coupling (or *Russel-Saunders* coupling) in which all orbital angular momenta combine to give a resultant orbital angular momentum vector L, irrespective of the spin angular momenta, and, similarly, all spin angular momenta combine to give a resultant spin angular momentum vector S, irrespective of the orbital angular momenta. The total angular momentum of the atom is thus represented by vector J = L + S, such that J must be an integer or an odd multiple of 1/2 according as S is an integer or an odd multiple of 1/2. So that, possible number of values that J can have if L > S is (2S + 1) and the number of values that it can have when L < S is (2L + 1). Of course, in case L = 0, J can have only a single value, viz., S.

SOLVED EXAMPLES

Example 1:

A neutron of energy 1 Me V passes a proton at such a distance that the angular momentum of the neutron relative to the proton approximately equals 10^{-26} erg-sec. What is the distance of closest approach ? (Neglect energy of interaction between the two particles).

Solution:

As we know, the angular momentum of the neutron is given by J = r × mv, where m is the mass of the neutron, v, its velocity and r, its position vector relative to the proton,

If r be equal to the *distance of closest approach,* r_A, we have J = r_A × mv.

And therefore, J = r_A mv sinθ.

Since at the distance of closest approach, the velocity vector is perpendicular to the position vector, θ = 90_o and, therefore, sinθ = 1.

Hence, J = $r_A$mv. Or, r_A = J/mv.

Now, K.E. *of the neutron,* T = 1/2 mv^2, whence, $v^2 = 2T/m$ and ∴ $v = \sqrt{2T/m}$.

Substituting this value of v in the expression for r_A above, we have

$$r_A = \frac{J}{m\sqrt{2T/m}} = \frac{J}{\sqrt{2Im}}$$

Since J = 10^{-26} erg-sec and T = 1 MeV = $1.6 \times 10^{-12} \times 10^6 = 1.6 \times 10^6$ erg and m, the mass of the neutron may be taken to be 1.67×10^{-24} gm, we have

$$\textit{distance of closest approach, rA} = \frac{10^{-26}}{(2\times1.6\times10^{-6}\times1.67\times10^{-24})^{\frac{1}{2}}}$$

$$= 4.326 \times 10^{-12} \text{ cm}$$

Example 2:

A vessel at rest explodes, breaking into three pieces. Two pieces, having equal masses, fly off perpendicular to one another with the same speed of 30 m/sec. The third piece has three times the mass of each other piece. What is the direction and magnitude of its velocity immediately after the explosion?

Solution:

Here, the masses of the three pieces of the vessel are given to be as 1 : 1 : 3. Let one of the smaller pieces move along the axis of x and the other, along the axis of y then clearly *momentum of the first small piece* = (1) (30i) = 30i,

and *momentum of the second small piece* = (1) (30j) = 30j.

If the speed of the third, bigger piece be v m/sec, we have

momentum of the bigger piece = (3) (v) = 3v.

Since initially the vessel was at rest and hence its linear momentum, zero, the total momentum of the three pieces, after explosion, should, in accordance with the law of conservation of momentum, be also zero, since no external force is acting on the vessel. We, therefore, have

30i + 30j + 3v = 0, whence, 3v = – 30i – 30j or, v = – (10i × 10j).

Thus, *the magnitude of the velocity of the bigger piece, i.e.*, v = √ $10^2 + 10^2$) = √200 = 10 √2 m/sec. And its direction with the axis of x or the direction of motion of the first piece, say θ = $\tan^{-1}$ = (–1/1), *i.e.*, $\tan^{-1}$ (–1) = 135°, the same as the angle that its direction makes with the axis of y or the direction of motion of the second piece.

Thus, *the velocity of the bigger piece immediately after the explosion is* 10√2 m/sec, *inclined at an angle of* 135° *to the direction of motion of either small piece.*

Example 3:

(a) The distance between the centres of the centres of the carbon and oxygen atoms in the carbon monoxide (CO) gas molecule is 1.130 × 10 [10] metres. Locate the centre of mass of the molecule relative to the carbon atom.

(b) Find the centre of mass of a homogenous semicircular plate of radius a.

Solution:

(a) The centre of mass of CO molecule will obviously lie on the line joining the C and 0 atoms. Let it be at a distance x from the carbon atom. Then, since the atomic weights of carbon and oxygen are respectively 12 and 16, we have

$$12x = 16\ (1.130 \times 10^{-10} - x) = 16 \times 1.130 \times 10^{-10} - 16x.$$

Or, $28x = 16 \times 1.130 \times 10^{-10}$,

whence, $x = 16 \times 1.130 \times 10^{-10}/28 = 6.457 \times 10^{-11}$ m.

Hence, *the centre of mass of the* CO *molecule lies at a distance of* 6.457×10^{-11} m *from the centre of the carbon atom on the line joining the carbon and oxygen atoms.*

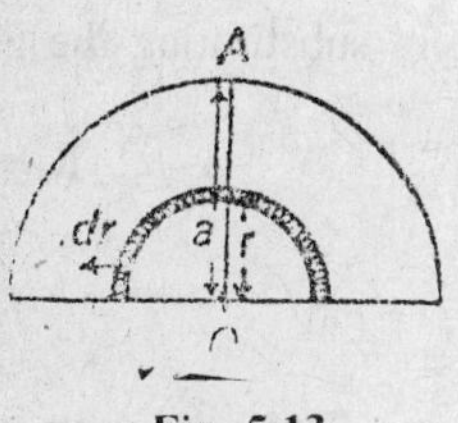

Fig. 5.13

(b) Here, too, the centre of mass obviously lie on one line of symmetry OA (Fig. 6.13) passing through the mid-point O of the case and perpendicular to it.

Since it is a homogeneous plate, the centre of mass will be at a distance R from O, given by the relation $R = \frac{1}{M}\int_0^a r.dm$, where dm is the mass of an element of it of radius r and M, the mass of the whole plate.

Clearly, *area* of the semicircular plate = $\pi a^2/2$. We, therefore, have

mass per unit area of the plate $= M / \frac{\pi a^2}{2} = \frac{2M}{\pi a^2}$.

And, *area of the element of the plate, of radius r* = (*length of shaded strip*) × (*its width dr*) = πrdr.

∴ *mass of the strip, i.e.,* dm = (πrdr) $(2M/\pi a^2)$

∴ *distance of centre of mass of the semicircular plate from 0, i.e.,*

$$R = \frac{1}{M}\int_0^a r\pi r dr \frac{2M}{\pi a^2} = \frac{2}{a^2}\int r^2 dr = \frac{2}{a^2}\left[\frac{r^3}{3}\right]_0^a = \frac{2}{a^2}\left(\frac{a^3}{3}\right) = \frac{2a}{3}.$$

Thus, *the centre of mass of the semicircular plate lies on the right bisector OA of its base at a distance of* 2a/3 *from its mid-point O.*

Example 4:

Two particles, of masses 100 and 300 gm, have at a given time positions 2i + 5j + 13k and – 6i + 4j – 2k cm respectively and velocities 10i – 7j – 3k and 7i – 9j + 6k cm/sec respectively. Deduce (i) the instantaneous position of the centre of mass, (ii) the velocity of the second particle in a frame of reference travelling with the centre of mass.

Solution:

(i) We know that the distance of the centre of mass from the chosen origin is given by $R = \dfrac{m_1 r_1 + m_2 r_2}{m_1 + m_2}$

∴ substituting the given values of m_1, m_2, r_1 and r_2, we have

$$R = \frac{100(2i+5j+13k)+300(-6i+4j-2k)}{100+300}$$

$$\frac{-16i+17j+7k}{4}\text{cm.}$$

Thus, the centre of mass lies at a distance of (–16i + 17j + 7k)/4 cm from the origin.

(ii) The *velocity of the centre of mass* is given by

$$V = \frac{m_1 v_1 + m_2 v_2}{m_1 + m_2} = \frac{100(10i-7j-3k)+300(7i-9j+6k)}{100+300}$$

$$= \frac{(31i-34j+15k)}{4}\text{cm / sec.}$$

∴ *velocity of the second particle in the frame of reference travelling with the centre of mass*

$$v_2 - V = (7i - 9j + 6k) - \frac{31i - 34j + 15k}{4} = \frac{-3i - 2j + 9k}{4} \text{cm/sec.}$$

Example 5:

Two particles P and Q are initially at rest, 1.0 metre apart. p has a mass of 0.10 kg and Q, a mass of 0.30 kg. P and Q attract each other with a constant force of 1.0 × 10^{-2} N. No external forces act on the system. Describe the motion of the centre of mass. At what distance from P's original position do the particles collide?

Solution:

Since no external forces act on the system, the question of motion of the centre of mass does not arise. It, therefore, continues to be at rest.

Let the distance at which the two particles collide be x metres from P's original position. Then, clearly, the two particle cover distances x *metres* and (1 – x) metres respectively, in the same interval of time t, say, that they take to come into contact with each other.

Now, acceleration of particle P $= \frac{\text{force}}{\text{mass}} = \frac{1 \times 10^{-2}}{0.1} = 10^{-1}$ m/sec^2

and that of particle Q $= \frac{1 \times 10^{-2}}{0.3} = 1/2 \times 10^{-1}$ m/sec^2.

∴ *distance x covered by particle P in time t* = $1/2 \times 10^{-1} \times t^2 = 1/20\ t^2$ *metres*.

and *distance* (1 – x) *covered by particle Q in the same time t* = $1/2 \times 1/3 \times 10^{-1} \times t^2 = 1/60\ t^2$ *metres*.

Clearly, therefore, $\frac{1}{20}t^2 \frac{1}{60}t^2 = (x) + (1 - x) = 1,$

whence, $t^2 = 15.$

Hence $x = \frac{1}{20}t^2 = \frac{1}{20} \times 15 = \frac{3}{4} = 0.75$ metres.

Thus, *the two particles collide at a distance of* 0.75 *m from the initial position of particle P.*

Example 6:

A bomb is flight explodes into two fragments when its velocity is 10i × 2j. If the smaller mass M flies with velocity 20i + 50j, deduce the velocity of the larger mass 3 M. Deduce also the velocities of the fragments in the centre of mass reference frame and show these in a diagram.

Solution:

Here, clearly, *initial momentum of the unexploded bomb* = (total mass) × (velocity) = (M + 3M) (10i + 2j) = 4M (10i + 2j)

After explosion, momentum of the smaller fragment = Mv_1 = M (20i + 50j)

and *momentum of the larger fragment* 3 Mv_2,

where v_1 and v_2 are the velocities of the two fragments respectively.

Since no external force acts on the system, its momentum must remain conserved. We, therefore, have *total momentum after explosion = total momentum before explosion.* Or,

M (20i + 50j) + $3mv_2$ = 4M (10i + 2j). Or, $3Mv_2$ = 4M (10i + 2j) – M (20i + 50j),

whence, v_2 = 1/2 (20i – 42j).

Thus, *the velocity of the larger mass* 3 M is 1/3 (20i – 42j).

Now, in the absence of any external force, the centre of mass continues to move with the same velocity as that of the unexploded bomb, viz.,

$$V = (10i + 2j).$$

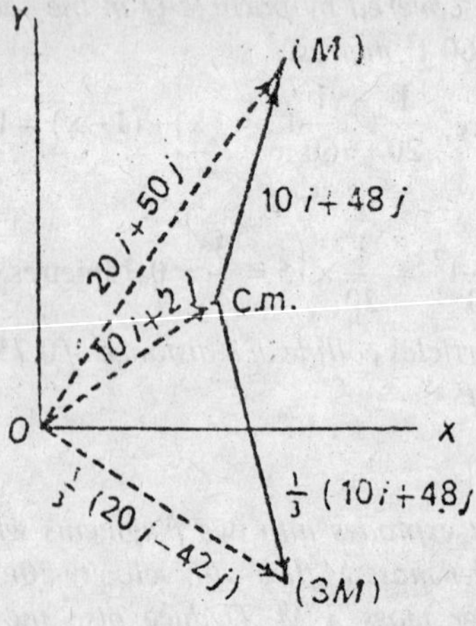

Fig. 5.14

∴ *velocity of the smaller mass in the centre of mass reference frame* = $v_2 - V$ = (20i + 50j) – (10i + 2j) = 10i + 48j and *velocity of the larger mass in the centre of mass reference frame* = $v_2 - V$ = 1/2 (20i – 42j) – (10i – 2j) = 1/2 (– 10i – 28j) = – 1/3 (10i + 48j)

All these velocities are shown in Fig. 6.14, both relative to the origin O and the centre of mass (cm), the former in dotted, and the latter, in full lines.

Example 7:

If the centre of mass of three particles of masses 1, 2 and 3 kg be at the point 3, 3, 3, where should a fourth mass of 4 kg be placed so that the centre of mass of the four particles be at the point 1, 1, 1?

Let $x_1, y_1, z_1, x_2, y_2, z_2$ and x_3, y_3, z_3 be the position of the three particles respectively.

Then, for the x – coordinate, $X = \dfrac{m_1x_1 + m_2x_2 + m_3x_3}{m_1 + m_2 + m_3}$

Or, $$3 = \frac{x_1 + 2x_2 + 3x_3}{1+2+3} = \frac{x_1 + 2x_2 + 3x_3}{6}$$

Or, $$x_1 + 2x_2 + 3x_3 = 18 \qquad \text{...(i)}$$

Now, suppose the fourth particle of mass 4 kg must be placed at the point x_4, y_4, z_4 so that the centre of mass of the combination of all the four particles lies at 1, 1, 1. Then, again, for the x-coordinate, we have

$$1 = \frac{x_1 + 2x_2 + 3x_3 + 4x_4}{1+2+3+4} = \frac{x_1 + 2x_2 + 3x_3 + 4x_4}{10}$$

Or, $$x_1 + 2x_2 + 3x_3 + 4x_4 = 10. \qquad \text{...(ii)}$$

Subtracting relation (i) from (ii). we have $4x_4$ = 10 – 18 = – 8, whence, x_4 = – 2.

Proceeding in the same manner for the y and z coordinates also, we obtain y_4 = – 2 and z_4 = – 2.

Thus, *the fourth mass of the* 4 *kg must be placed at the point* – 2, – 2, – 2, *in order that the centre of mass of the combination of all the four particles may be at the point* 1, 1, 1.

Example 8:

(a) A particle of mass m_1, moving with a velocity u_1, collides head-on with a particle of mass m_2 at rest, such that, after collision, they travel

with velocities v_1 and v_2 respectively. If the collision be an elastic one, show that $v_2 = 2x_1\left(1+\frac{m_2}{m_1}\right)$.

(b) If the collision in (a) be assumed to be an inelastic one, calculate the common velocity of the two particles after collision and the loss of kinetic energy due to collision.

Solution:

(a) A *'head-on'* collision means that, after the collision, the two particles travel along the same line.

Now, in accordance with the law of conservation of momentum, we have

$$m_1u_1 + m_2 \times 0 = m_1v_1 + m_2v_2.$$

Or
$$m_1u_1 = m_1v_1 + m_2v_2, \quad \text{...(i)}$$

whence, $v_1 = (m_1u_1 - m_2v_2)m_1 = u_1 - \frac{m_2}{m_1}v_2.$ ...(ii)

And, in accordance with the law of conservation of energy, we have

$$1/2m_1u_1^2 + 0 = 1/2m_1v_1^2 + 1/2m_2v_2^2.$$

$$\text{Or, } 1/2m_1u_1^2 + 1/2m_2v_2^2 \quad \text{...(iii)}$$

This gives
$$u_1^2 = v_1^2 + \frac{m_2}{m_1}v_2^2 \quad \text{...(iv)}$$

Or, substituting the value of v_1 from relation (ii) above, we have

$$u_1^2 = u_1^2 + \frac{m_2^2}{m_1^2}v_2^2 - \frac{2m_2u_1v_2}{m_1} + \left(\frac{m_2}{m_1}v_2\right)^2$$

$$\text{Or,} \quad 0 = \frac{m_2}{m_1}\left(v_2^2 + \frac{m_2}{m_1}v_2^2 - 2u_1v_2\right)$$

$$\text{Or,} \quad 0 = \frac{m_2}{m_1}\left(v_2 + \frac{m_2}{m_1}v_2 - 2u_1\right)$$

$$\text{Or,} \quad v_2 + \frac{m_2}{m_1}v_2 - 2u_1 = 0.$$

$$\text{Or,} \quad v_2\left(1+\frac{m_2}{m_1}\right)2u_1, \text{ whence, } v_2 = \frac{2u_1}{1+\frac{m_2}{m_1}}$$

(b) Here, the collision being *inelastic,* the two particles stick together on collision and move with a common velocity v, say.

Then, *for conservation of momentum,* we have $m_1u_1 + 0 (m_1 + m_2)v$. whence, $v = m_1u_1/(m_1 + m_2)$.

And, clearly, *loss of kinetic energy due to collision* = $1/2m_1u_1^2 - 1/2\ (m_1 + m_2)v^2$.

Or, substituting the value of v_1, we have

loss of kinetic energy = $1/2m_1u_1^2 - 1/2\ (m_1 + m_2)$

$$\left(\frac{m_1u_1}{m_1+m_2}\right)^2 = \frac{1}{2}m_1u_1^2 - \frac{1}{2}\frac{m_1^2u_1^2}{m_1+m_2}$$

$$= \frac{1}{2}m_1u_1^2\left(1-\frac{m_1}{m_1+m_2}\right) = \frac{1}{2}m_1u_1^2\left(\frac{m_2}{m_1+m_2}\right)$$

Example 9:

A steel ball weighing 1 lb is fastened to a cord 27 inches long and is released when the cord is horizontal. At the bottom of its path the ball strikes a 5.0 lb-block initially at rest on a frictionless surface (Fig. 5.15). The collision is elastic. Find the speed of the ball and the speed of the block just after collision.

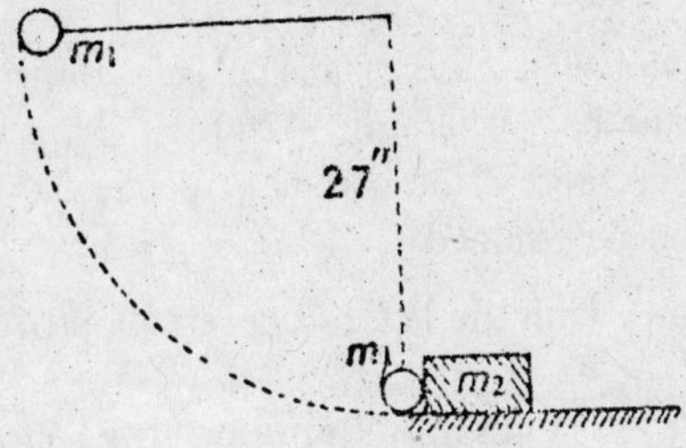

Fig. 5.15

Solution:

At the ball comes down and hits the block, it is clearly a head-on collision. If, therefore, m_1 and m_2 be the masses of the ball and the block respectively, u_1, the velocity acquired by the ball on falling through 27", with which it hits the block, v_1, the *velocity of recoil of the ball* and v_2, *the forward velocity of the block,* we have, for conservation of momentum,

$$m_1u_1 + 0 = m_1v_1 + m_2v_2 \qquad ...(i)$$

K.E. acquired by the ball on falling through 27" or 9.4 ft = $1/2m_1u_1^2$ = m_1gh, whence, $u_1^2 = 2gh = 2 \times 32 \times 9/4 = 144$.

$$\text{Or, } u_1 = \sqrt{144} = 12 \text{ ft/sec.}$$

Now, proceeding as in example 9 above, we have

$$v_2 = \frac{2u_1}{1 + m_2/m_1} = \frac{2 \times 12}{1 + 5/1} = \frac{24}{6} = 4 \text{ ft/sec.}$$

Substituting v_2 = 4 ft/sec in the momentum relation (i) above, we have

$$1 \times 12 = -1 \times v_1 + 5 \times 4,$$

Or $\quad v_1 = 20 - 12 = 8$ ft/sec.

Thus, *the ball recoils with a velocity of* 8 ft/sec *and the block moves forward with a velocity of* 4 ft/sec.

Example 10:

A sand bag of mass 10 kg is suspended with a 3 metre long weightless string. A bullet of mass 200 gm is fired with a speed 20 m/sec into the bag and stays in the bag. Calculate (i) the speed acquired by the bag. (ii) the maximum displacement of the bag, (iii) energy converted to heat in the collision.

Solution:

(i) This is obviously a case of an inelastic collision between the bullet and the bag, where m_1 = 200 gm = 0.2 kg, m_2 = 10 kg, u_1 = 20 m/sec and v, the common speed of the bullet and the bag is to be determined.

In accordance with the law of conservation of momentum, we have

$$m_1u_1 = (m_1 + m_2)v, \text{ whence,}$$

$$v = \frac{m_1u_1}{m_1 + m_2} = \frac{0.2 \times 20}{0.2 + 10} = \frac{4}{10.2} = 0.3922 \text{ m/sec.}$$

(ii) The bag will oscillate like a simple pendulum of length 3 metres about its point of suspension under a *restoring force* F = – mg sin θ = Cx, say, (Fig. 6.16), where x is the *amplitude* or the *maximum displacement* of the bag and the bullet lodged into it.

If l be the length of the pendulum, we have – mg sin θ = – mgx/*l* = – Cx, where C = mg/*l*. since m (the mass of the bag and the bullet) = 10.2 kg, g=9.8 m/sec², we have

$$C = 10.2 \times 9.8/3.$$

The *total work done in this displacement of the bag and the bullet* $\int_0^x Cxdx = \frac{1}{2}Cx^2$. This work is obviously done at the expense of the K.E. acquired by the bag and the bullet, equal to $1/2mv^2$. We, therefore, have $1/2\ Cx^2 = 1/2mv^2 = 1/2\ (10.2)\ (0.3922)^2$.

Or $\quad Cx^2 = (10.2)\ (0.3922)^2.$

$$\text{Or,}\quad x^2 = \frac{(10.2)\ (0.3922)^2\ (3)}{(10.2)\ (9.8)}$$

$$\text{whence,}\quad x = \sqrt{\frac{(0.3922)^2 \times 3}{(9.8)}} = 0.217 \text{ metre.}$$

The maximum displacement of the bag is thus 0.217 *metre.*

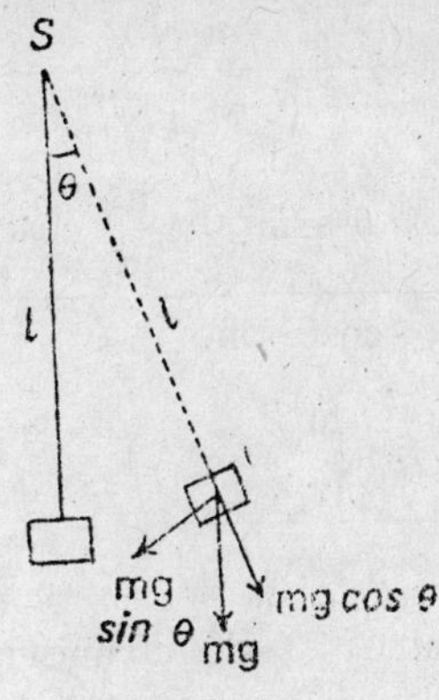

Fig. 5.16

(iii) The *energy converted into heat* is obviously equal to the difference between the initial K.E. of the bullet and the final K.E. of the bullet and the bag *i.e.*,

$$= 1/2\ m_1u_1^2 - 1/2\ (m_1+m_2)v^2$$

$= 1/2\ (0.2)\ (20)^2 - 1/2\ (10.2)\ (0.3922)^2 = 40 - 0.7845 = 39.2$ *joules.*

Example 11:

Show that in the case of an elastic and glancing collision between two particles of masses m_1 and m_2 respectively, the maximum value of the scattering angle θ_1 in the laboratory frame corresponds to the scattering angle θ in the centre of mass reference frame, where $\theta = \cos^{-1}(-m_2/m_1)$. Also show that this maximum value of the scattering angle

$$\theta_1 = \tan^{-1}\left(\frac{m_1^2}{m_2^2}-1\right)^{-\frac{1}{2}}.$$

Solution:

We know that the scattering angle θ_1 in the laboratory reference frame is related to the scattering angle θ in the centre of mass reference frame by the relation $\tan\theta_1 = \dfrac{\sin\theta}{\cos\theta + (m_1/m_2)}$ ...(i)

Differentiating with respect to θ, therefore, we have

$$\sec^2\theta_1 \frac{d\theta_1}{d\theta} = \frac{\left(\cos\theta + \frac{m_1}{m_2}\right) - \sin\theta\,(-\sin\theta)}{(\cos\theta + m_1/m_2)^2}$$

$$= \frac{\cos^2\theta + \sin^2\theta + \frac{m_1}{m_2}\cos\theta}{(\cos\theta + m_1/m_2)^2}$$

$$= \frac{1 + \frac{m_1}{m_2}\cos\theta}{(\cos\theta + m_1/m_2)^2}$$

For maximum value of 0_1, $d\theta_1/d\theta$ must be equal to zero. This means, in other words, that $1 + \dfrac{m_1}{m_2}\cos\theta = 0$,

or that $\cos\theta = -\dfrac{m_2}{m_1}$,

whence, $\theta = \cos^{-1}\left(-\dfrac{m_2}{m_1}\right)$

Now, $\sin\theta = (1 - \cos^2\theta)^{1/2} = (1 - m_2^2/m_1^2)^{1/2}$. So that, substituting the values of inθ and $\cos\theta$ in relation (i) above, we have

$$\tan\theta_1 = \frac{(1 - m_2^2/m_1^2)^{1/2}}{-\frac{m_2}{m_1} + \frac{m_1}{m_2}}$$

$$= \frac{(m_1^2 - m_2^2)^{1/2}/m_1}{(m_1^2 - m_2^2)/m_1 m_2} = \frac{m_2}{(m_1^2 - m_2^2)^{1/2}}$$

$$= \left(\frac{m_2^2}{(m_1^2 - m_2^2)}\right)^{1/2} = \left(\frac{m_1^2 - m_2^2}{m_2^2}\right)^{-1/2}$$

$$\text{Or, } \theta_1 = \tan^{-1}\left(\frac{m_1^2}{m_2^2}\right)^{-1/2}.$$

Example 12:

A rocket motor consumes 100 kg of fuel per second, exhausting it with a speed of 5 × 10^3 m/sec. What force is exerted on the rocket ? What will be the velocity of the rocket at the instant that its mass is reduced to 1/20th of its initial mass, its initial velocity being zero ? (Neglect gravitational and viscous forces etc.)

Solution:

We have *force or thrust on the rocket* given by $F = -\frac{dM}{dt}v.$

Here, *rate of consumption of fuel,* dM/dt = 100 kg/sec and *exhaust velocity* v = 5 × 10^3 m/sec. So that, *thrust on the rocket* = – 100 × 5 × 10^3 = –5 × 10^5 *newtons.*

Now, as we know, *velocity of the rocket at an instant t* is given by

$$V = V_o + v\log_e\left(\frac{M_o}{M}\right).$$

where V_o is its *initial velocity* at t = 0, M_o, its *initial mass* and M, its *mass at instant t.*

Here, initial velocity V_o, is given to be zero. So that, $V = v\log_e \frac{M_o}{M}$.

Since the mass of the rocket is reduced to 1/20th of its initial mass, $M_o/M = 20$.

We, therefore, have $V = v \log_e 20 = v \times 2.3026 \log_{10} 20 = v \times 2.3026 \times 1.3010 = 2.995v = 3v$, *i.e.*, very nearly 3 times the exhaust velocity.

Since the exhaust velocity v = 5 × 10^3 m/sec, we have

velocity of the rocket at the instant its mass is reduced to 1/20th *of its initial mass* = 3 × 5 × 10^3 = 15 × 10^3 *metres per second* very nearly.

Example 13:

If the maximum possible exhaust velocity of a rocket be 2 km/sec, calculate the ration M_0/M for it if it is to attain the escape velocity 11.2 km/sec. How long will it take the rocket (starting from rest) to attain this velocity if its rate of change of mass in terms of its initial mass, i.e., $\beta = 1/10^2$.

Solution:

we have velocity of the rocket, starting from rest (*i.e.*, $V_o = 0$), given by

$$V = v \log_e \frac{M_o}{M} \text{ So that, } \log_e \frac{M_o}{M} = \frac{V}{v} = \frac{11.2}{2} = 5.6.$$

Or, $$2.3026 \log_{10} \frac{M_o}{M} = 5.6.$$

Or, $$\log_{10} \frac{M_o}{M} = \frac{5.6}{2.3026} = 2.432,$$

whence, $$M_0/M = 270.$$

Thus, *the ration* M_0/M *for the rocket must be* 270.

Now, as we know, $M = M_o(1 - \beta t)$. So that, $M/M_o = 1 - \beta t$. Or, $\beta t = (1 - M/M_0)$,

whence, $$t = \frac{1}{\beta}\left(1 - \frac{M}{M_o}\right) = 10\left(1 - \frac{1}{270}\right) = \frac{10 \times 269}{270} = 9.96 \text{ sec.}$$

Thus, *the rocket will, in this case, attain the escape velocity in* 9.96 sec.

Example 14:

A rocket starts vertically upward with speed v_0. Show that its speed v at a height h is given by $v_0^2 - v^2 = 2gh/(1 + h/R)$, where R is the radius of the earth and g the acceleration due to gravity at the earth's surface. Hence deduce an expression for the maximum height reached by a rocket fired with a speed 90% of the escape velocity.

Solution:

For the velocity of a body going vertically upward with an initial velocity u, we have the relation $v^2 = u^2 - 2gh$, where v is the acquired by it on covering a height h and g, the acceleration due to gravity, assumed to be constant over comparatively small heights above the surface of the earth.

In the case of a rocket, large heights being involved, g cannot be presumed to remain constant throughout. So that, at the surface of the earth g being equal to MG/R^2 (where M is the mass and R, the radius of the earth, and G the gravitational constant), we have at height h above the surface of the earth, the value of acceleration due to gravity given by

$$g' = MG/(R + h)^2 = MG/R^2 (1 + h/R)^2 = g/(1 + h/R)^2$$

$$[\because MG/R^2 = g.$$

So that, $$v^2 = v_o^2 - 2\int_0^h \frac{g}{(1+h/R)^2} dh = v_o^2 - 2g\left[-\frac{R}{1+h/R}\right]_0^h$$

$$= v_o + 2g\left[\frac{R}{1+h/R} - R\right]$$

Or, $$v^2 = v_o^2 - \frac{2gh}{1+h/R}. \quad v_o^2 - v^2 = \frac{2gh}{1+h/R}.$$

Now, as we know, the *escape velocity* is equal to $\sqrt{2gR}$ (see chapter VIII) and therefore, 90% of this velocity is $0.9\sqrt{2gR}$. Hence, the *initial velocity of the rocket in the second case* is $v_o = 0.9\sqrt{2gR}$. And, at the maximum height h_m, obviously, $v = 0$. So that, substituting these values in the expression above, we have

$$(0.9\sqrt{2gR})^2 - 0 = \frac{2gh}{1+h_m/R}.$$

Or $$0.81 \times 2gR = \frac{2gh_m R}{R+h_m}.$$

Or $$0.81 = \frac{h_m}{R+h_m},$$

whence, $$h_m = \frac{0.81}{0.19}R = 4.263R.$$

Thus, *the maximum height attained by the rocket in this case is* 4.26 R.

Example 15:

(a) A rocket, set for vertical firing, weighs 50 kg and contains 450 kg of fuel. It can have a maximum exhaust velocity of 2km/sec. What should be its minimum rate of fuel consumption (i) to just lift it off the launching pad, (ii) to give it an acceleration of 20 m/sec² ?

(b) What will be the speed of the rocket when the rate of consumption of fuel is (i) 10 kg.sec (ii) 50 kg/sec?

Solution:

(a) We know that, at any instant, the thrust on the rocket is given by

$$F = M\frac{dV}{dt} = -\frac{dM}{dt}v - Mg.$$

(i) when the rocket is still on the launching pad. $M = M_o = 50 + 450 = 500$ kg and *to just lift it off the pad with minimum fuel consumption,* therefore,

$$-\frac{dM_o}{dt}v = M_o g.$$

So that, $$-\frac{dM_o}{dt} = \frac{M_o g}{v} = \frac{500 \times 9.8}{2 \times 10^3} = 2.45 \text{kg/sec}.$$

[∵ $M_o = 500$ kg and $v = 2$ km/sec $= 2 \times 10^3$ m/sec.

Or, *the minimum rate of consumption of fuel to just lift the rocket off the launching should be 2.45 kg/sec.*

(ii) Again, $F = MdV/dt = Ma = -vdM/dt - Mg$, where a is the acceleration of the rocket

Since the rocket is still on the launching pad, $M = M_o = 500$ kg. We, therefore,

have $$M_o a = -\frac{dM_o}{dt}v - M_o g,$$

whence, $$-\frac{dM_o}{dt} = \frac{M_o(g+a)}{v}$$

Or, $$-\frac{dM_o}{dt} = \frac{500(9.8+20)}{2 \times 10^3} = 7.45 \text{ kg/sec}.$$

Or, *the rate of the fuel consumption, in the case, should be* 7.45 kg/sec.

(b) Obviously, the rocket will acquire its maximum speed when its entire fuel has been consumed.

(i) Here, the rate of fuel consumption is 10 kg/sec. So that the time for the consumption of the entire fuel = 450/10 = 45 sec.

Now, velocity acquired by the rocket, starting from rest, in time t, is given be $V = v\log_6 (M_o/M) - gt$.

Here, $M_o = 500$ kg and $M = 50$ kg (the entire fuel 450 kg) bing burnt up)

$\therefore V = 2 \times 10^3 \log_e (500/50) - 9.8 \times 45 = 2 \times 10^3 \times 2.3026 \log_{10} 10 - 9.8 \times 45$

$= 2 \times 10^3 \times 2.3026 - 9.8 \times 45$ km/sec.

Thus *the speed of the rocket, in the case, will be* 4.164 km/sec.

(ii) Here, the entire fuel is burnt up in 450/50 = 9 sec. And, therefore, *velocity of the rocket,* $V = 2 \times 10^3 \log_e (500/50) - 9.8 \times 9 = 2 \times 10^3 \times 2.3026 - 9.8 \times 9 = 4517$ m/sec = 4.517 km/sec.

Thus, *the maximum speed of the rocket here will be* 4.517 km/sec.

Example 16:

A particle of mass M travels with velocity v = vi along the axis of y. At t = 0, the particle is at the point (0, y_1). Calculate (a) the angular momentum of the particle about the origin, (b) the total angular momentum about the point (0, y_2) where $y_2 > 0$.

Solution:

(a) Here, clearly, the *position vector* of the particle, *i.e.*, $r = y_1 j$ and its velocity vector v = vi. Therefore, its *angular momentum about the origin, i.e.,* $J = r \times mv$.

Or, $$J = y_1 j + Mvi = -Mvy_1 k$$

(b) In this case, the position vector of the particle, $r = (y_2 - y_1)\, j$ and its *velocity vector, the same as before, i.e.,* v = vi.

$\therefore$ *total angular momentum about the point* 0, y_2 *i.e.,* $J = r \times Mv$.

$$= (y_2 - y_1)\, j \times Mvi = -Mv(y_2 - y_1)\, k.$$

Example 17:

(a) A satellite of mass Ms is going round the earth in a circular orbit or radius Rs. Obtain an expression for its angular momentum about the cenre of its orbit (i.e., the centre of the earth).

(b) Express the total energy of the satellite in terms of its angular momentum.

Solution:

(a) Let v be the speed of the satellite in its circular orbit around the earth. Then, *its angular momentum about the centre of its orbit, i.e., about the centre of the earth* is given by J = Rs. × M_Sv. And its magnitude is

$$J = RsM_Sv \sin\theta. \qquad ...(i)$$

in a direction perpendicular to the plane of its motion.

Since the orbit is a circular one, $\theta = \pi/2$ and, therefore, $\sin\theta = 1$ .So that, *angular momentum of the satellite about the centre of its orbit*

$$= RsM_Sv \qquad ...(ii)$$

To obtain the value of v we note that as the satellite goes round the earth, the outward pull on it due to centrifugal force, M_Sv^2/R_S is just balanced by the inward gravitational pull M_EM_SG/R_S^2 due to the earth (where M_E is the mass of the earth).

We, therefore, have $M_Sv^2/RS = M_EM_SG/R_S^2$,

whence, $v = (M_EG/R_S)^{1/2}$.

Substituting this value of v in expression II above, we have

angular momentum of the satellite about the centre of its orbit, i.e.,

$$J = (R_SM_S^2M_EG)^{1/2} \qquad ...(iii)$$

(b) The *kinetic energy of the satellite* is given by $T = 1/2\ M_Sv^2 = 1/2\ M_SM_EG/R_S$

From relation III above, M_E = J2/RSMS2G. So that,

$$T = \frac{1}{2}M_S.\frac{J^2}{R_SM_S^2G}\frac{G}{R_S} = \frac{J^2}{2R_S^2M_S}$$

And, *potential energy of the satellite* is given by $U = -M_SM_EG/R_S$

Again, *substituting the value of* M_E, we have

$$U = -\frac{M_S G}{R_S}\left(\frac{J^2}{R_S M_S{}^2 G}\right)$$

$$= -\frac{J^2}{R_S{}^2 M_S}.$$

∴ *total energy of the satellite,*

$$E = T + U = \frac{J^2}{2R_S{}^2 M_S} - \frac{J^2}{R_S{}^2 M_S} = -\frac{J^2}{2R_S{}^2 M_S}.$$

Example 18:

The moon revolves about the earth so we always see the same face of the moon. (a) How are spin and orbital parts of the angular momentum of the moon with respect to the earth related ? (b) By how much would its spin angular momentum have to change if we were to be able to see all the moon's surface during the course of a month ? (Mass of the moon, $M_m = 7.35 \times 10^{25}$ gm, radius of the moon, R_m 1.74×10^8 cm, radius of moon's orbit around the earth, $R_E = 3.84 \times 10^{10}$ cm and time of one revolution or the moon round the earth, $T = 2.36 \times 10^6$ sec).

Solution:

(a) As we know (see next chapter), the angular momentum of a body rotating about an axis is given by $J = I\omega$, where I is the *moment of inertia* and ω, the *angular velocity* of the body about that axis.

For a sphere M.I. about its diameter = (2/5) MR^2. Therefore, *spin angular momentum of the moon, due to rotation about its own axis, i.e., about its own diameter, say,*

$$S = I\omega = (2/5)M_m R_m{}^2.$$

Now, the moon revolves around the earth so as to always present the same face towards it, showing clearly that its rate of rotation about its own axis is the same as its rate of revolution about the earth, so that $\omega = 2\pi/2.36 \times 10^6$.

We, therefore, have $S \frac{2}{5} M_m R_m{}^2 \times \frac{2\pi}{2.36 \times 10^6}$

$$= \frac{2}{5} \times 7.35 \times 10^{25} \times (1.74 \times 10^8)^2 \times \frac{2\pi}{2.36 \times 10^6}$$

$$= 2.369 \times 10^{36} \text{ erg-sec.}$$

And, *orbital angular momentum of the moon about the earth,* say, $L = R_E M_m v$, where v is the speed of the moon in its orbit around the earth.

Since $v = R_E \omega$, we have $L = R_E^2 M_m \omega$

$= (3.84 \times 10^{10})^2 \times 7.35 \times 10^{25} \times 2\pi/2.36 \times 10^6 = 2.885 \times 10^{41}$ ergsec

$\therefore$ *ratio of the orbital to spin angular momentum of the moon. i.e.,* L/S

$= 2.885 \times 10^{41}/2.369 \times 10^{36} = 1.217 \times 10^5$.

(b) Obviously, the moon must spin about its axis half as fact or half as slow as it actually does now, in order that both faces of it may be visible on the earth during the course of a moon. *There should thus be an increase or a decrease in its spin angular momentum of half its present value.*

Example 19:

(a) The maximum and minimum distances of a comet from the sun are 1.4 × 10^{12}m and 7 × 10^{10}m. If its velocity nearest to the sun is 6 × 10^4 m/sec, what is its velocity when farthest ? Assume in both positions that the comet is moving in a circular orbit.

(b) A wheel is rotating with an angular speed of 500 rev/min on a shaft whose rotational inertia is negligible. A second identical wheel, initially at rest, is suddenly coupled to the same shaft. What is the angular speed of the resultant combination of the shaft and the two wheels?

Solution:

(a) Let m be the mass of the comet, r_1, r_2, v_1 and v_2, the radii of its orbits and its velocities when nearest to the sun and farthest from it respectively.

The law of conservation of angular momentum demands that $mv_1r_1 = mv_2r_2$.

Since r_1 and r_2 are always perpendicular to the velocity vectors v_1 and v_2 respectively (the orbits being circular), we have $mv_1r_1 = mv_2r_2$.

Or, $v_2 = v_1r_1/r_2 = 6 \times 10^1 \times 7 \times 10^{10}/1.4 \times 10^{10}$

$= 3 \times 10^8 = 300$ m/sec.

Thus, *the speed of the comet when farthest from the sun*

= 300 m/sec.

(b) If I be the M.I. of the wheel and ω, its angular velocity about the shaft, we have

angular momentum of the wheel about the shaft = $I\omega$.

When a second identical wheel is coupled with the first, the M.I. of the combination of the two wheels about the shaft = 2I.

If ω' be the angular velocity of the combination about the shaft, we have

Its angular momentum about the shaft = $2I\omega'$

Since no external torque has been applied to the system, its angular momentum must remain conserved. And we, therefore, have

$$2\,I\omega' = I\omega, \text{ whence, } \omega' = \omega/2.$$

Since ω = 500 rev/min, we have ω' = 500/2 = 250 rev/min,

i.e., the angular speed of the resultant combination is 250 rev/min.

Example 20:

A 500 gm, mass is whirled round in a circle at the end of a string 40 cm. long, the other end of which is held in the hand if the mass makes 5 rev/sec, what is its angular momentum ? If the number of revolutions is reduced to just one, after 20 seconds, calculate the mean value of the torque acting on the mass.

Solution:

Here, angular momentum of the mass is given by $J = r \times mv$, in a direction perpendicular to the plane of the circle.

Or, $J = r \times mv \sin\theta$.

Since r is at any instant perpendicular to v (Fig. 6.17), $\theta = 90°$ and $\therefore$ sin 0 = I.

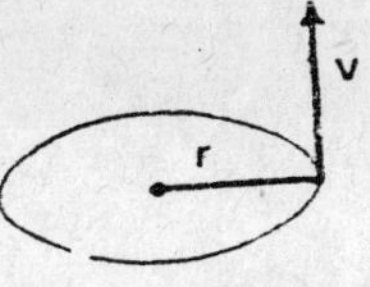

Fig. 5.17

we, therefore, have $J = rmv = rm\,(r\omega) = mr^2\omega$, where ω is the *angular velocity of the mass* = $5 \times 2\pi$/sec.

So that, substituting the values of m, r and ω, we have *angular momentum,* $J = 500(40)^2\,(5 \times 2\pi) = 8 \times 10^6\pi = 2.5 \times 10^7$ erg-sec.

Now, *torque* $\tau = dJ/dt = mr^2 d\omega/dt = 500\ (40)^2\ (2\pi/20) = 2.5 \times 10^5$ dyne-cm.

Example 21:

A string of length l, tied to the top of a pole, carries ball at its other end, as shown in Fig 6.18. On giving the ball a single hard blow, it acquires an initial velocity v_0 in the horizontal plane and moves in a spiral of decreasing radius, with the sting winding itself around the pole whose radius to neglisible compared with the length of the string. (i) What is the instantaneous centre of revolution ? (ii) Is the angular momentum conserved ? (iii) Assuming the kinetic energy to be conserved, calculate the angular velocity of the ball after it has made five complete revolutions.

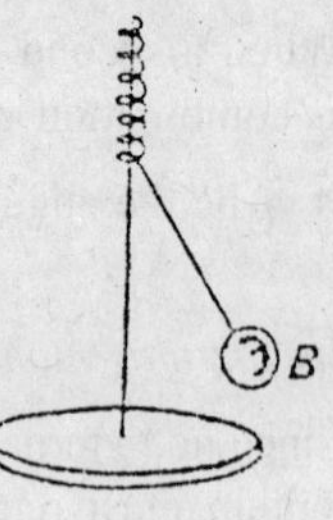

Fig. 5.18

Solution:

(i) The instantaneous centre of the ball is obviously the point of contact of the string with the circumference of the pole at the instant considered.

(ii) Let r be the position vector of the ball measured from the axis of the pole passing through 0 to the centre B of the ball (Fig. 6.19) and let F be the force applied to the ball, direct towards the point P on the circumference of the pole, which is the centre of revolution at the instant, and not towards O, So that F is not parallel to r and *the torque on the ball is not, therefore, zero.* It thus follows as a natural consequence that *the angular momentum of the system is not conserved.*

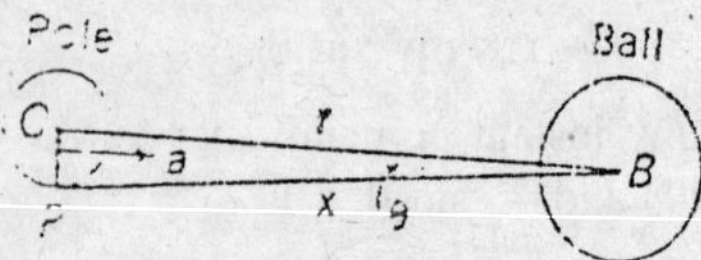

Fig. 5.19

(iii) Since the K.E. of the system is assumed to be conserved, it is clear that the velocity of the ball remains constant at its initial value v_o. The component of the velocity perpendicular to the radius vector OB is $v_o \cos\theta = v_o r/r$ and the angular velocity of the ball about the axis of the pole is thus $\omega = \frac{v_0 x/r}{r} = v_0 x/r^2$

After 5 revolutions, the length of the string will decrease to l – 5 (2π a) = l – 10πa, where a is the radius of the pole. So that, the length of the string will now be (l – 10 πa). And, since $r^2 = OP^2 + PB^2 = a^2 + x^2 = a^2 + (l - 10\pi a)^2$, the angular velocity will now be given by $\omega'' v_0 = \frac{l-10\pi a}{a^2 + (l-10\pi a)^2}$. Or, Since a < < l, we have $\omega' = \frac{v_1}{l - 10\pi a}$.

This will than be the angular velocity of the ball about the point on the circumference if the pole after five complete revolutions of the ball.

Example 22:

The distance of closest approach of a positive particle to a nucleus of atomic number is 2 × 10 ¹² cm. If the charge on the particle be equal to e and its initial kinetic energy equal to 2 MeV, obtain the value of the impact parameter for it.

Solution:

We have the relation

$$\frac{Ze^2}{rA} = \frac{1}{2} m v_o^2 \left[1 - \left(\frac{b}{rA}\right)^2\right]$$

where rA is the *distance of closest approach* of the particle, b, the *impact parameter* for it and 1/2 mv_o^2, its *initial* K.E.

Since initial K.E. of the particle is given to be 2 MeV, we have

$1/2\, Mv_o^2 = 2 \times 1.6 \times 10^{-12} \times 10^{-6} = 3.2 \times 10^6$ ergs.

So that,

$$\frac{Ze^2}{rA} = 3.2\times10^{-6}\left[1 - \left(\frac{b}{rA}\right)^2\right], \text{ whence, } 1 - \left(\frac{b}{rA}\right)^2 = \frac{Ze^2}{3.2\times10^{-6}\, rA}.$$

$$\text{Or,} \quad \left(\frac{b}{rA}\right)^2 = 1 - \frac{Ze^2}{3.2\times10^{-6}\times rA} = 1 - \frac{80\left(4.8\times10^{-10}\right)^2}{3.2\times10^{-6}\times2\times10^{-11}}$$

$\therefore\ b^2 = 0.712 \times (rA)^2$ Or, $b = rA\sqrt{0.712} = 2 \times 10^{-11} \times 0.8437$
$= 1.6874 \times 10^{-11} = 1.7 \times 10^{-11}$ cm.

Thus, the *impact parameter for the particle* $= 1.7 \times 10^{-11}$ cm.

Example 23:

Given that the orbital velocity of the sun about the centre of our galaxy is 3×10^7 cm/sec and its distance nearly equal to 3×10^{22} cm from the axis of the galaxy, estimate the mass of the galaxy ($G = 6.67 \times 10^{-8}$ C.G.S. units).

Solution:

Obviously, the inward pull on the sun due to gravitational attraction between it and the galaxy just balances the outward pull on it due to centrifugal force. If, therefore, M_S and M be the masses of the sun and the galaxy respectively, we have

$$\frac{MM_S}{r^2}G = \frac{M_S v^2}{r}, \quad \text{whence,} \ M = \frac{v^2 r}{G}.$$

Substituting the values of v, r and G, therefore, we have

mass of the galaxy,

$$M = \frac{\left(3\times10^7\right)^2 \times 3\times10^{22}}{6.67\times10^{-8}} = \frac{27\times10^{44}}{6.67} = 4.05\times10^{44}\ \text{gm.}$$

Example 24:

A hunter has rifle that can fire 0.06 kg bullets with a muzzle velocity of 900 m/sec. A 40 kg leopard springs at him at a speed of 10 m/sec. How many bullets must the hunter fire into the leopard in order to stop it in its tracks?

Solution:

Here, clearly, *momentum of the leopard* = (mass) (velocity) = (40) (10) kg-m/sec. The momentum of the bullets too that must be fired into the leopard must be at least this much in order to stop it in its tracks. Let this number of bullets be n. Then, *momentum of the bullets (in the opposite direction to that of the leopard* = *n* (0.06) (900) *kg-m/sec.* Equating the two, therefore, we have

$$n\ (.06)\ (900) = (40)\ (10),$$

whence $n = \frac{40 \times 10}{.06 \times 900} = \frac{400}{54} = 8.$

Thus, *the hunter must fire* 8 *bullets into the leopard in order to stop it.*

Example 25:

(a) What is the momentum of an electron of kinetic energy 100 electron volts ? (mass of the electron may be taken to be 9×10^{-28} gm).

(b) At what velocity will a 10, 000 kg truck have (i) the same momentum, (ii) the same kinetic energy as a 4, 000 kg car at 30 m/sec?

Solution:

(a) We have K.E. *of the electron* = $1/2mv^2 = 100eV = 100 \times 1.6 \times 10^{-12}$ ergs.

$\therefore$ *momentum of the electron* = $mv = \sqrt{m^2v^2} = \sqrt{2m \times 1/2mv^2}$

$= \sqrt{2 \times 9 \times 10^{-28} \times 100 \times 1.6 \times 10^{-12}}$

$= \sqrt{18 \times 1.6 \times 10^{-38}} = 5.37 \times 10^{-19}$ gm-cm/sec.

(b) (i) Let the truck have the same momentum as the car at a velocity of v m/sec. Then, clearly,

$10{,}000\ v = 4{,}000 \times 30$, whence, $v = 4{,}000 \times 30/10{,}000 = 12$m/sec.

(ii) If the truck has the same K.E. as the car at velocity v', we have

$1/2 \times 10{,}000 \times v'^2 = 1/2 \times 4{,}000 \times 30^2$,

whence, $v' = \sqrt{360} = 18.97$ m/sec.

Thus, *the truck will have the same momentum as the car at a velocity of* 12 m/sec *and the same* K.E. *as the car at a velocity of* 18.97 m/sec.

EXERCISES

1. A bomb of mass 40.0 kg explodes in flight at the instant when its velocity is $v_o = 10i + 5j$ m/sec. It splits in two fragments of mass ration 1 : 3 and the larger mass is observed to have velocity $v_2 = 100i + 50j + 10k$ m/sec just after explosion. Deduce (i) velocity v_1 of the smaller fragment just after explosion, (ii) kinetic energy produced in the explosion.

Ans. (i) – (260i + 130j + 30k) m/sec.
(ii) 6.135×10^2 joules.

[**Hint.** If M be the mass of the unexploded bomb, *its momentum before explosion* = Mv_o. And if m_1 and m_2 be the masses of the two fragments after explosion and v_1 and v_2, their respective velocities, we have $Mv_o = m_1v_1 + m_2v_2$.

Or, 40 (10i + 5j) = $10v_1$ + 30 (100i + 5j + 10k),

whence, v_1 = – (260in + 130j + 30k) m/sec.

And K.E. *produced in the explosion* = $1/2m_1v_1^2 + m_2v_2^2 - 1/2Mv_0^2$ = 1/2 × 10($260^2 + 130^2 + 30^2$) 1/2 × 30 ($100^2 + 50^2 + 10^2$) – 1/2 × 40($10^2 + 5^2$) = 6.135 × 10^2 joules].

2. Three particles of masses 20 gm, 30 gm and 40 gm are initially moving along the positive direction of the three coordinate axes respectively with the same velocity of 20 cm/sec, when due to their mutual interaction, the first particle comes to rest, the second acquires a velocity acquires a velocity 10j + 20k. What is then the velocity of the third particle?

Ans. 10i + 7.5j + 5k.

[**Hint:** *Initial momentum of the system* = 20 (20i) + 30 (20j) + 40(20k) and *final momentum of the system* = 20 (0) + 30 (10j + 20k) + 40v, where v is the velocity of the third particle. Equating the two, v can be evaluated]

3. Two blocks A and B of masses m_A and m_B are connected together by means of a spring and are resting on a horizontal frictionless table. The blocks are then pulled apart so as to stretch the spring and then released. Show that the kinetic energies of the blocks are, at any instant, inversely proportional to their respective masses.

4. A projectile is fired from a gm at an angle of 45° with the horizontal and with a muzzle speed of 1500 ft/sec. At the highest point in its flight the projectile explodes into two fragments of equal mass. One fragment; whose initial speed is zero, falls vertically. How far from the gun does the other fragment land, assuming a level terrian?

Ans. 1.1 × 10^5 ft.

5. The wire cage of a bird is suspended from a spring balance. How does the reading on the balance differ when the bird flies about from that when it just sits quietly?

6. (a) What is centre of mass? Show, that, in the absence of external force, the velocity of the centre of mass remains constant.

   (b) Show that the position of the centre of mass of a system of particles is quite independent of the frame of reference used.

7. Show that the centre of mass of two particles is on the line joining them at a point whose distance from each particle is inversely proportional to the mass of that particle.

8. If only an external force can change the state of motion of the centre of mass of a body, how does it happen that the internal force of the brakes can bring a car to rest?

9. (a) Locate the centre of mass of three particle of 2 kg. 3kg and 4 kg placed at the three corners of an equilateral triangle of 1 metre side. **Ans.** x = 0.56 m, y = 0.385 m.

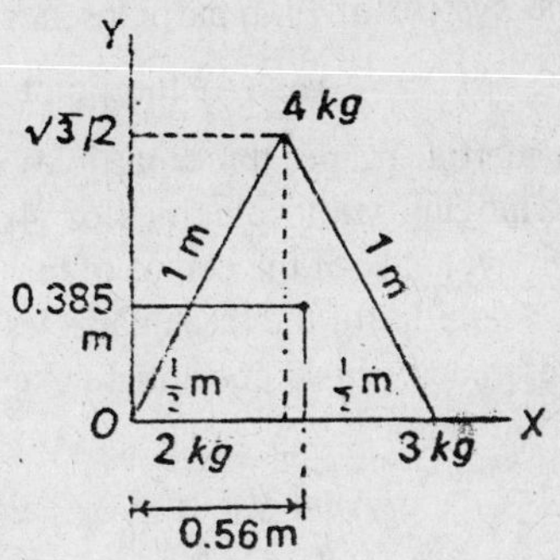

Fig. 5.20

[**Hint:** Imagine the triangle to lie in X – Y plane, with one side along the axis of x, as shown in Fig. 6.20. Then, we have

$$x = \frac{\sum m_k x_k}{\sum m_k} = \frac{2(0)+3(1)+4\left(\frac{1}{2}\right)}{2+3+4}$$

$$= \frac{5}{9} = 0.56 \text{ m} \quad \text{and} \quad y = \frac{\sum m_k y_k}{\sum m_k}$$

$$= \frac{2(0)+3(0)+4\left(\sqrt{3/2}\right)}{2+3+4} = \frac{2\sqrt{3}}{9} = 0.385 \text{ m.}]$$

(b) Where does the centre of mass of a triangular plate lie?

Ans. *At the point of intersection of the three mediation.*

10. (a) What is meant by a centre of mass frame of reference? What is its particular advantage?

(b) Must there necessarily be any mass at the centre of mass in the case of a solid body and must it always lie within the body?

11. Find the centre of mass of (i) a solid hemisphere (ii) a thin hemispherical shell, of radius r.

Ans. (i) 3/8 r *above the centre of the plane face*
(ii) 1/2 r *above the mid-point of open face.*

12. The centre of mass of three particles of masses 4 gm, 5 gm and 6 gm respectively lies at the point (2, 2, 2). Where should a fourth particle of mass 12 gm be placed so that the centre of mass of the system of four particles lies at (0, 0, 0) ?

Ans. At the point (– 2.5, – 2.5, – 2.5)

13. In a given inertial frame, three particles of masses 10 gm and 20 gm are moving with velocities 6i, 4j and 9k respectively. Calculate the velocity of the centre of mass of the system. What will be the velocities of the three particles in the centre-of-mass frame of reference whose coordinate axes are parallel to those of the inertial frame?

Ans. $V = i + 2j + 3k$, $v_1 = 5i - 2j - 3k$,
$v_2 = -i + 2j - 3k$, $v_3 = -i - 2j + 6k$.

[**Hint:** $V = \dfrac{\sum m_k V_k}{\sum m_k} = \dfrac{10(6i)+30(4j)+20(9k)}{10+30+20} = i+2j+3k.$

And, velocities in the centre-of-mass frame of reference, are $v_1 = (6i) - V$, $v_2 = (4j) + V$ and $v_2 = (9k) - V$.]

14. The respective positions of three particles A, B and C, of masses 10 kg, 5 kg and 5 kg are given by (4i + 2j), (–2i + 2j) and (2i – 4j) metres. If forces 20j, – 10i and 25i newtons act on them respectively, calculate the acceleration of the centre of mass of the system. Illustrate your answer by means of a proper diagram.

Ans. 1.25 m/sec^2 at an angle of 53° with the axis of x.

[**Hint:** Here, the x-component of the resultant force F acting on the centre of mass is $F_z = 25i - 10i = 15i$ and its y-component, $F_y = 20j$.

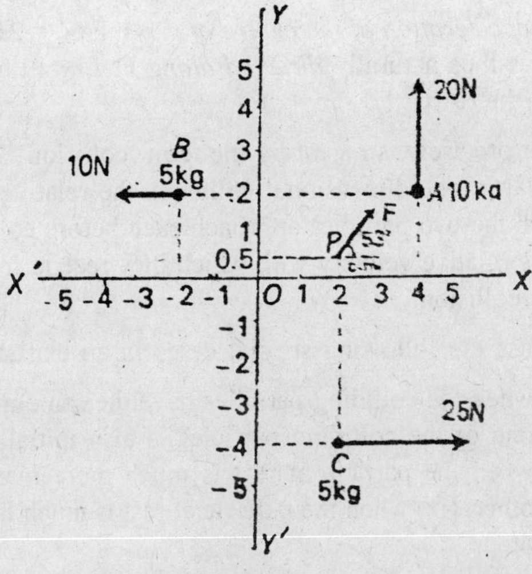

Fig. 5.21

∴ *magnitude of the resultant force acting at the centre of mass* $= \sqrt{15^2 + 20^2} = \sqrt{0.25} = 25N$. If θ be the angle that this force makes with the x-axis, we have $\theta = \tan^{-1}(20/15) = \tan^{-1}(4/3) = 53°$.

Hence *acceleration of the centre of mass* =

$$\frac{\text{force}}{\text{total mass}} = \frac{25}{10+5+5} = \frac{25}{20}$$

$= 1.25$ m/sec^2 *at an angle of* 53° *with the x-axis.*

To illustrate this by means of a diagram, we must first indicate the positions of the particles A, B and C, as shown in Fig. 6.21. Then, the x and y coordinates of the centre of mass are obtained as usual, *i.e.*,

$$x = \frac{10(4)+5(-2)+5(2)}{10+5+5} = 2\text{m} \text{ and } y$$

$$= \frac{10(2)+5(2)+5(-4)}{10+5+5} = 0.5\text{m}.$$

Or, the position of the centre of mass P is (2i + 0.5j).

Thus, the c.m. lies at the point P, as shown, and the force F 25 N acts at this point P at an angle of 53° with the x-axis.

The *acceleration of the centre of mass,* F/M = 25/20 = 1.25 m/sec^2 is thus naturally *directed along* F, *i.e.*, at an angle of 53° with the x-axis.]

15. What precisely is meant by the term 'collision' ? Show that in an elastic one-dimensional collision, the relative velocity with which the two particles approach each before collision is equal to the relative velocity with which they recede from each other after collision.

16. Discuss the following special cases in an elastic collision:

    (i) when the colliding particles have the same mass, (ii) when one of the colliding particles is also initially at rest, (iii) when the particle at rest is much more massive than the other, (iv) when the particle at rest is much lighter than the other.

17. Show that in a head-on elastic collision between two particles, the transference of energy is the maximum when their mass ration is unity.

18. A neutron of mass 1.67×10^{-27} kg and speed 10^5 m/sec collides with a stationary deuteron of mass 3.34×10^{-27} kg. The particles do not stick together and no kinetic energy is lost in collision. What is the subsequent speed of each particle ?

    Ans. *Speed of neutron* = 3.33×10^4 m/sec;
    Speed of deuteron = 6.67×10^4m/sec in opposite direction.

19. Obtain an expression for the fractional decrease in kinetic energy of a neutron of mass m_1 when it makes a head-on collision with an atomic nucleus of mass m_2 initially at rest and calculate the percentage decrease when the nucleus concerned is the one of (i) lead, (iii) hydrogen.

    Ans. $\frac{4m_1m_2}{(m_1+m_2)^2}$, (i) 2%, (ii) 28%, (iii) 100%.

    [**Hint:** If u_1 and v_1 be the velocities of the neutron before and after collision respectively, we have K.E. of *the neutron before* *collision,* $T_1 = 1/2 m_1 u_1^2$ and its K.E. *after collision,*

$T_2 = 1/2 m_1 v_1^2$

$\therefore$ fractional loss of energy on collision

$$= \frac{T_1 - T_2}{T_1} = \frac{\frac{1}{2}m_1 u_1^2 - \frac{1}{2}m_1 v_1^2}{\frac{1}{2}m_1 u_1^2}$$

$$= \frac{u_1^2 - v_1^2}{u_1^2} = 1 - \frac{v_1^2}{u_1^2}. \qquad ...(i)$$

But, as we have (b) above, in a collision of this type,

$$v_1 = \left(\frac{m_1 - m_2}{m_1 + m_2}\right) u_1.$$

Substituting this value of v_1 in relation I above, we have

$$\frac{T_1 - T_2}{T_1} = 1 - \left(\frac{m_1 - m_2}{m_1 + m_2}\right)^2 u_1^2 / u_1^2$$

$$= 1 - \left(\frac{m_1 - m_2}{m_1 + m_2}\right)^2 = \frac{4 m_1 m_2}{(m_1 + m_3)^2}.$$

$\therefore$ (i) For *lead*, $m_2 \backslash m_1 = 206$ and $\therefore$ $m_2 = 206 m_1$. So that,

$$\frac{T_1 - T_2}{T_1} = \frac{4 \times 206}{(1+206)^2} = .01912 \text{ or } 1.91\% \approx 2\%$$

(ii) For *carbon*, $m_2/m_1 = 12$, or, $m_2 = 12 m_1$. And $\therefore$

$$\frac{T_1 - T_2}{T_1} = \frac{4 \times 12}{(1+12)^2} = .02841 \text{ or } 28\%.$$

And (iii) for *hydrogen*, $m_2/m_1 = 1$ or, $m_2 = m_1$. And $\therefore$

$$\frac{T_1 - T_2}{T_1} = \frac{4 \times 1}{(1+1)^2} = 1 \text{ or } 100\%.]$$

20. Show that in a one-dimensional elastic collision the speed of the centre of mass of two particles, m_1 moving with initial speed u_1, and m_2 moving with initial speed u_2 is given by

$$V = \left(\frac{m_1}{m_1 + m_2}\right) u_1 + \left(\frac{m_2}{m_1 + m_2}\right) u_2$$

both before and after collision.

[**Hint:** *Before collision,* $(m_1 + m_2)V = m_1u_1 + m_2u_2$, whence,

$$V = \left(\frac{m_1}{m_1 + m_2}\right)u_1 + \left(\frac{m_2}{m_1 + m_2}\right)u_2$$. After collision, using the values of v_1 and v_2, as given under § 6.6 and substituting in the relation $(m_1 + m_2)V = m_1v_1 + m_2v_2$ we obtain the same value of V as before collision.]

21. Two objects of masses $m_1 = 2$ gm and $m_2 = 5$ gm possess velocities $u_1 = 10i$ cm/sec and $u_2 = 3i + 5j$ cm/sec just prior to a collision during which they become permanently attached to each other. Obtain the values of (i) the velocity of the centre of mass of the system, (ii) final momentum of the system in (a) the laboratory (b) the fraction of the initial total kinetic energy associated with the motion after collision.

Ans. (i) 5i + 25/7j cm/sec,
(ii) 35i + 25j gm-cm/sec, (b) 0, (iii) 0.72.

[**Hint:** (i) $(m_1 + m_2)V = m_1u_1 + m_2u_2$. (ii) (a) Since momentum is conserved in both elastic and inelastic collisions, the final momentum is equal to the initial momentum $m_1u_1 + m_2u_2$ or equal to $(m_1 + m_2)V$. (b) Since in the centre-of-mass reference frame, the centre of mass remains at rest, V = 0 and hence the momentum of the system = $(m_1 + m_2)V = 0$. (iii) Initial K.E., $T_1 = 1/2m_1u_1^2 + 1/2m_2u_2^2$ and final K.E., $T_2 = 1/2\,(m_1 + m_2)V^2$, where v is the velocity of the combination after collision, equal to $(m_1u_1 + m_2u_2)/(m_1 + m_2)$, *i.e.*, the same as the velocity of the centre of mass. The fraction of the total K.E., associated with the motion after collision is thus T_2/T_1].

22. The amount of energy transferred to a stationary body of mass m_2 by a collision with a moving body of mass m_1 can be increased by interposing another body of mass m_2 between them, so that m_1 strikes m_2 which, in turn, strikes m_2. Show that the energy transferred to m_2 is a maximum when $m_3 = \sqrt{m_1m_2}$.

23. A bullet of mass 20 gm and moving with a speed of 20m/sec hits the bob of a simple pendulum of mass 0.5 kg and lodges into it. If the length of the pendulum be 100 cm and g = 980 cm/sec^2, find the angular amplitude of the swing of the pendulum.

Ans. 14°15'.

24. Show that the total angular momentum of a system of particles about a fixed or a reference point is given by the relation $J = R \times P + J_{c.m.},$ where $(R \times P)$ is the angular momentum of the centre of mass about that point and $J_{c.m.},$ the angular momentum of the system about the centre of mass.

    What are spin and orbital angular momenta ?

25. (a) State the law of conservation of angular momentum. Give one example of its application.

    (b) show that for a centre force, the; angular momentum is conserved.

26. (a) Define angular momentum J and torque N, Using Newton's second law in an inertial frame of reference, show that dJ/dt = N.

    (b) Show that for a particle subjected to a central force, the angular momentum is a constant of motion.

27. (a) A man turns on a rotating table with an angular speed ω. He is holding two equal masses at arm's length. Without moving his arms, he drops the two masses. What change, if any, is there in his angular speed? Is the angular momentum conserved ? Explain.

    (b) Show that the conservation of the angular momentum of a system is a consequence of the rotational invariance of its potential energy.

28. Show that the conservation of angular momentum applied to planetary motion leads to the law of constant areal velocity.

    Why is the velocity of a satellite the maximum when it is closest to the sun and the minimum when it is farthest from it in its orbit around it?

29. State the law of conservation of angular momentum. Show that the sum of all internal torques is zero.

    Derive an expression for the distance of closest approach of a proton projected into the Coulomb field of a heavy nucleus.

30. A proton of speed u travels towards an atomic nucleus such that if it is undeflected, the nearest approach would be b. Show that the actual nearest approach b' is given by $Ze^2/b' = 1/2\ mu^2 [1 - (b/b')^2]$, where Z is the atomic number of the nucleus and m is the mass of the proton.

31. (a) For a particle of mass m = 10.0 gm. position r = 10i + 6j cm and velocity v = 5i cm/sec, calculate the angular momentum about the origin. **Ans.** 300k.

(b) Show that the angular momentum of a particle about a fixed point is equal to the product of its mass and double the are a(1/2 r^2 Example 20ω) described by the rotating line in unit time and has a direction perpendicular to this area.

32. What is the angular momentum of a neutron of kinetic energy 1 MeV about a nucleus it the impact parameter be 1 A.U. (Mass of neutron may be taken to be 1.67×10^{-24} gm; 1 A.U. = 10^{-8} cm.) **Ans.** 2.31 × 10–23 gm-cm^2/sec.

33. Explain why due to atmospheric friction, the velocity of a satellite in its circular (or nearly circular) orbit around the earth increases. Does it not violate the law of conservation of energy? What happens to its angular momentum?

[**Hint:** Due to atmospheric friction, the satellite certainly loses some energy in the form of heat. Since, however, its total energy E = its kinetic energy T + its potential energy U we have

$$E = T + U = 1/2mv^2 - Mm\,G/r,$$

34. (a) State and explain principle of conservation of linear momentum. Show that in the absence of any external force acting on it, the linear momentum of a system of particles remains constant.

(b) A light and a heavy body have equal kinetic energies of translation. Which one has the larger momentum?

35. Explain whether (a) it is possible for a body to have energy without having momentum, and vice versa, (b) a sail boat can be propelled by air blown at the sails from a fan attached to the boat itself.

36. Write a short note on the law of conservation of momentum and its importance in physics Does the law also hold good in nuclear and relativistic physics?

37. (a) A 60 lb and a 90 lb boy are at rest on frictionless roller skates. The larger boy pushes the other so that the latter rolls away at a speed of 5 mi/hr. What is the effect of this action on the larger boy himself ?

(b) A 200 kg bomb falling freely explodes and splits into two fragments, a larger on of 160 kg and a smaller one of 40 kg. The former moves away with a speed $v = i + 2j + 3k$. What is the speed of the latter? **Ans.** $-4\ (i + 2j + 3k)$

38. Show that when the vector sum of the external forces facing upon a system of particles equals zero, the total linear momentum of the system remains constant.

39. Show that the law of conservation of linear momentum of a system is a direct consequence of the transitional invariance of the potential energy of the system.

40. A radioactive nucleus, initially at rest, decays by emitting an electron and a neutrino at right angles to one another. The momentum of the electron of the electron is 1.2×10^{-22} kg-m/sec and that of the neutrino is 6.4×10^{-23} kg-m/sec. (a) Find the direction and magnitude of the momentum of the recoiling nucleus. (b) The mass of the residual nucleus is 5.8×10^{-26} kg. What is the kinetic energy of recoil?

Ans. (a) 1.36×10^{-22} kg-m/sec at $\angle\ 28°4'$ *with the direction of the electron.*

(b) 1.595×10^{-19} joule or 1 eV.

[**Hint:** Assuming the direction of motion of the electron to be along the axis of x, *momentum of the electron* = 1.2×10^{-22}, i kg-m/sec and *momentum of neutrino* = $6.4 \times 10^{-23}\,j = 0.64 \times 10^{-22}\,j$ kg-m/sec.

And, *momentum of the residual nucleus* = mv, where m is its mass and v, its *velocity*.

Since, initially, the undecayed nucleus was at rest, its momentum was zero, Hence, in accordance with the law of conservation of momentum,

$$1.2 \times 10^{-22}\,i + 0.64 \times 10^{-22}\,j + mv = 0,$$

whence, $mv = -(1.2 \times 10^{-22}\,i + 0.64 \times 10^{-22}\,j)$

Hence, *magnitude of momentum of recoiling nucleus, mv*

$= \sqrt{(1.2 \times 10^{-22})^2 + (0.64 \times 10^{-22})^2}$

$= 1.36 \times 10^{-22}$ kg-m/sec.

If its direction of motion be inclined at $< \theta$ with the x-axis or direction of motion of the electron, we have $\tan\theta = 0.64/1.2 = 0.5333$, whence, $\theta = \tan^{-1}(0.5333) \approx 28°4'$.

And, *velocity of the recoiling nucleus*

$$= \frac{\text{momentum (mv)}}{\text{mass (m)}} = \frac{1.36\times10^{-22}}{5.8\times10^{-26}}\,\text{m/sec.}$$

$$\therefore \text{K.E. } \textit{of recoil} = 1/2mv^2 = 1/2 \times 5.8 \times 10^{-26} \times \left(\frac{1.36\times10^{-22}}{5.8\times10^{-26}}\right)^2$$

$$= 1.595 \times 10^{-19}\text{ joule} = \frac{1.595\times10^{-4}}{1.60\times10^{-19}} = 1\text{ eV.}]$$

41. A railroad flat car of weight W can roll without friction along a straight horizontal track. Initially, a man of weight ω is standing on the car which is moving to the right with speed v_0. What is the change in velocity of the car if the man runs to the left so that his speed relative to the car is v just before he jumps off at the left end? **Ans.** wv/(W + w)